Exercise Workbook of Engineering Drawing (Chinese-English)

工程制图习题集（中英双语）

Editors in Chief by Yingna LIANG, Jingli LU
主 编 梁瑛娜 陆景丽
Associate Editors by Cheng GUO, Xiaoxia YU, Mingchao DING, Xinbo CHEN
副主编 郭 成 于晓霞 丁明超 陈新博
Auditors in Chief by Zhikui DONG
主 审 董志奎

燕山大学出版社
·秦皇岛·

图书在版编目（CIP）数据

工程制图习题集=Exercise Workbook of Engineering Drawing (Chinese-English)：汉、英 / 梁瑛娜，陆景丽主编．—秦皇岛：燕山大学出版社，2024.1
ISBN 978-7-5761-0557-5

I．①工… II．①梁… ②陆… III．①工程制图—习题集—汉、英 IV．①TB23-44

中国国家版本馆 CIP 数据核字（2023）第 162342 号

工程制图习题集（中英双语）

Exercise Workbook of Engineering Drawing (Chinese-English)

梁瑛娜 陆景丽 主编

出 版 人：陈 玉
责任编辑：孙志强　　**责任印制**：吴 波　　**封面设计**：刘馨泽
出版发行：燕山大学出版社 YANSHAN UNIVERSITY PRESS　　**电 话**：0335-8387555　　**地 址**：河北省秦皇岛市河北大街西段 438 号
邮政编码：066004　　**印 刷**：涿州市般润文化传播有限公司　　**经 销**：全国新华书店

开 本：787mm×1092mm 1/16　　**印 张**：13
版 次：2024 年 1 月第 1 版　　**印 次**：2024 年 1 月第 1 次印刷
书 号：ISBN 978-7-5761-0557-5　　**字 数**：140 千字
定 价：78.00 元

Foreword

With the continuous improvement of China's competitiveness and influence in the world, it is an inevitable choice for the higher education in China to carry out multi-level and wide-ranging educational exchanges and cooperation, and cultivate international talents who have an international vision, understand international rules and can participate in international affairs and competition.

Under Sino-foreign cooperative education mode in Yanshan University, this workbook is designed to cultivate high-quality innovative talents in the 21st century based on the excellent disciplinary and professional foundation of Yanshan University and combined with the high-quality educational resources of Silesian University of Technology. The workbook can be used as a bilingual teaching textbook for undergraduate mechanical and near-mechanical majors in ordinary universities. It can also be used for reference by other professionals and relevant engineering and technical personnel.

This workbook is written according to the "*Basic Requirements for Engineering Graphics Teaching Courses in Colleges and Universities*" adopted by the Engineering Graphics Teaching Steering Committee of the Ministry of Education and the latest national standards of "*Technical Drawing*" and "*Mechanical Drawing*", and combined with the practical experience of teaching reform over the years and the new requirements of engineering graphics development.

The workbook has the following features,

1. Bilingual arrangement in Chinese and English to cultivate students' language ability and academic communication ability.

2. The questions are written from easy to deep, step by step, and the combination of difficult and easy.

3. Strengthen the diversity and practical characteristics of the exercises of composite objects, machine part expression and part drawing to cultivate students' engineering design and expression ability.

4. Develop the question types of National Advanced Drawing Competition, such as configuration design and part expression, to cultivate students' divergent thinking and innovative design consciousness.

The chief editors of this workbook are Yingna Liang from Yanshan University and Jingli Lu from Qinhuangdao Vocational and Technical College. The associate editors are Cheng Guo, Xiaoxia Yu, Mingchao Ding and Xinbo Chen, all from Yanshan University. The chief auditors is Zhikui Dong from Yanshan University. The specific division of labor is as follows: Yingna Liang wrote Chapter 1, Chapter 2, Chapter 3, Chapter 8, Chapter 12; Jingli Lu wrote Chapter 10 and Chapter 11; Cheng Guo wrote Chapter 6 and Chapter 7; Xiaoxia Yu wrote Chapter 4 and Chapter 5; Mingchao Ding wrote Chapter 13; Xinbo Chen wrote Chapter 9. The whole workbook was compiled by Yingna Liang and reviewed by Zhikui Dong.

The publication of this workbook is supported by the Yanshan University Excellent Academic Works and Textbook Publishing Fund. In addition, in the process of writing, postgraduate students Zhepeng Zhang, Wei Wang, Ziliang Liu, Jiahao Yang, Huan Wang, Yuqin Guo and Shenghua Gao participated in the drawing work. Hereby express our sincere thanks.

This workbook has referred to some relevant books and textbooks domestic and overseas, and great appreciation is extended to the authors concerned.

Due to various limitations, the workbook inevitably has some mistakes, please criticize and correct.

Editors
August, 2023

前　　言

随着中国在世界上竞争力和影响力的不断提升，开展多层次、宽领域的教育交流与合作，培养具有国际视野、通晓国际规则、能够参与国际事务和竞争的国际化人才，是中国高等教育面临的必然选择。

本习题集是燕山大学中外合作办学模式下，为培养21世纪高素质的创新型人才，立足燕山大学优良的学科专业基础，同时结合西里西亚技术大学优质的教育资源编写而成的，可用作普通高等院校本科机械类和近机械类各专业开展双语教学的教材，也可供其他专业和有关工程技术人员参考。

本习题集根据教育部工程图学教学指导委员会通过的《普通高等院校工程图学教学课程基本要求》，以及最新《技术制图》和《机械制图》国家标准，结合多年来的教学改革实践经验和工程图学发展的新要求编写而成。

本习题集具有以下特点：

1. 中英文双语对照编排，培养学生语言能力和学术交流能力。
2. 题目编写由浅入深，循序渐进，难易结合。
3. 强化组合体、机件表达和零件图习题的多样性和实践性特色，培养学生工程设计和表达能力。
4. 开拓构型设计、零件表达等全国先进成图大赛题型，培养学生发散思维和创新设计意识。

本习题集由燕山大学梁瑛娜、秦皇岛职业技术学院陆景丽任主编；燕山大学郭成、于晓霞、丁明超、陈新博任副主编；燕山大学董志奎任主审。具体分工如下：梁瑛娜编写第1章、第2章、第3章、第8章、第12章；陆景丽编写第10章、第11章；郭成编写第6章、第7章；于晓霞编写第4章、第5章；丁明超编写第13章；陈新博编写第9章。全习题集由梁瑛娜统稿，董志奎审稿。

本习题集的出版获得了燕山大学优秀学术著作及教材出版基金资助。另外，在编写过程中，硕士研究生张喆鹏、王威、刘梓良、杨家豪、王欢、郭雨芹、高升华参加了绘图工作，在此表示诚挚的感谢。

本习题集参考了一些国内外相关著作和教材，在此特向有关作者表示感谢。

由于编者水平有限，本习题集难免存在不足之处，敬请读者批评指正。

编者

2023 年 8 月

Contents 目 录

Contents 目 录

1-1 Font practice in accordance with the national standard on long Song characters.
按照关于长仿宋体字的国家标准规定进行字体练习。

序 其 余 技 术 要 求 未 注 圆 角 铸 件 清 砂 车 铣 磨 刨 钻 轴

套 盘 盖 座 叉 架 箱 体 螺 栓 母 垫 圈 滚 动 轴 承 键 销 齿 轮

0 1 2 3 4 5 6 7 8 9 Φ 0 1 2 3 4 5 6 7 8 9 Φ R M 0 1 2 3 4 5 6 7 8 9

1-2 Linetype practice
线型练习

粗实线

细虚线

细点画线

细实线

细波浪线

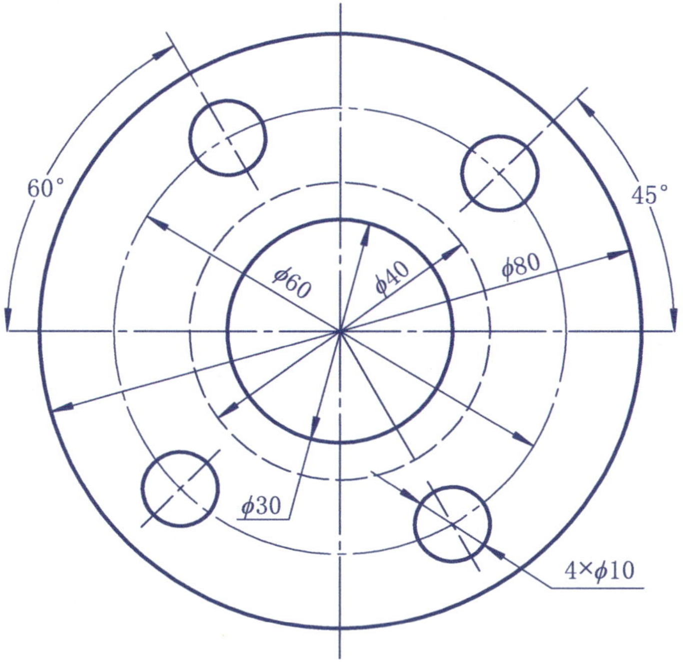

Class 班级：　　Name 姓名：　　ID Number 学号：

1-3 Practice drawing lines, slope, and taper. Draw the following figures at a scale of 1∶1 according to the size shown.
图线、斜度、锥度练习。按图示尺寸以1∶1比例画出下列图形。

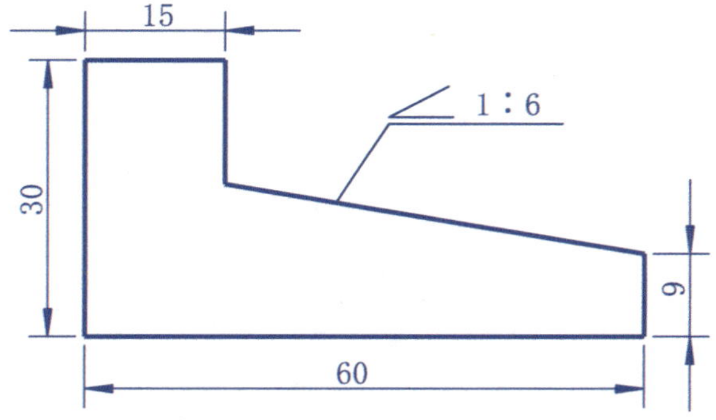

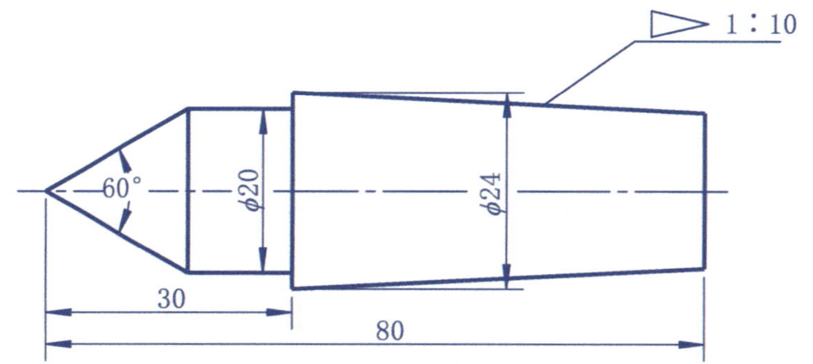

1-4 Practice drawing straight lines and arcs. Draw the following figures at a scale of 1∶1 according to the size shown.
直线、弧线练习。按图示尺寸以1∶1比例画出下列图形。

(1) Clevis joint
拖钩

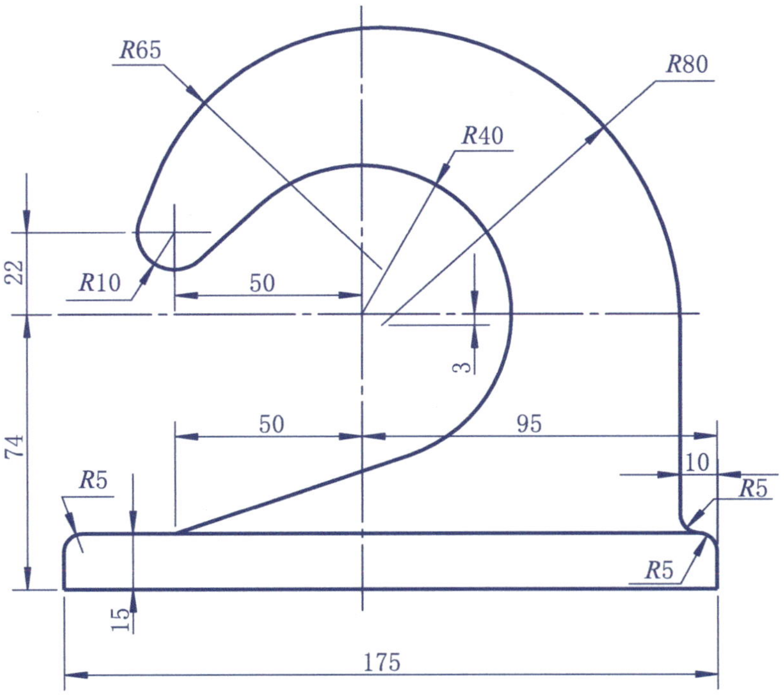

(2) Load hook
吊钩

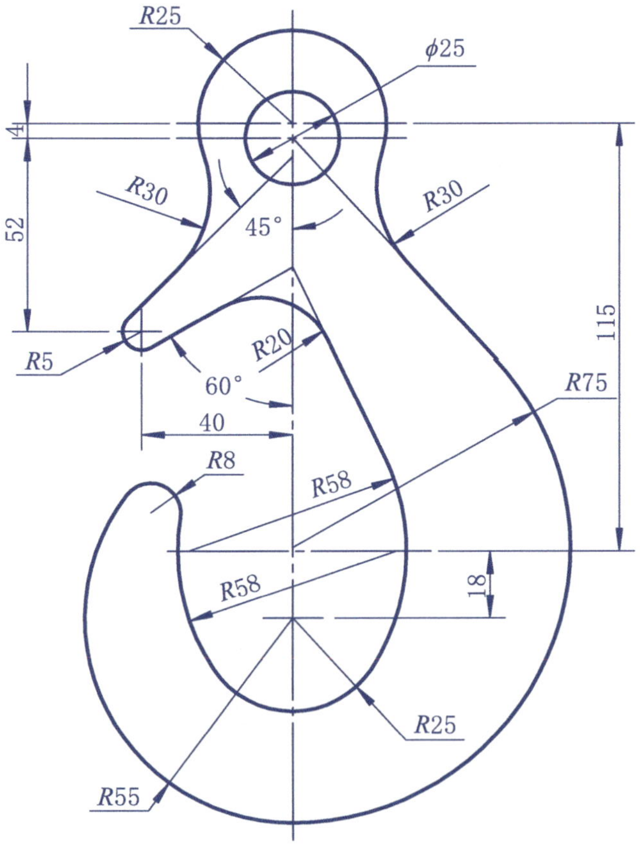

 Class 班级： Name 姓名： ID Number 学号：

1-5 Dimensioning exercise: Fill in the dimensions in the following figures and the numbers are measured from the figures (in mm, rounded).

尺寸标注练习：填注下列图形中的尺寸，数字从图中量取（以mm为单位，取整）。

（1） Linear dimension 线性尺寸

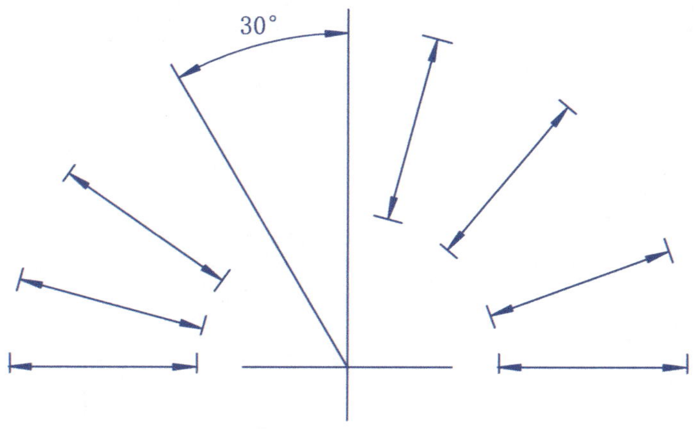

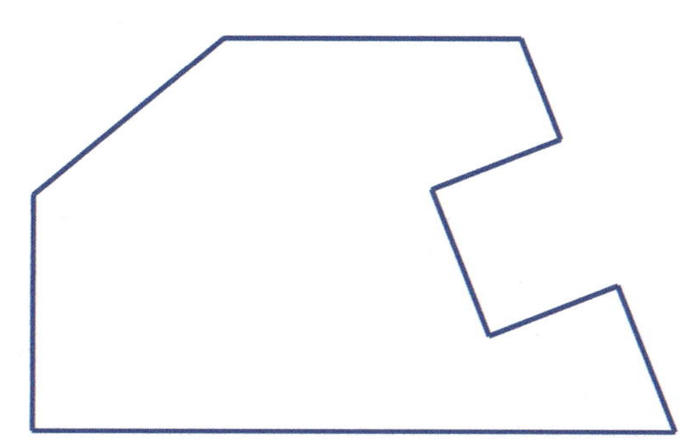

（2）The diameter of the circle and the radius of the circle

圆的直径和圆弧的半径

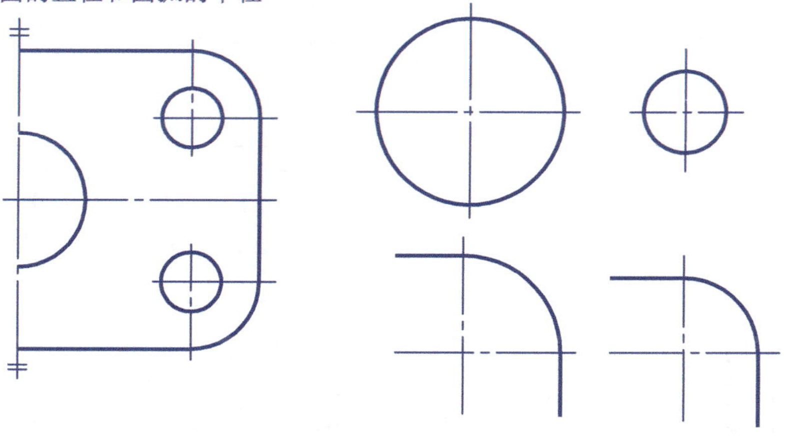

（3）Angular dimension

角度尺寸

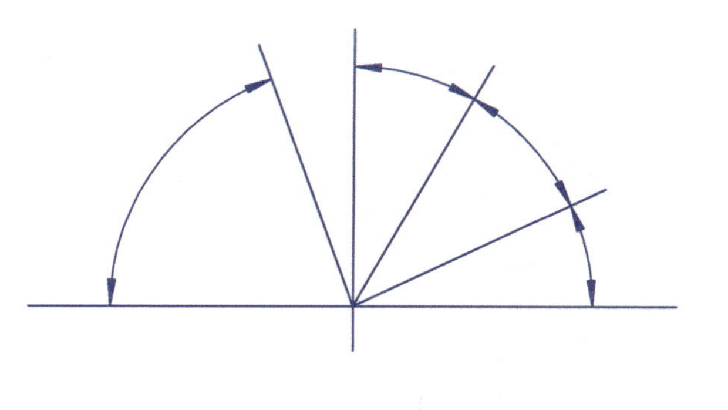

Indicate the axonometric projection to the three views. 指出三视图所对应的立体图。

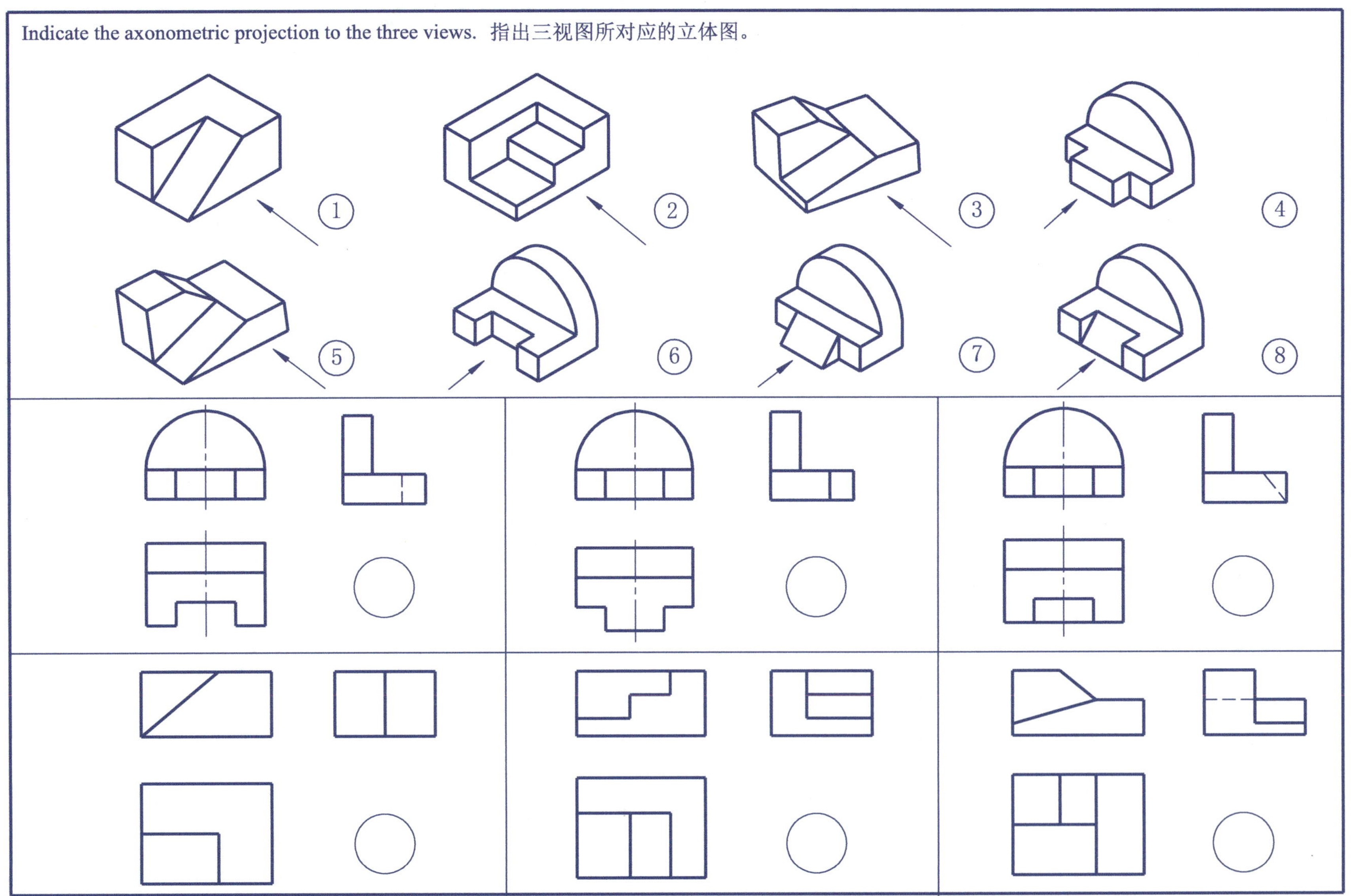

Class 班级： Name 姓名： ID Number 学号：

3-1 Given the spatial position of each point, draw its three projections (The size is measured in the stereogram and rounded).
已知各点的空间位置，画出其三面投影图（尺寸由立体图量取，并取整）。

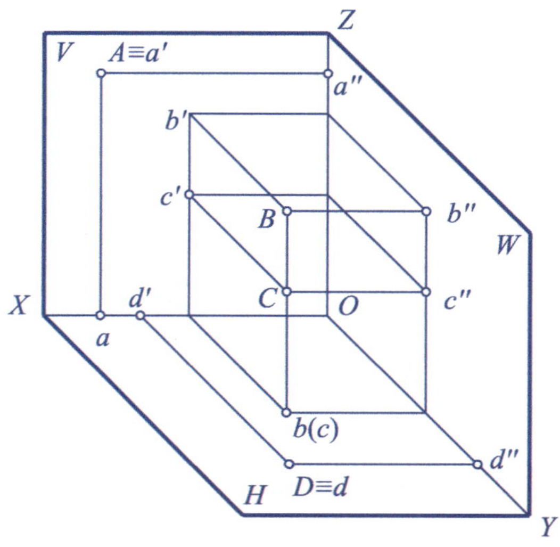

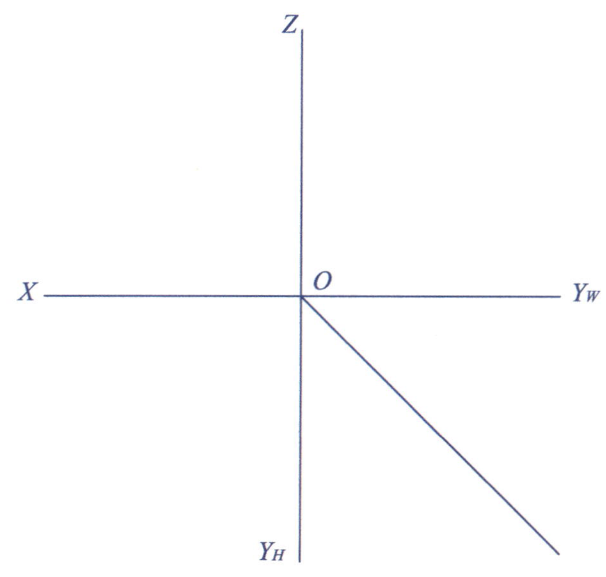

Point A is on the ____ plane.
Point D is on the ____ plane.
Points B and C are coincident points relative to _____ plane.
点A在____面上。
点D在____面上。
点B与C是相对于____面的重影点。

3-2 Given the two projections of each point, draw the third projection.
已知各点的两面投影，画出其第三面投影。

3-3 Given points *A* (40, 30, 35), *B* (25, 0, 20), *C* (0, 20, 15), draw the three projections of points *A*, *B* and *C*.
已知点*A* (40, 30, 35)，*B* (25, 0, 20)，*C* (0, 20, 15)，求作*A*、*B*、*C*三点的三面投影。

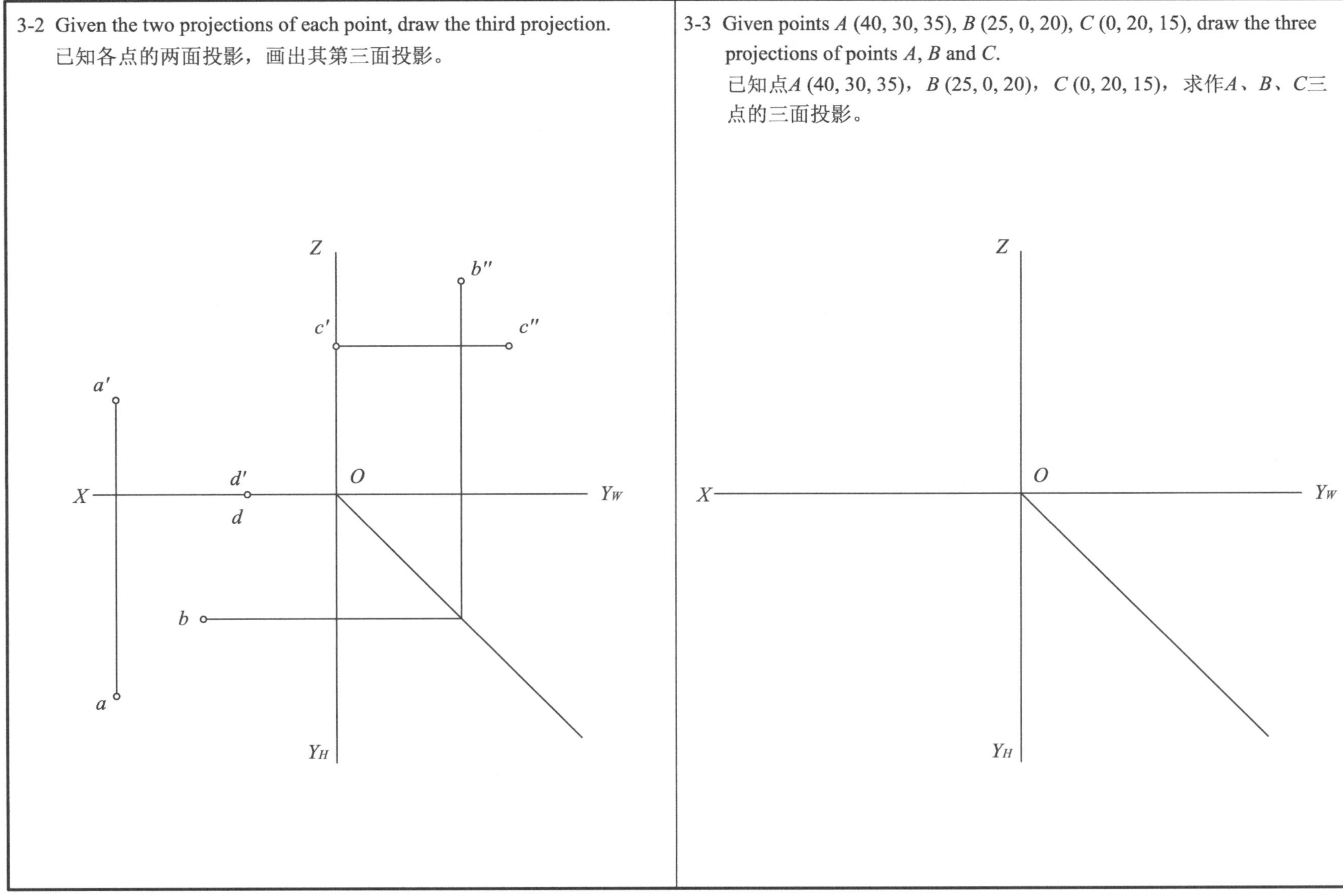

 Class 班级： Name 姓名： ID Number 学号：

3-4 Given point A (10, 10, 15), point B is 25, 18 and 5 away from the projection plane W, V and H, and point C is 15mm left to point A, 15mm in front of point A and 10mm above point A, draw the three projections of the three points A, B and C.

已知点A (10, 10, 15)，点B距离投影面W、V、H分别为25、18、5，点C在点A左方15mm，前方15mm，上方10mm，作出A、B、C三点的三面投影。

Z

X

O

Y_W

Y_H

3-5 Given the frontal projection of points A and B, the distance from point A to plane H and plane V is equal, and the distance from point B to plane V and plane W is equal. Find the other two projections of points A and B.

已知点A和点B的正面投影，点A到H面、V面的距离相等，点B到V面、W面的距离相等，求点A和点B的其他两面投影。

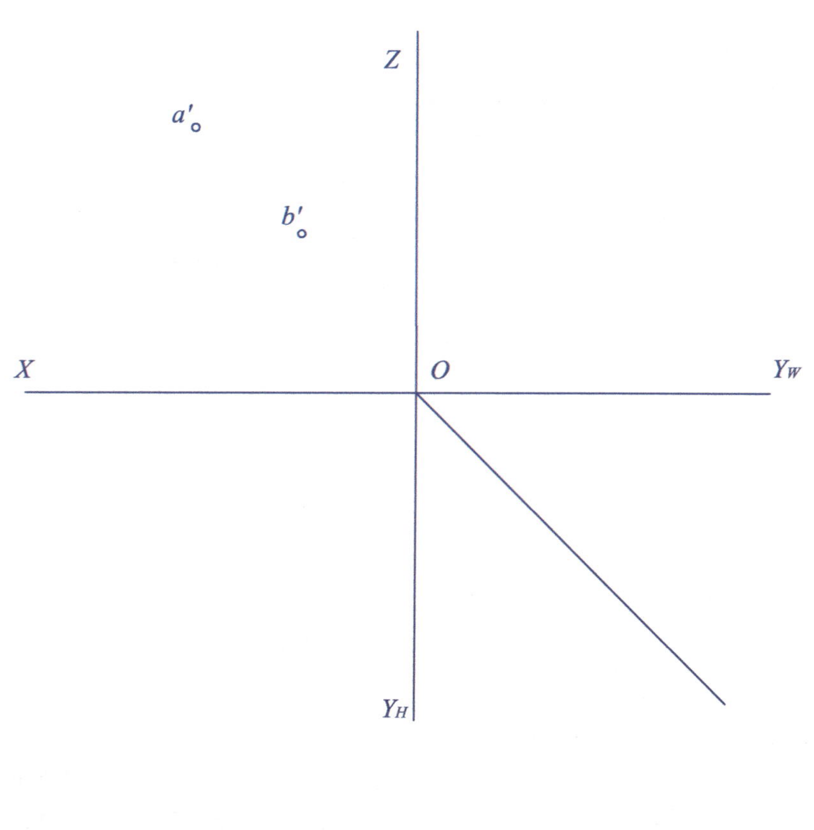

3-6 Draw the third projection of each of the following lines and determine the relative position to the projection plane.
画出下列各直线的第三投影，并判断对投影面的相对位置。

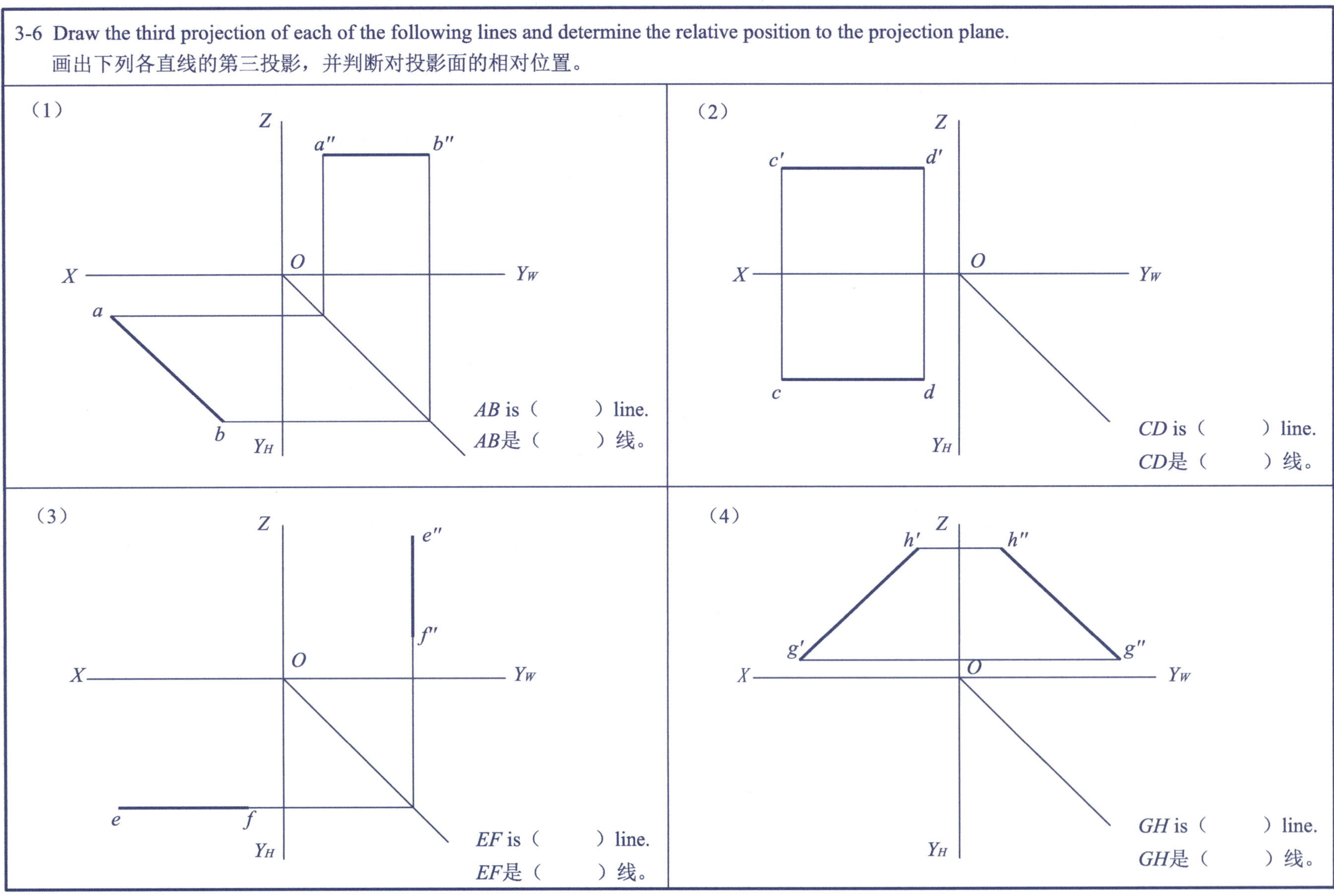

 Class 班级： Name 姓名： ID Number 学号：

3-7 Draw a line through point A, satisfying the following conditions. (Think: How many solutions to each problem? Only draw one solution.)
过点A作线段，使其满足下列各条件。（思考：下列各题有几解，只作一个解。）

（1）Draw a horizontal line AB, making AB=30mm and β=30°.
作水平线AB，使AB=30mm，β=30°。

（2）Draw a frontal line AB, making AB=35mm and γ=45°.
作正平线AB，使AB=35mm，γ=45°。

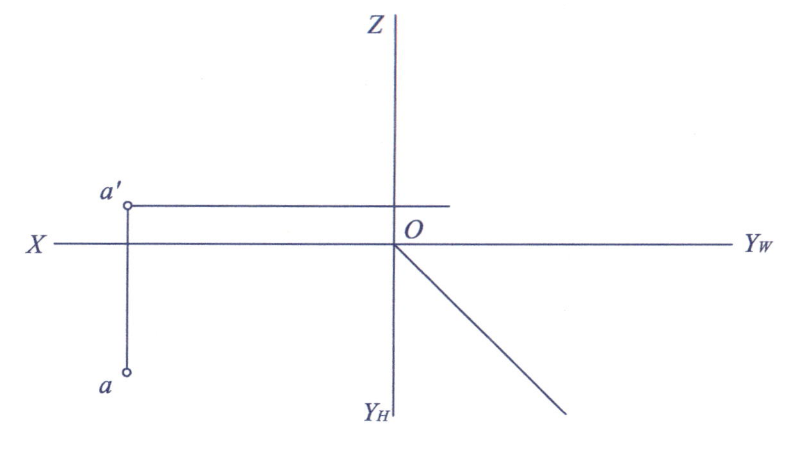

（3）Draw a profile line AB, making AB=25mm and α=60°.
作侧平线AB，使AB=25mm，α=60°。

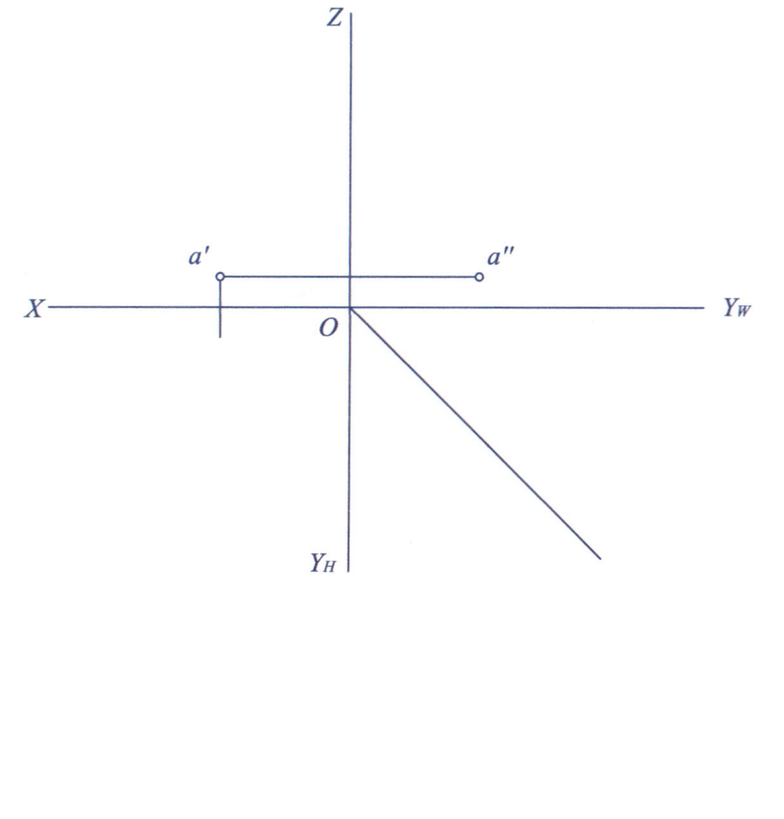

3-8 Find the projection of point *C* on the given line *AB*, making that *AC*∶*CB*=2∶3.

在已知线段*AB*上求一点*C*，使*AC*∶*CB*=2∶3, 求出点*C*的投影。

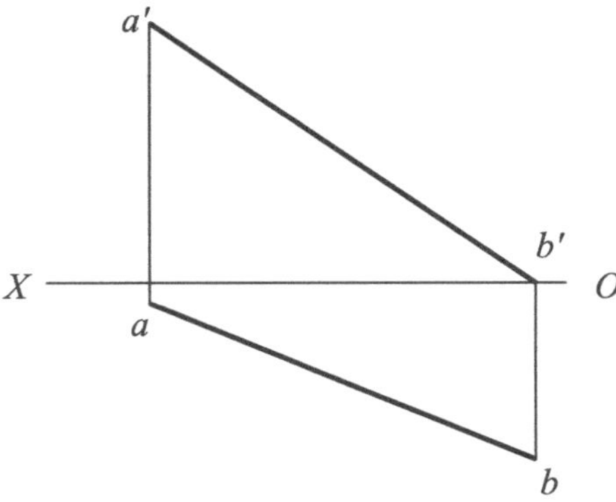

3-9 The two lines *AB* and *CD* intersect, draw another projection of line *CD*, and mark the projection of the intersection point.

*AB*和*CD*两直线相交，画出直线*CD*的另一投影，并标出交点的投影。

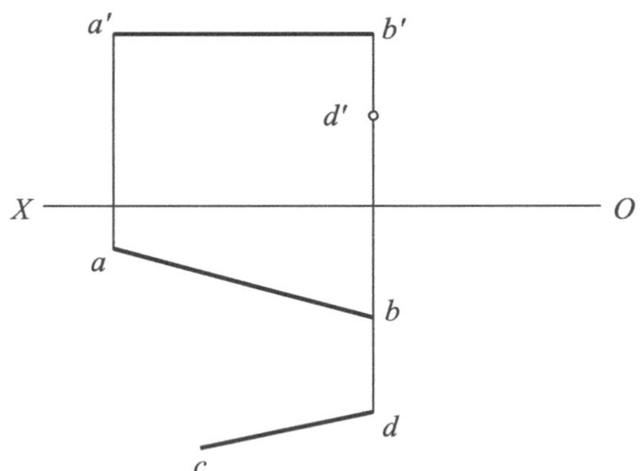

 Class 班级： Name 姓名： ID Number 学号：

3-10 Determine the relative position of two lines *AB* and *CD* (parallel, intersect or cross).
判断两直线*AB*与*CD*的相对位置（平行、相交、交叉）。

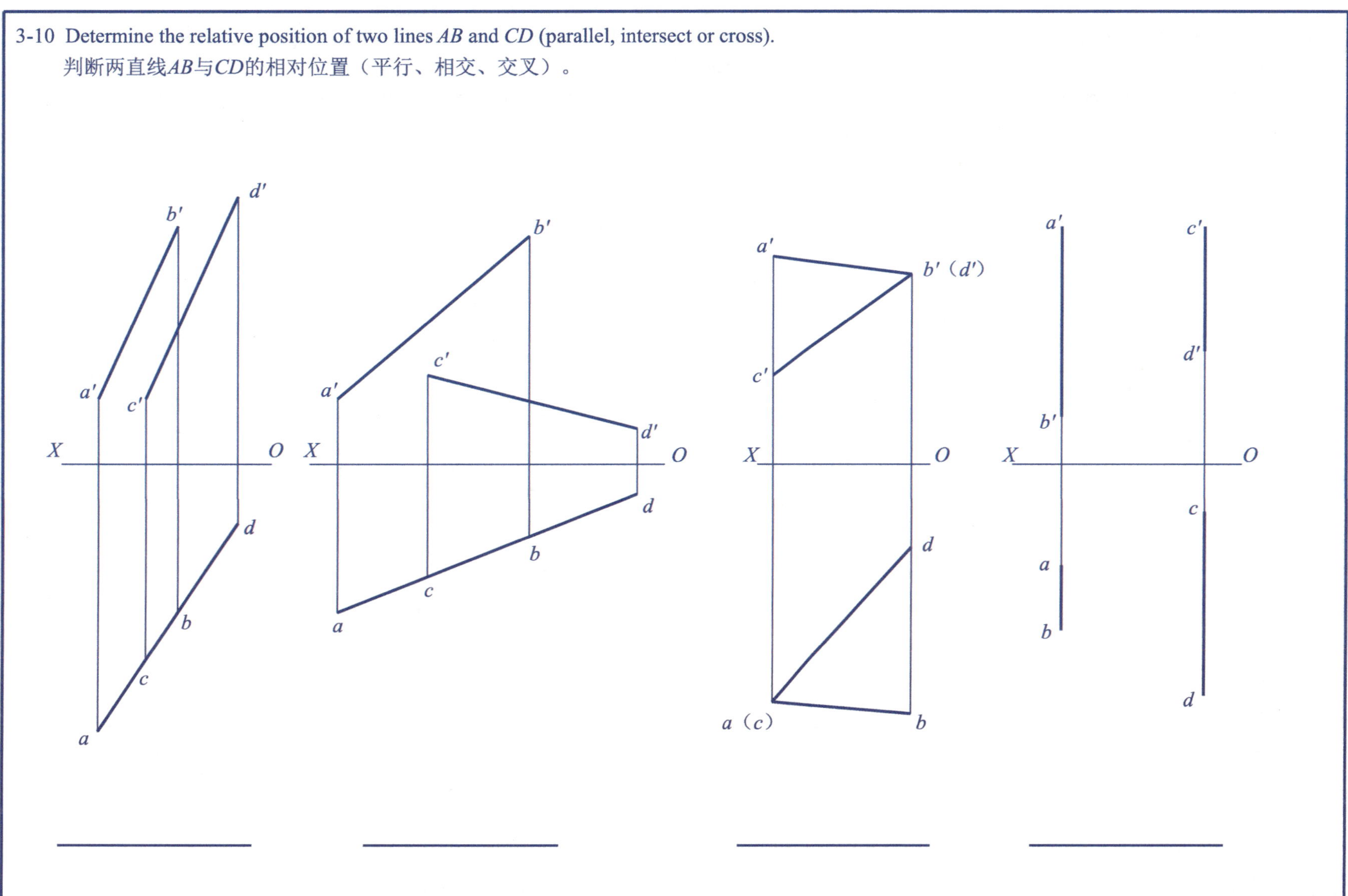

Class 班级： Name 姓名： ID Number 学号：

3-11 Draw a line *CD* through point *C*, parallel to line *AB*, and intersecting line *EF* at point *D*. Find the frontal projection and horizontal projection of *CD* and the frontal projection of *EF*.

过*C*点作一直线*CD*平行于直线*AB*，且与直线*EF*相交于点*D*，求*CD*的正面、水平投影及*EF*的正面投影。

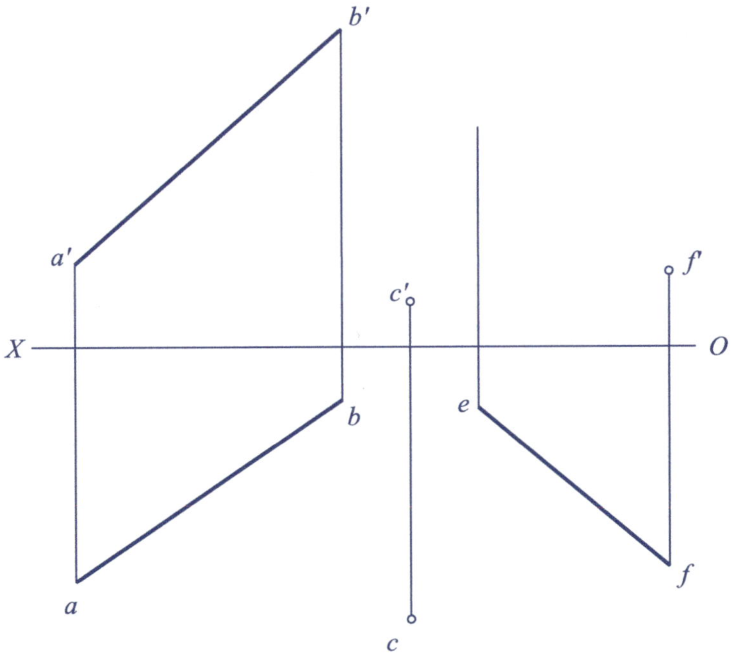

3-12 Draw a line intersecting with two lines *AB* and *CD* and parallel to line *EF*.

作一直线使其与两直线*AB*、*CD*相交，且平行于直线*EF*。

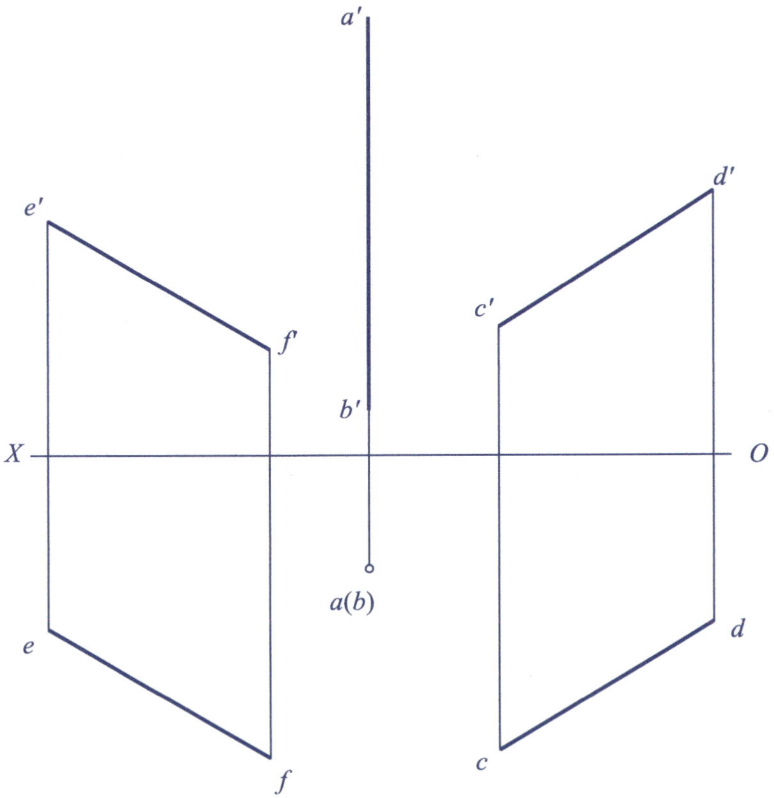

3-13 Given lines *AB*, *CD* and *EF*, draw a frontal line *MN* which intersects with lines *AB*, *CD* and *EF* at points *M*, *S* and *T* respectively, and point *N* is 10mm above plane *H*.
已知直线*AB*、*CD*、*EF*。作正平线*MN*，与*AB*、*CD*、*EF*分别交于点*M*、*S*、*T*，点*N*在*H*面之上10mm。

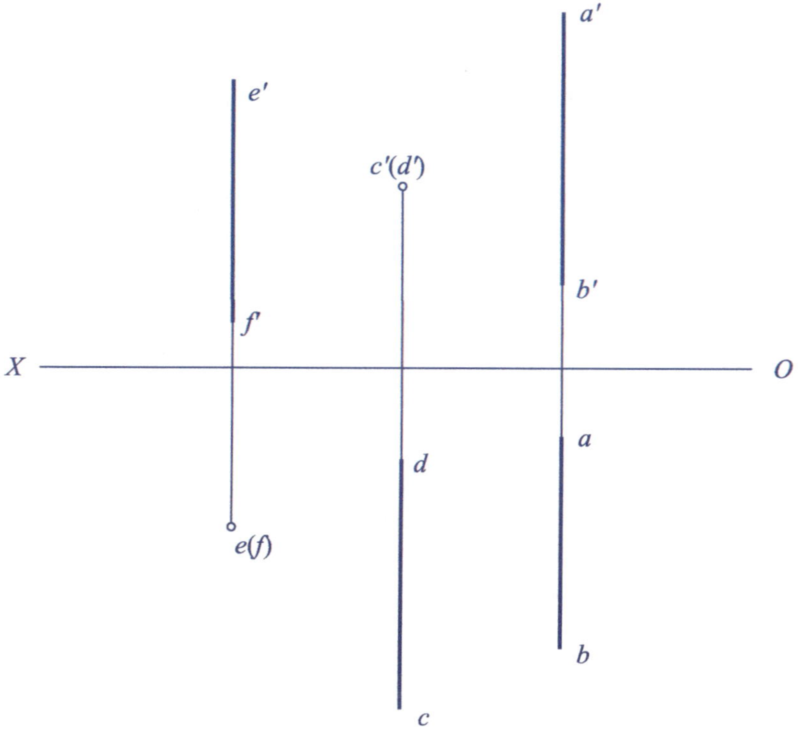

a′
e′
c′(d′)
b′
f′
X
O
a
d
e(f)
b
c

3-14 Draw a line *EF* through point *E* and intersecting with lines *AB* and *CD*.
过已知点*E*作直线*EF*与直线*AB*、*CD*都相交。

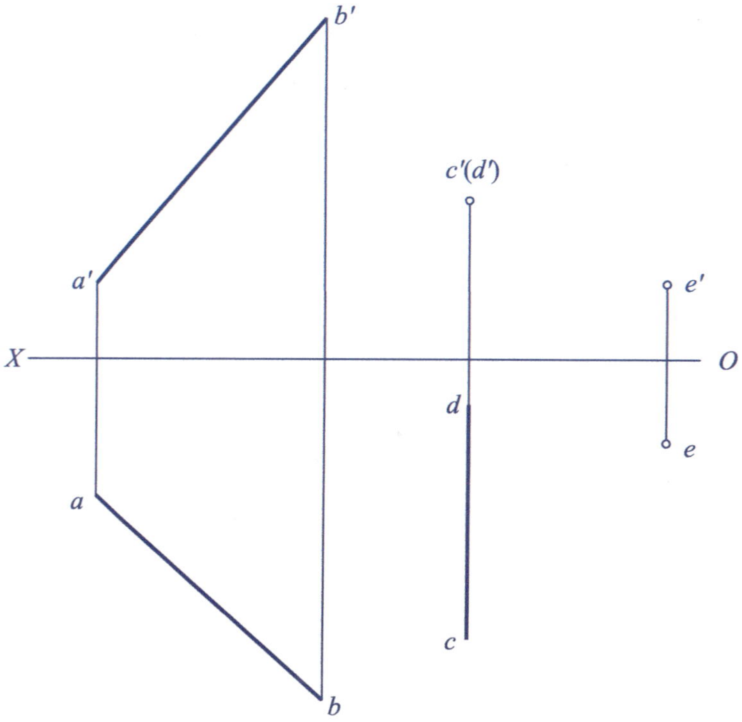

b′
c′(d′)
a′
e′
X
O
d
e
a
c
b

3-15 Draw a horizontal line MN that intersects with the two lines AB and CD and the distance from plane H is 20mm.

作水平线MN，使其与两已知直线AB、CD都相交，且距H面的距离为20mm。

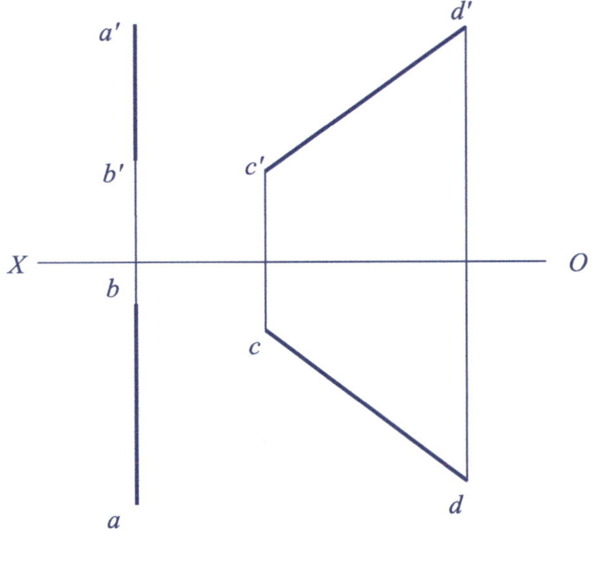

*3-16 Find the real length and the dip angles α and β with plane H and plane V of line AB.

求线段AB的实长及其与H、V面的倾角α、β。

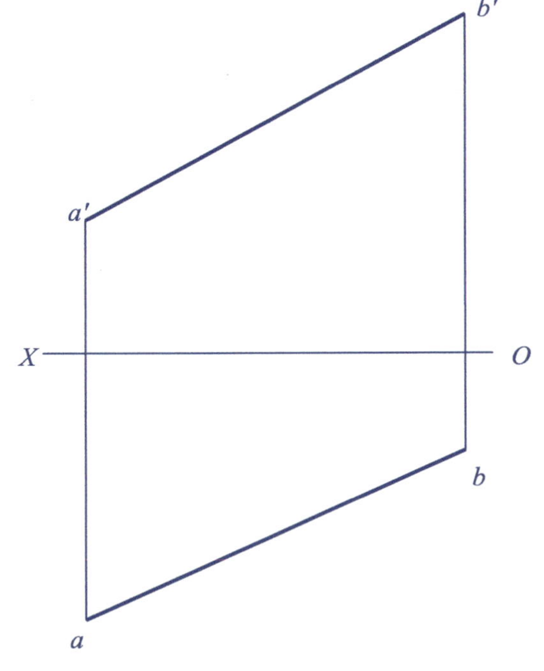

* 3-17 Given that the angle β between line AB and plane V is 30°, find its horizontal projection.

已知线段AB与V面的倾角β=30°，求其水平投影。

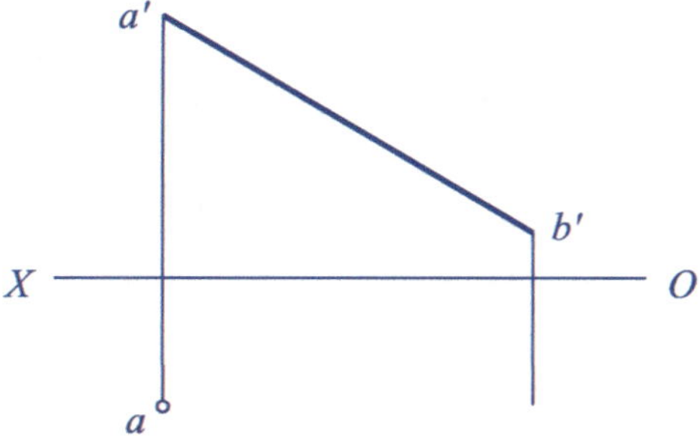

* 3-18 Given that point C is on line AB and AC=25mm, draw the two projections of point C.

已知点C在直线AB上，且AC=25mm，作出点C的两面投影。

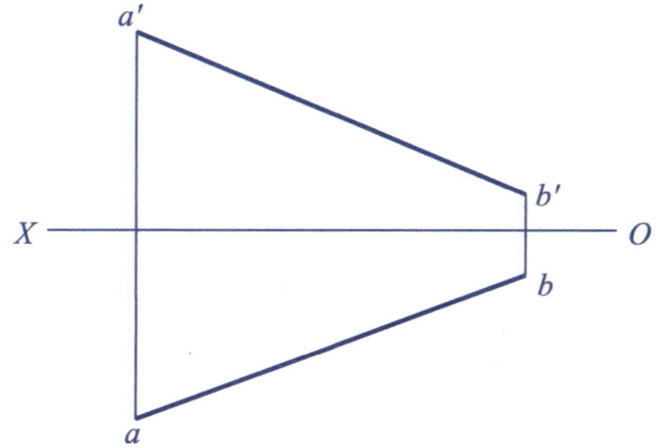

*3-19 Known △*ABC*, try to find the projection of angle bisector *AD* of ∠ *BAC*.

已知△*ABC*，试求∠*BAC*上的角平分线*AD*的投影。

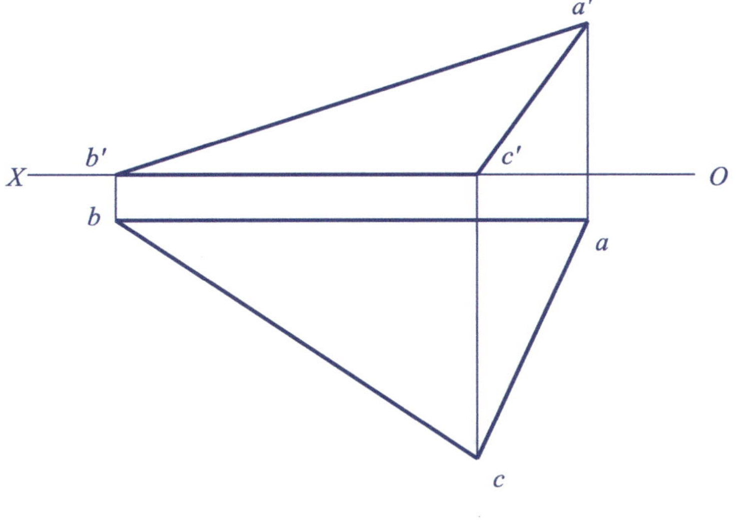

*3-20 Given a diamond *ABCD* and its diagonal *AC*, try to complete the two projections of the diamond.

已知菱形*ABCD*，*AC*为一对角线，试完成该菱形的两面投影。

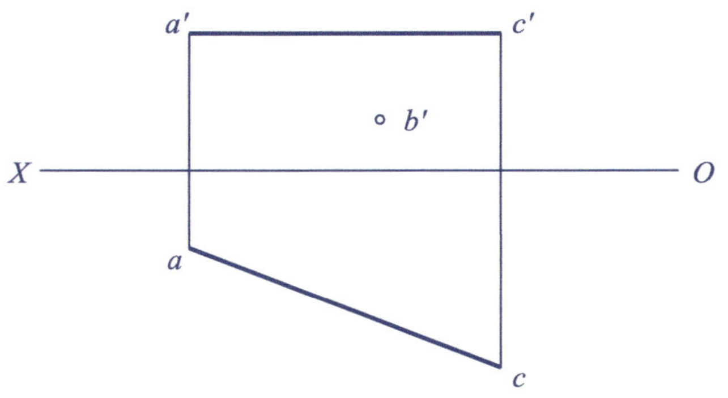

*3-21 The horizontal line AD is the height of side BC of equilateral $\triangle ABC$. Point C is in plane H. Find the two projections of equilateral $\triangle ABC$.

水平线AD是等边$\triangle ABC$的边BC上的高，点C在H面内，求等边$\triangle ABC$的两面投影。

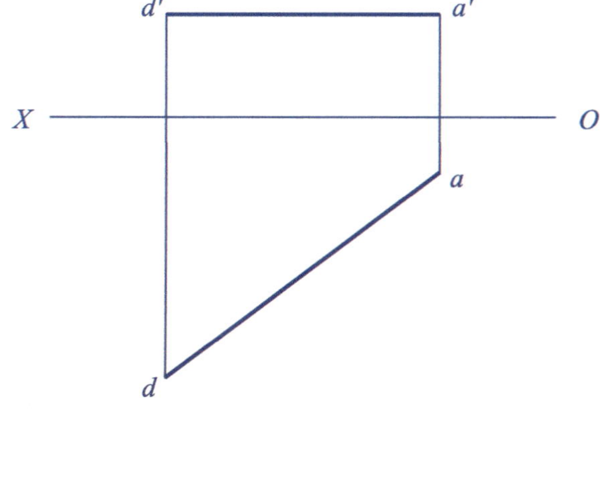

*3-22 Given that the shortest distance of lines AB and CD is 20mm, find $c'd'$ and the two projections of the shortest distance (common perpendicular).

已知直线AB、CD的最短距离为20mm，求$c'd'$及最短距离（公垂线）的两面投影。

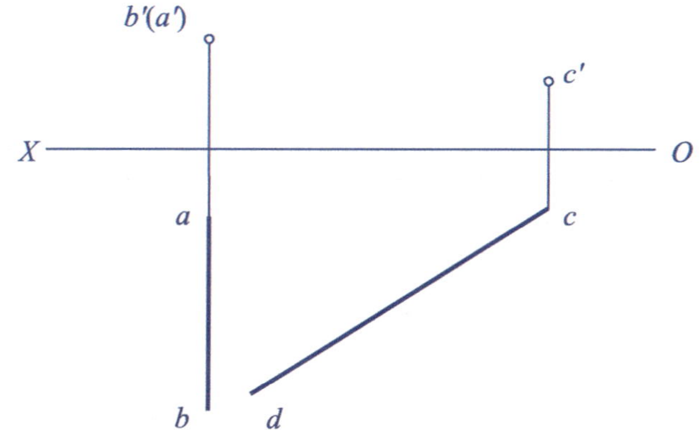

3-23 Draw the three projections of the plane with *AB* as one side.
以*AB*为一边作平面的三面投影图。

（1） Make equilateral △*ABC* as a horizontal plane.
作等边△*ABC*为水平面。

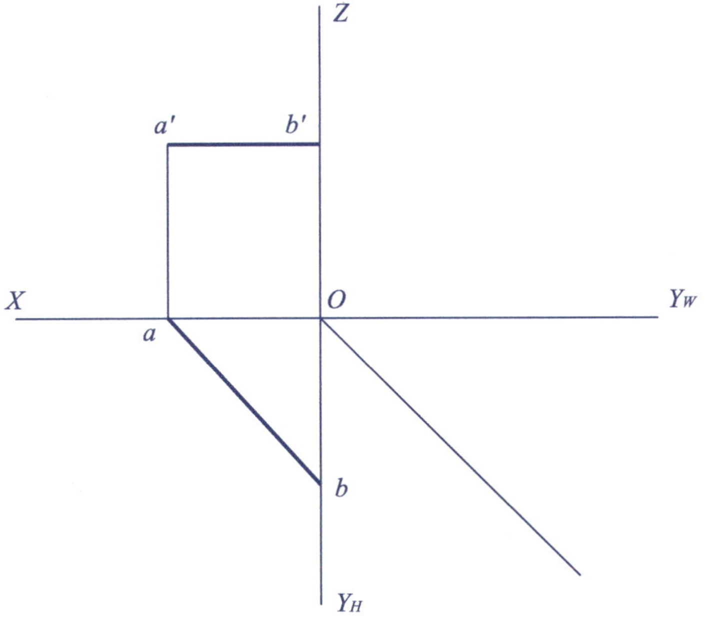

Z
a′ *b′*
X *O* *Yw*
a
b
YH

（2） Make square *ABCD* as a *H*-perpendicular plane.
作正方形*ABCD* 为铅垂面。

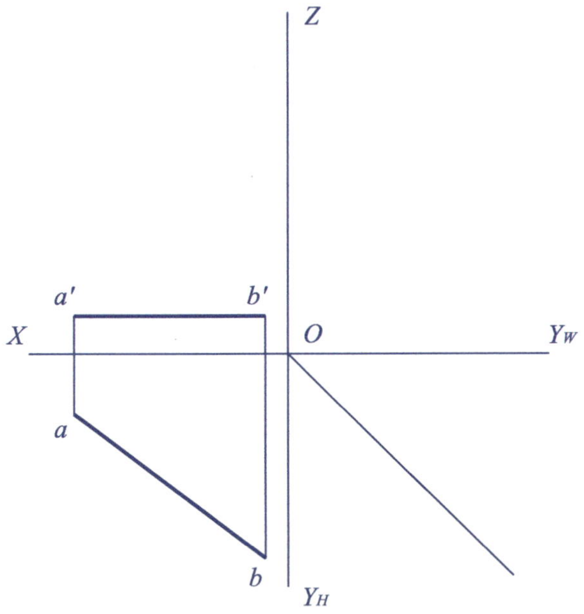

Z
a′ *b′*
X *O* *Yw*
a
b *YH*

 Class 班级： Name 姓名： ID Number 学号：

3-24 Determine what position plane the following planes are.

判断下列平面是什么位置平面。

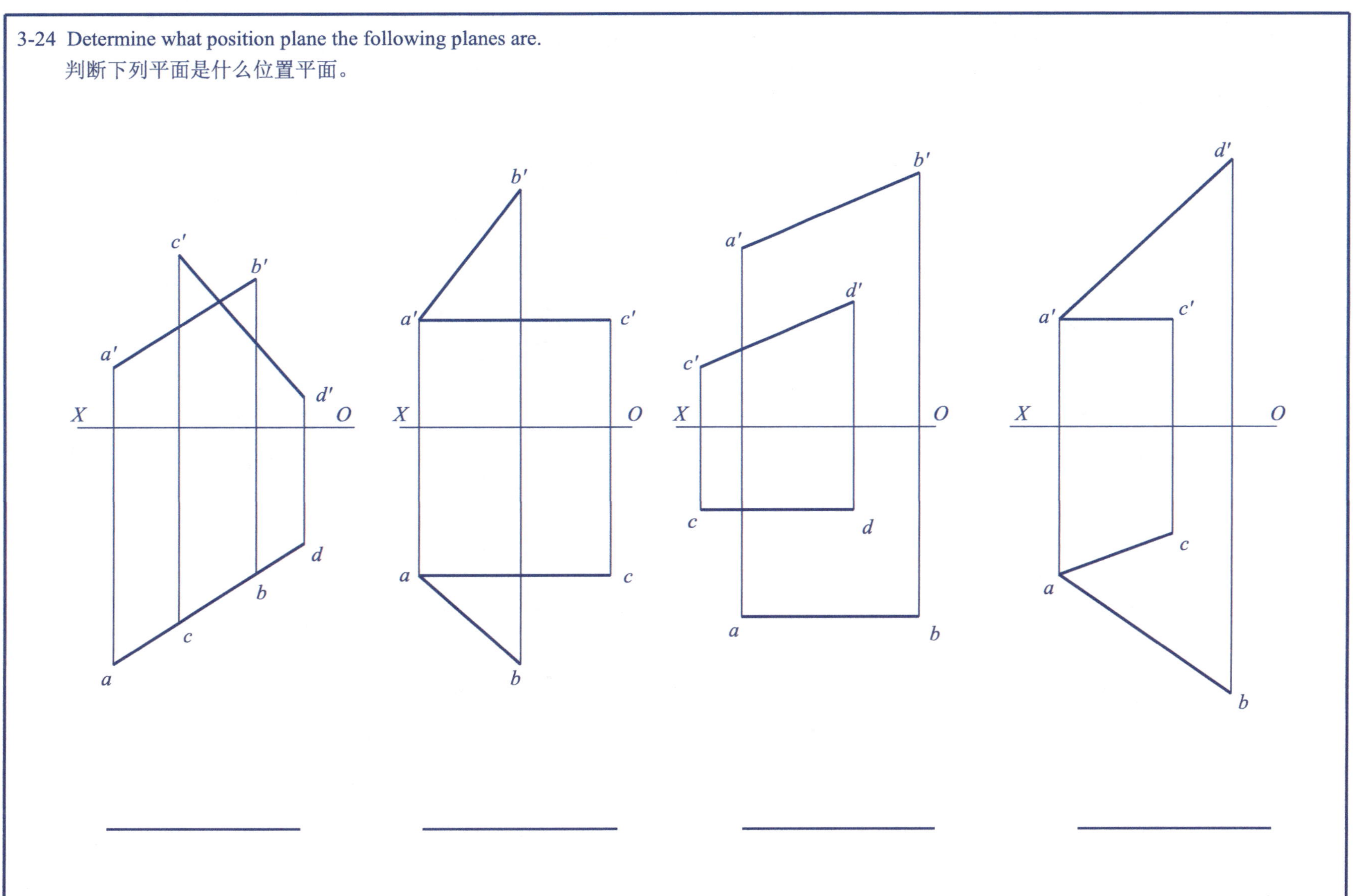

3-25 Draw the two projections of a frontal line *MN* in the plane △*ABC*, making that the distance between *MN* and plane *V* is 25mm.
在△*ABC*平面内求作一条正平线*MN*的两面投影，使*MN*到*V*面的距离为25mm。

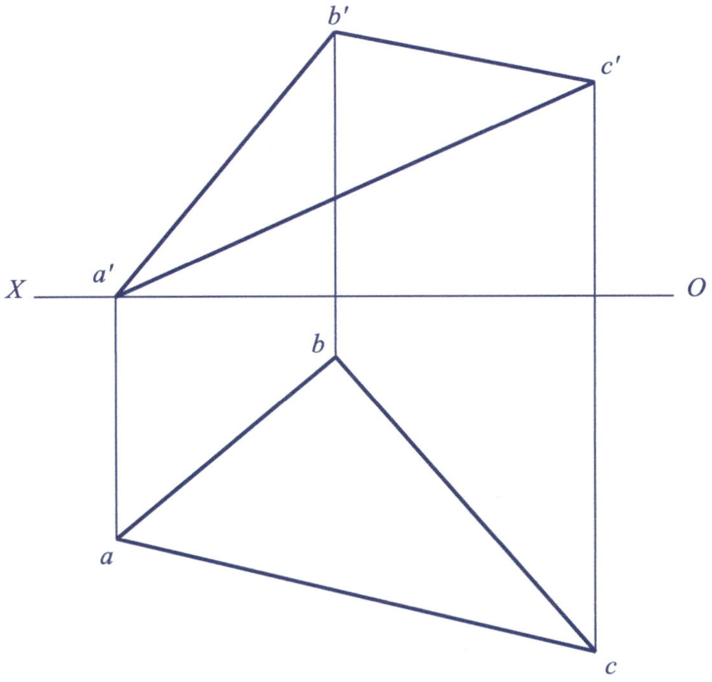

b′
c′
X
a′
O
b
a
c

3-26 Complete the horizontal projection of the planar pentagon.
完成平面五边形的水平投影。

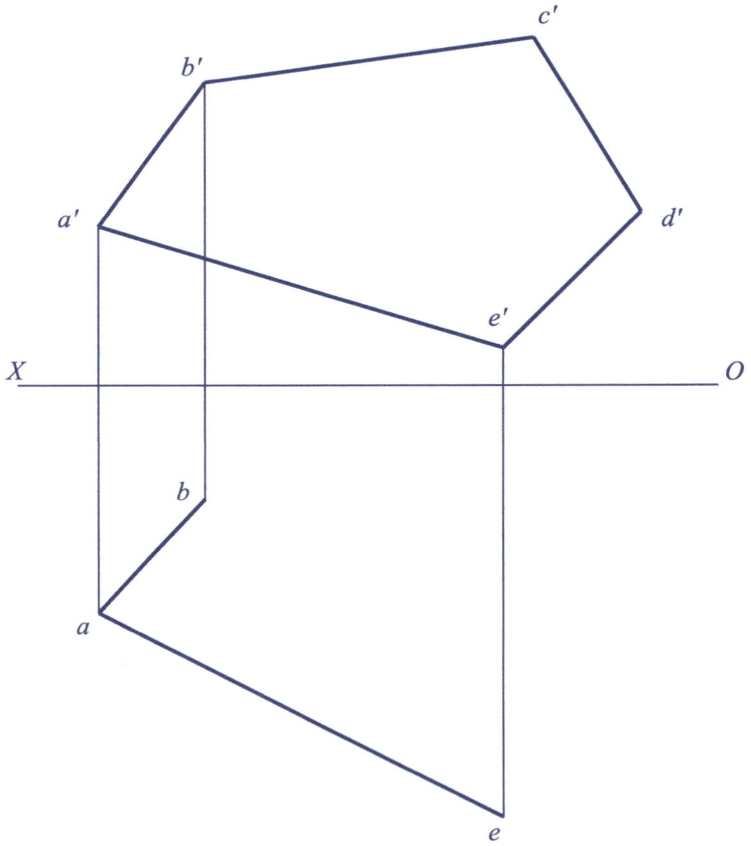

c′
b′
a′
d′
e′
X
O
b
a
e

 Class 班级： Name 姓名： ID Number 学号：

3-27 Known that the *CD* edge of a planar quadrilateral *ABCD* is a horizontal line, complete the frontal projection of *ABCD*.
已知平面四边形*ABCD*，*CD*边为水平线，完成*ABCD*的正面投影。

3-28 In planar quadrilateral *ABCD*, point *D* is in plane *V* and 35mm away from plane *H*. Complete the projection of quadrilateral *ABCD*.
平面四边形*ABCD*中，点*D*在*V*面内，距*H*面为35mm，完成四边形*ABCD*的投影图。

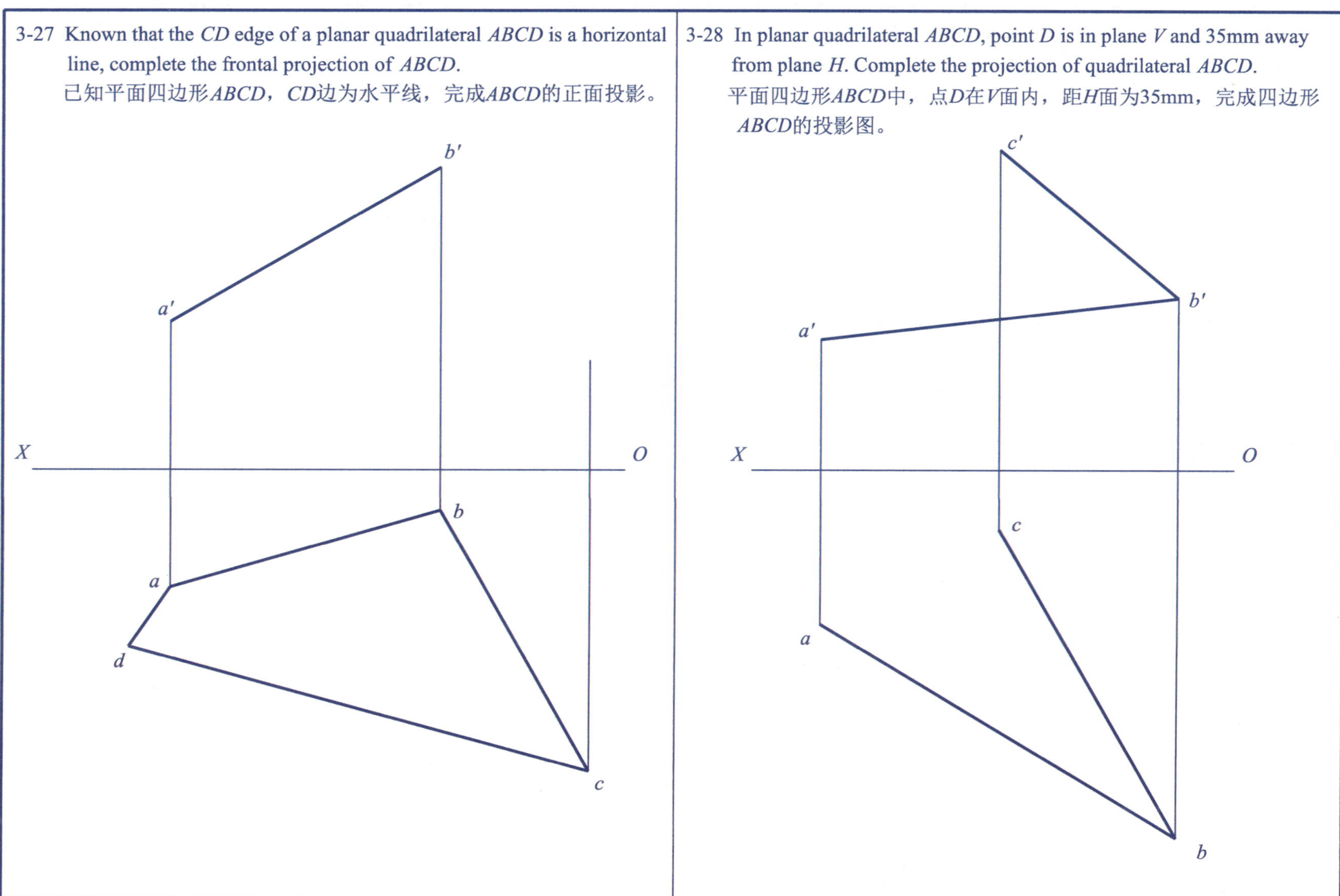

*3-29 Please draw to show whether the ball goes into the net.
请作图说明小球是否进入球网。

*3-30 Find the intersection point of the line and the plane and judge the visibility.
求作直线与平面的交点，并判断可见性。

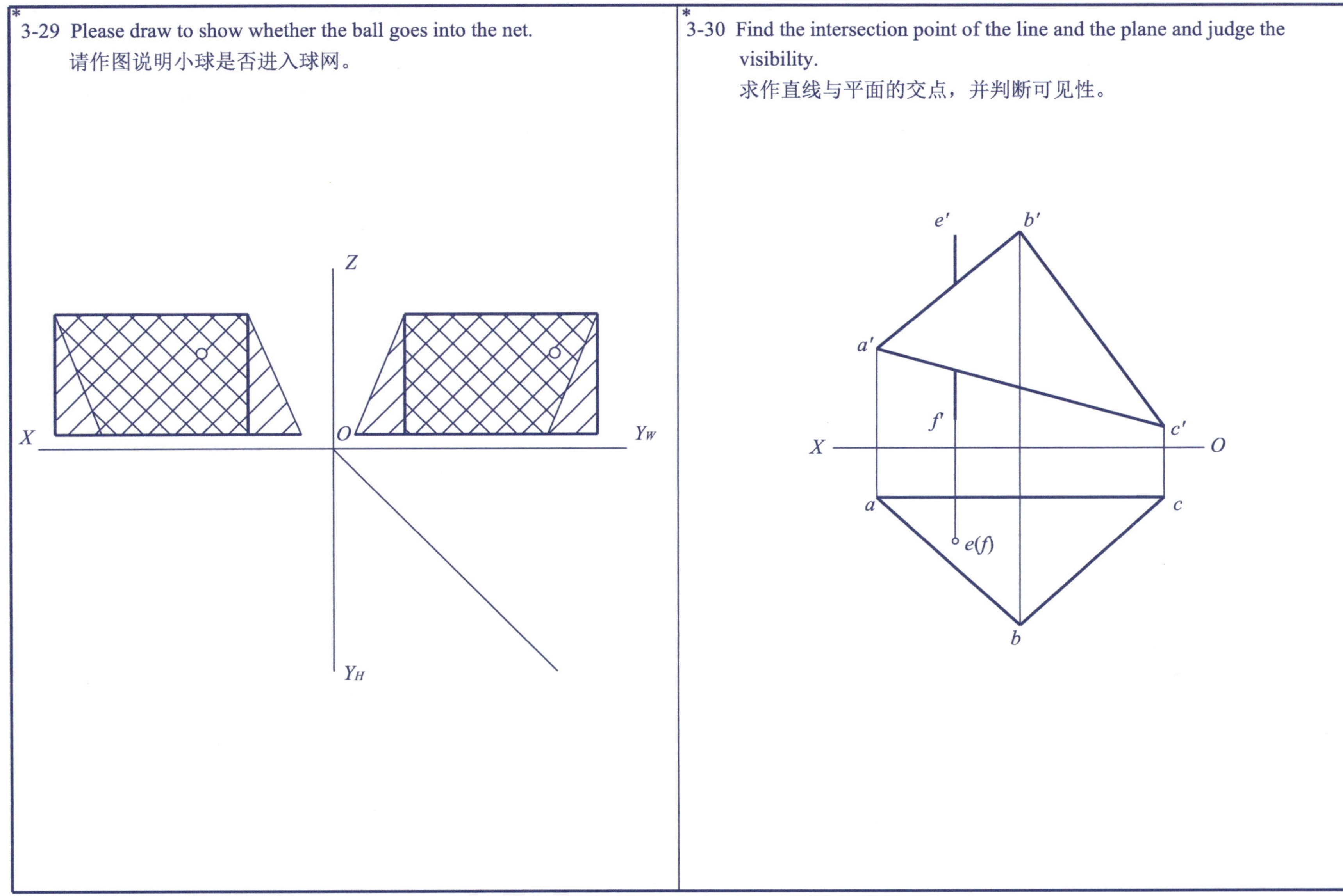

Class 班级： Name 姓名： ID Number 学号：

*3-31 Find the intersection point of the line and the plane and judge the visibility.
求作直线和平面的交点，并判断可见性。

*3-32 Find the intersection line of two planes and judge the visibility.
求作两平面的交线，并判断可见性。

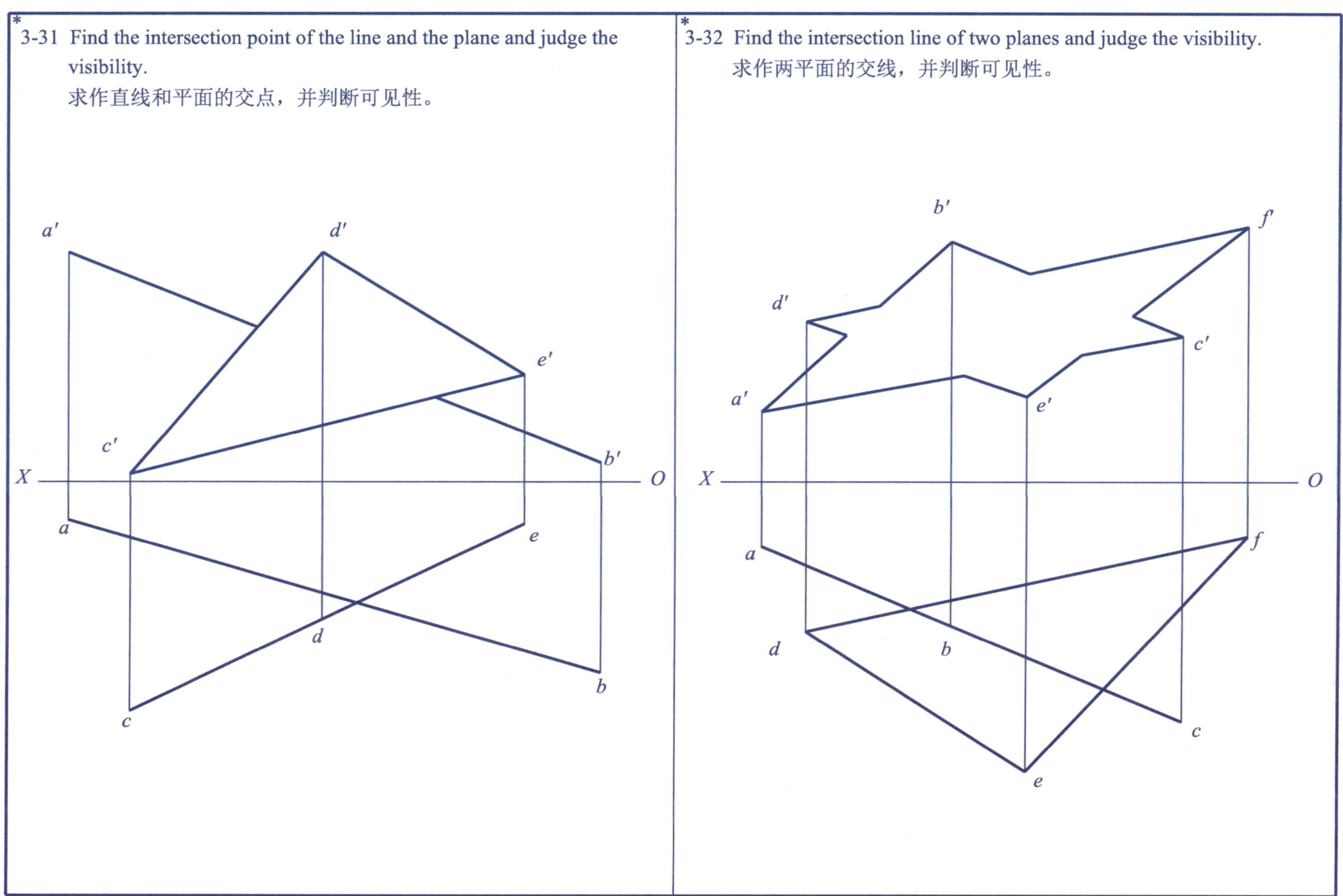

Class 班级： Name 姓名： ID Number 学号：

*3-33 Find the intersection line of two planes and judge the visibility.
求作两平面的交线，并判断可见性。

*3-34 Find the intersection point of the line and the plane and judge the visibility.
求作直线与平面的交点，并判断可见性。

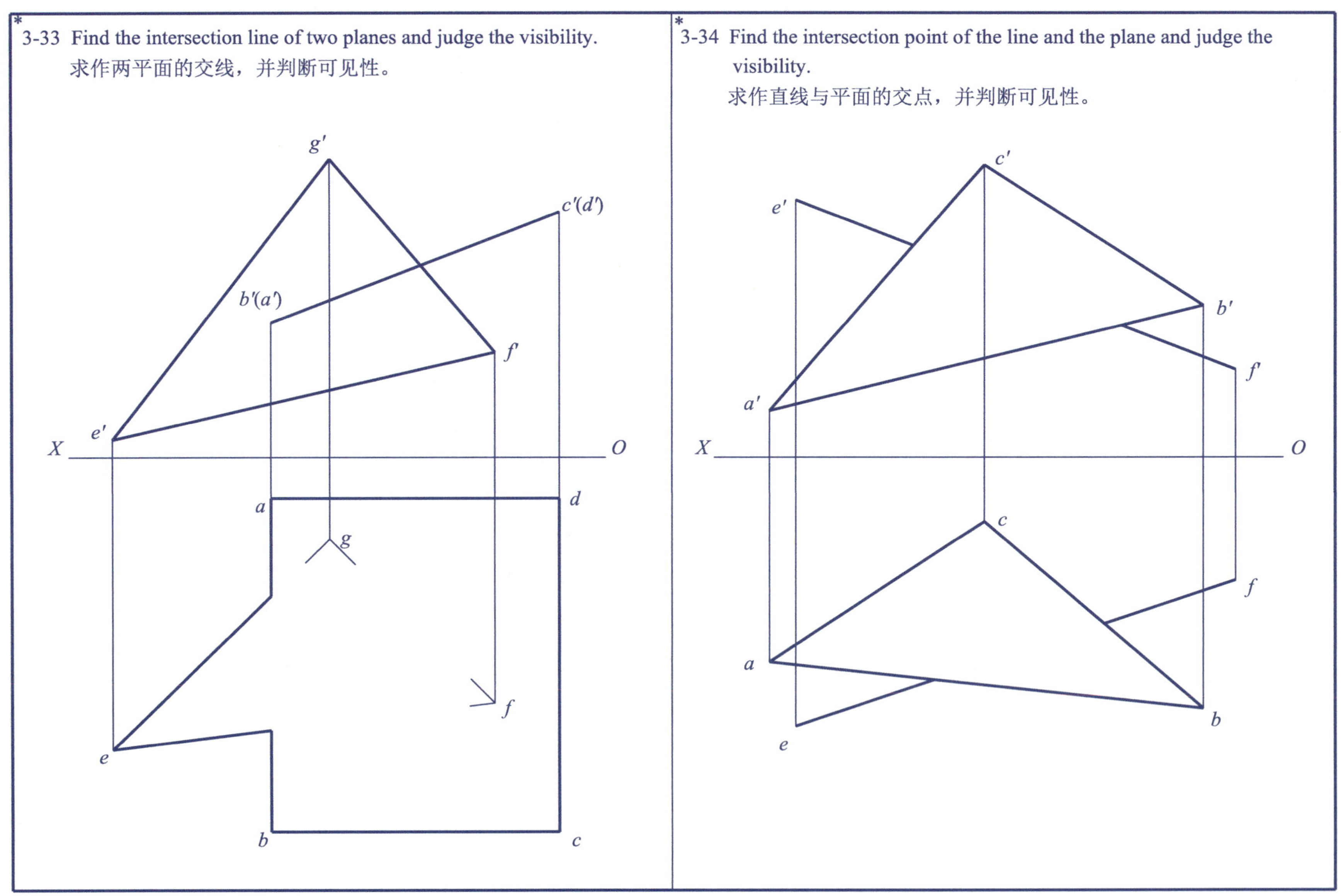

* 3-35 Find the intersection point of the line and the plane and judge the visibility.
求作直线与平面的交点，并判断可见性。

* 3-36 Find the intersection line of two planes and judge the visibility.
求作两平面的交线，并判断可见性。

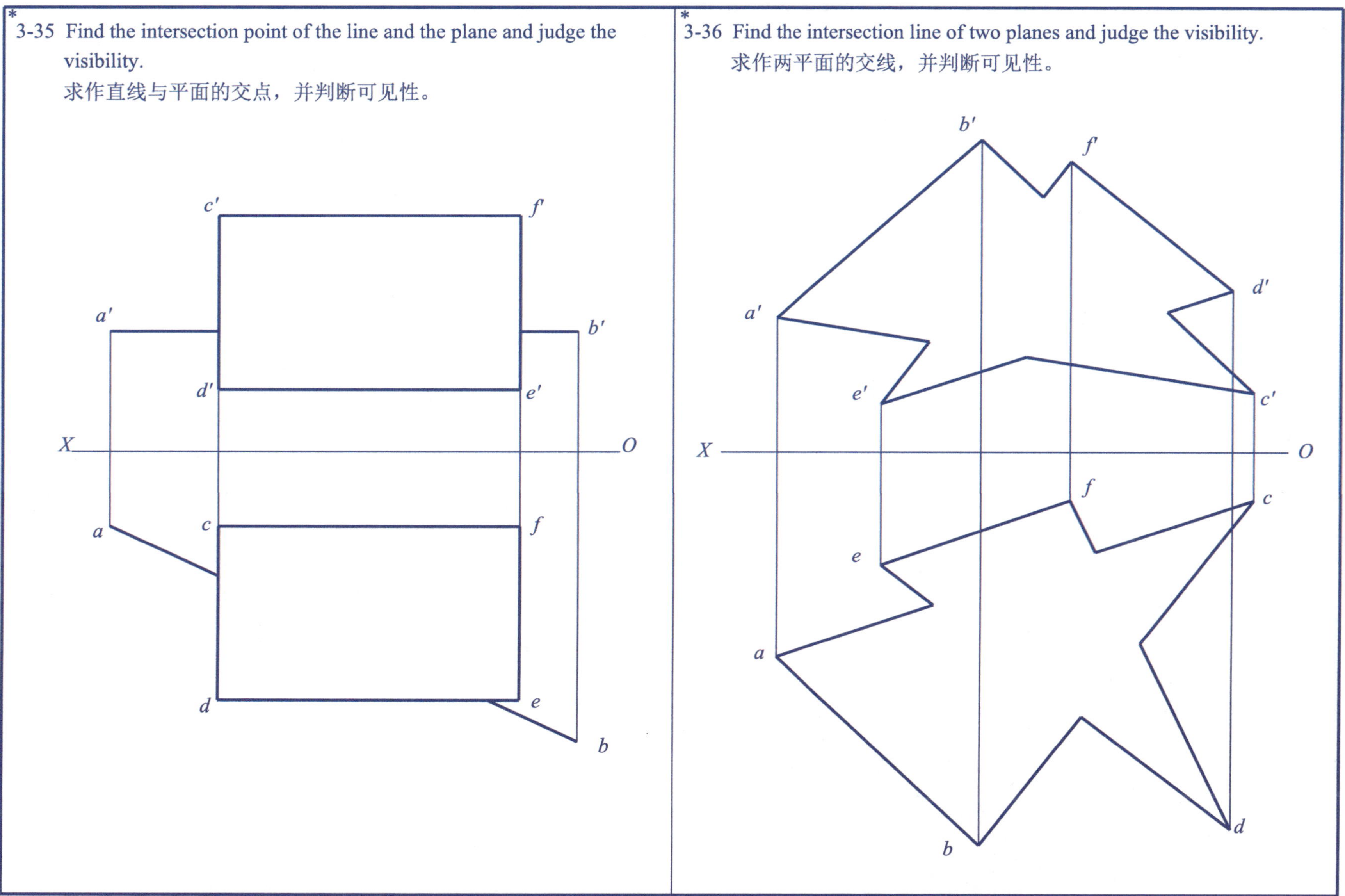

*3-37 Find the intersection line of two planes.
求作两平面的交线。

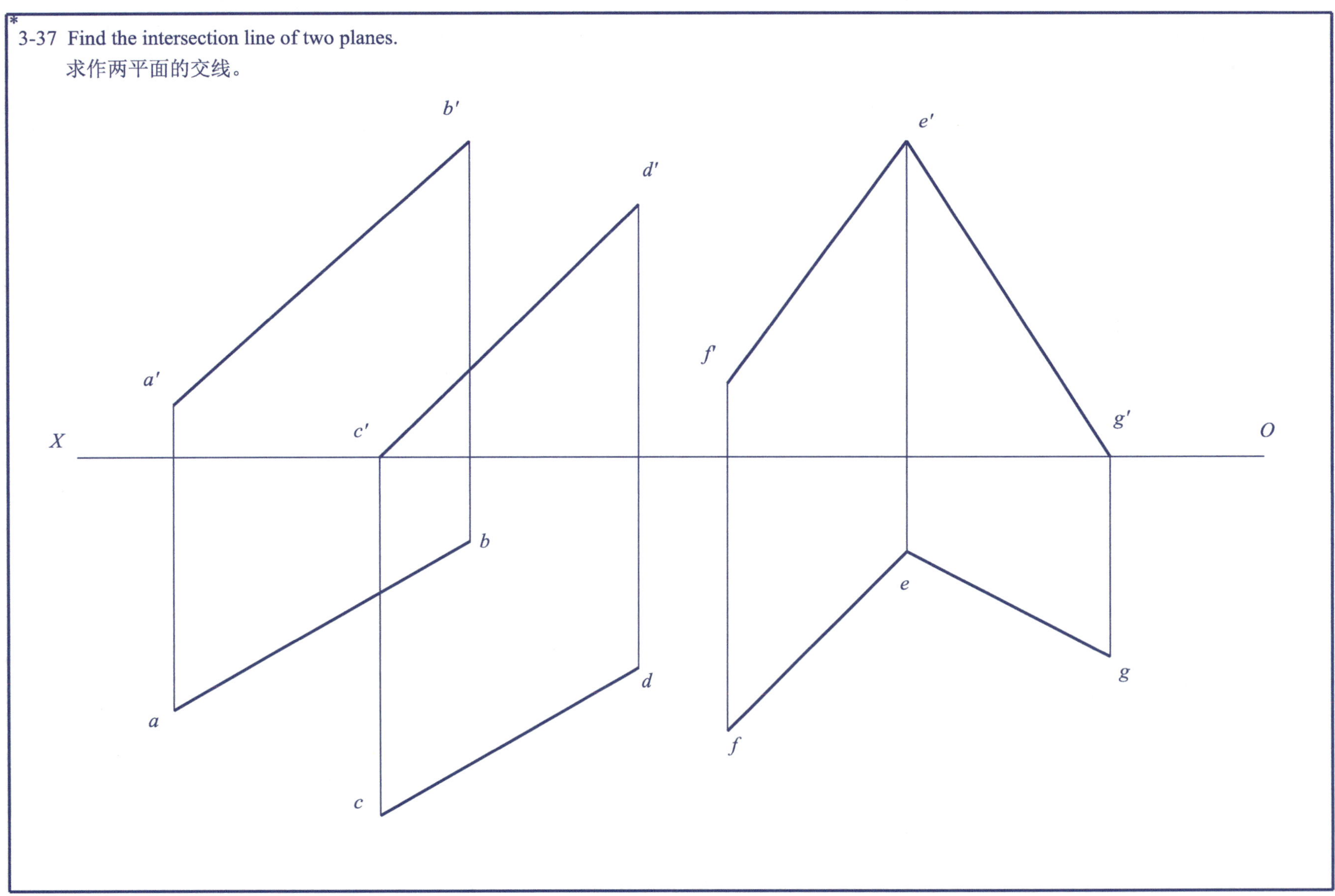

Class 班级：   Name 姓名：   ID Number 学号：

* 3-38 Draw a line through point *K* and intersecting with lines *AB* and *CD*.
过点*K*作直线与交叉两直线*AB*和*CD*相交。

* 3-39 Draw a line that intersects with lines *AB* and *CD* and is parallel to line *EF*.
作一直线与直线*AB*和*CD*都相交，且平行于直线*EF*。

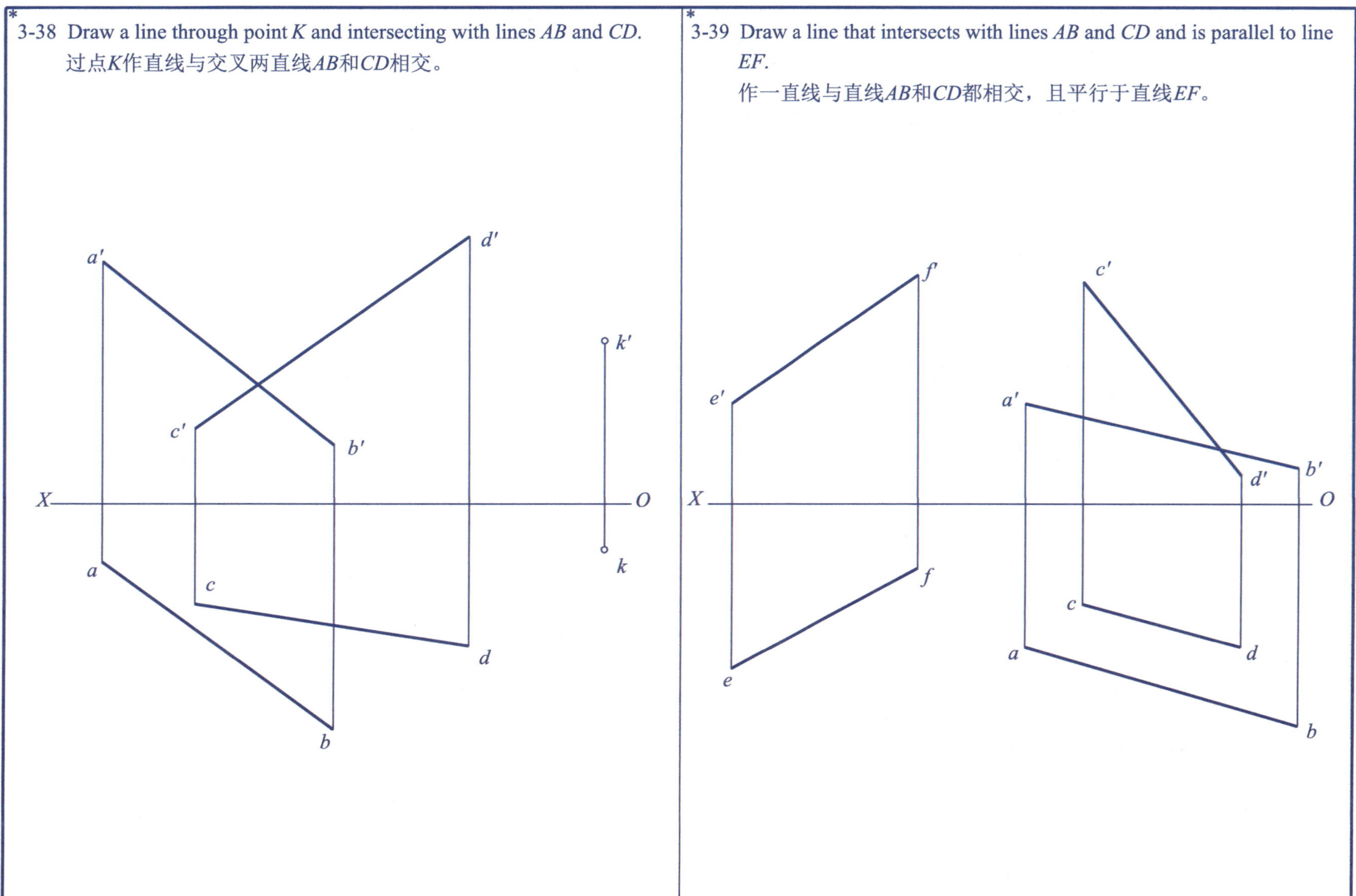

4-1 Given the frontal projection of lines *ABC* and point *D* on the surface of a pyramid, find their horizontal projection. Given the horizontal projection of point *E*, find its frontal projection.

已知四棱锥表面上线*ABC*和点*D*的正面投影，求作其水平投影；已知点*E*的水平投影，求作其正面投影。

4-2 Make the other two projections of points *A*, *B* and *C* on the surface of the pentagonal prism.

作出五棱柱表面上点*A*、*B*和*C*的另外两个投影。

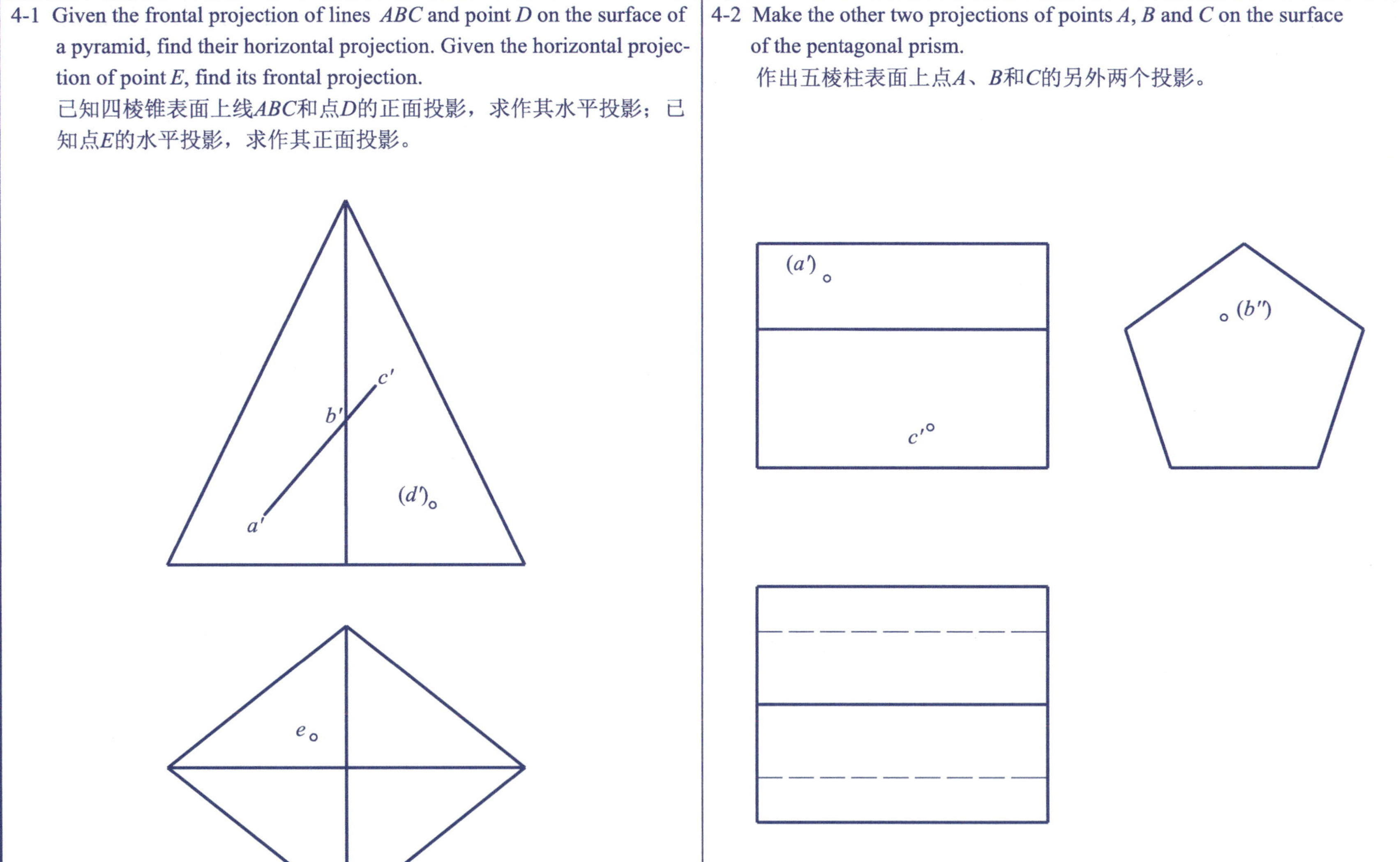

4-3 Draw the profile projection of the cylinder and find the other two projectins of line *ABC* and point *D* on the surface of cylinder.
补画出圆柱的侧面投影，并作出其表面上线*ABC*及点*D*的其他两面投影。

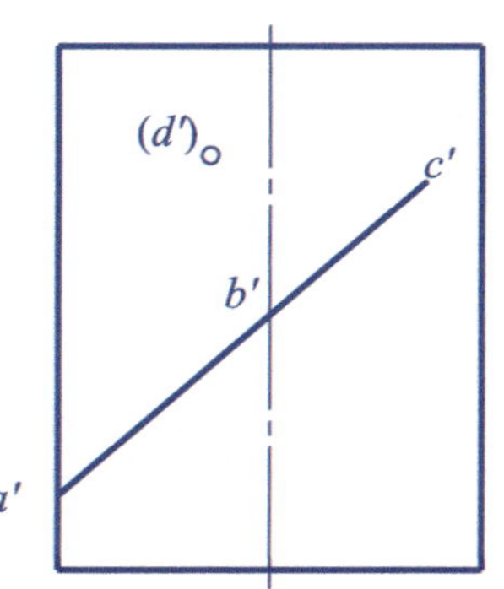

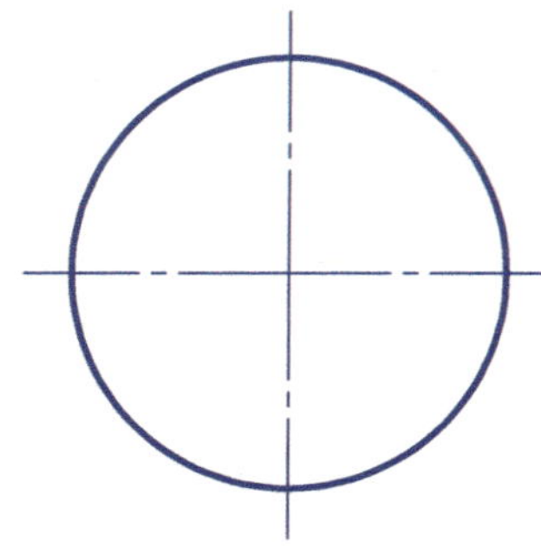

4-4 Draw the horizontal projection of the cylinder, and find the other two projections of lines *AB* and *BC* on its surface.
补画出圆柱的水平投影，并作出其表面上的线*AB*、*BC*的其他两面投影。

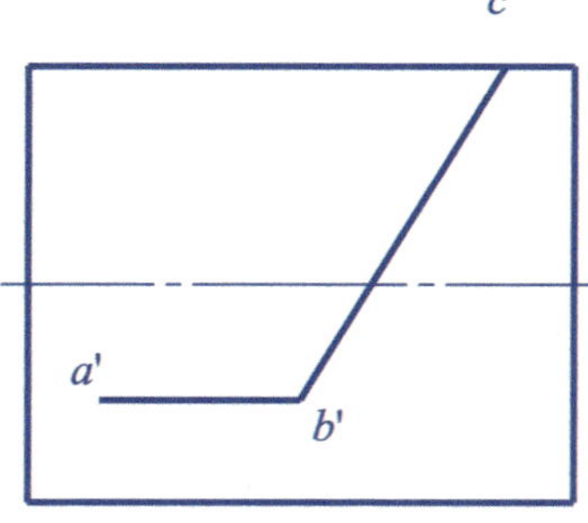

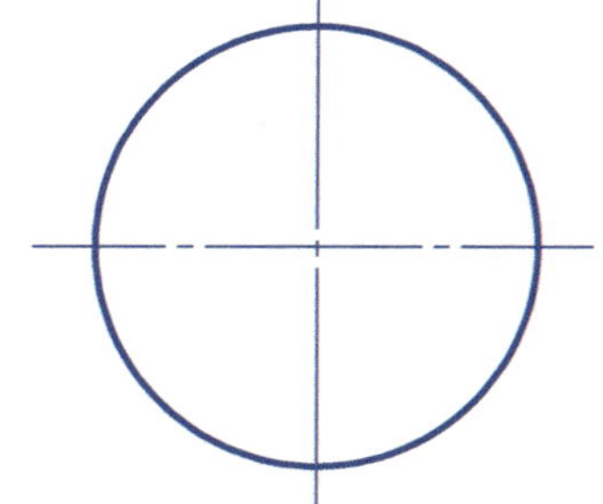

4-5 Draw the profile projection of the circular truncated cone and find the other two projections of line *ABC* and arc *DE* on its surface.
补画出圆台的侧面投影，并作出其表面上的线*ABC*和圆弧*DE*的其他两面投影。

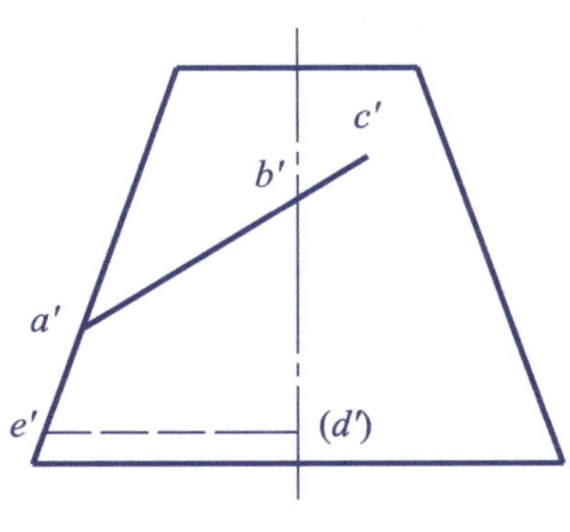

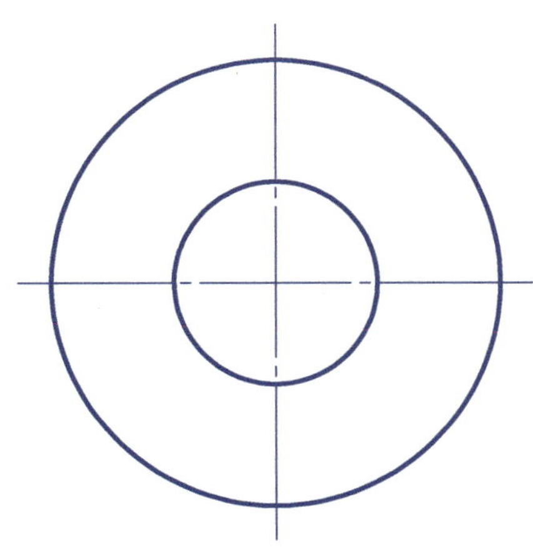

4-6 Known one projection of lines *AB* and *CEFG* on the spherical surface, find the other two projections.
已知球表面上线*AB*、*CEFG*的一个投影，分别作出其他两面投影。

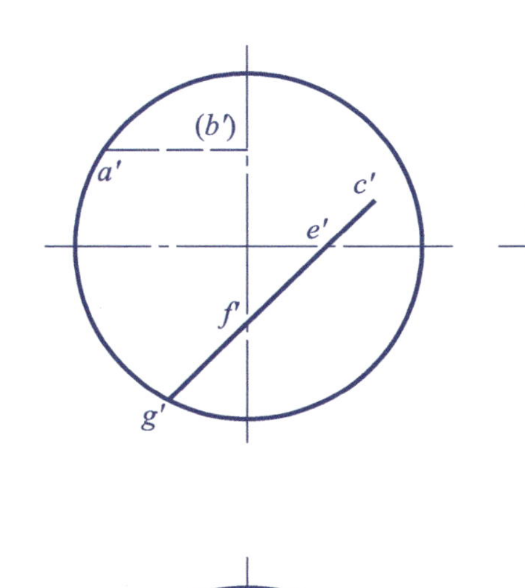

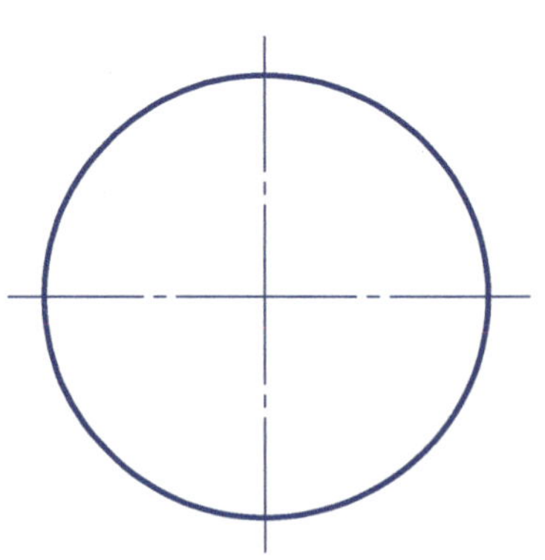

5-1 Draw the horizontal projection and profile projection of the truncated triangular pyramid.

补画三棱锥被截切后的水平投影和侧面投影。

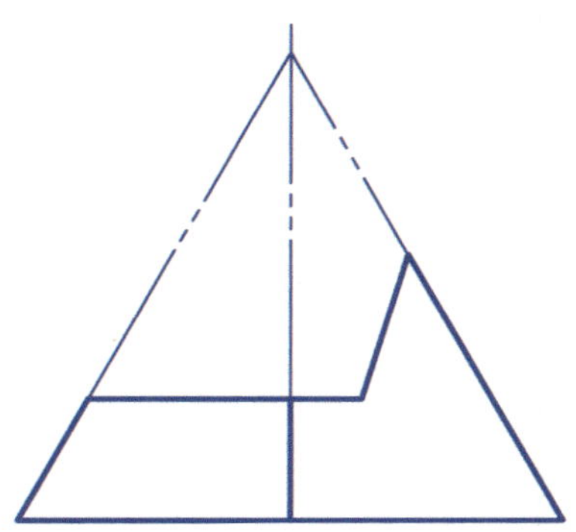

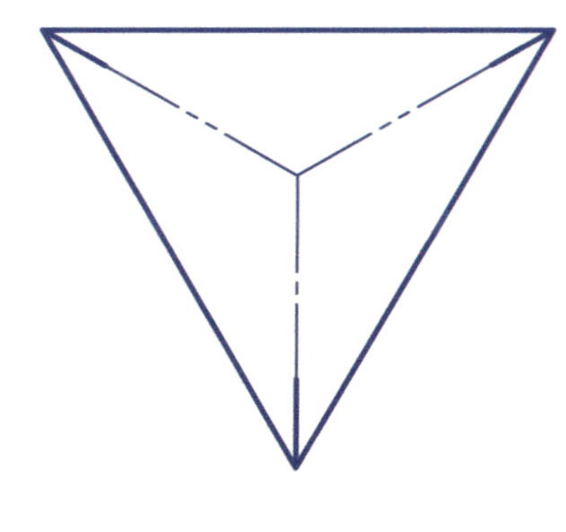

5-2 Draw the profile projection of the truncated quadrangular prism.

补画四棱柱被截切后的侧面投影。

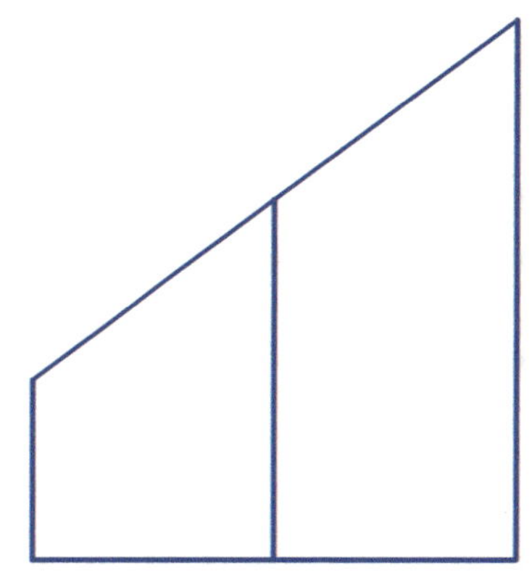

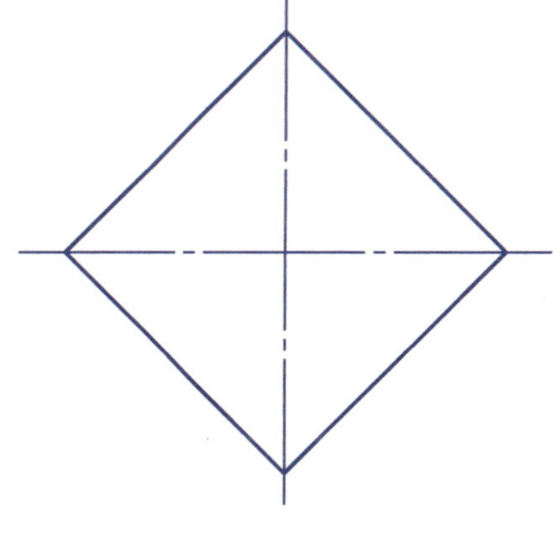

Class 班级： Name 姓名： ID Number 学号：

5-3 Draw the profile projection of the truncated pentagonal prism.
补画五棱柱被截切后的侧面投影。

5-4 Draw the profile projection of the truncated hexagonal prism.
补画六棱柱被截切后的侧面投影。

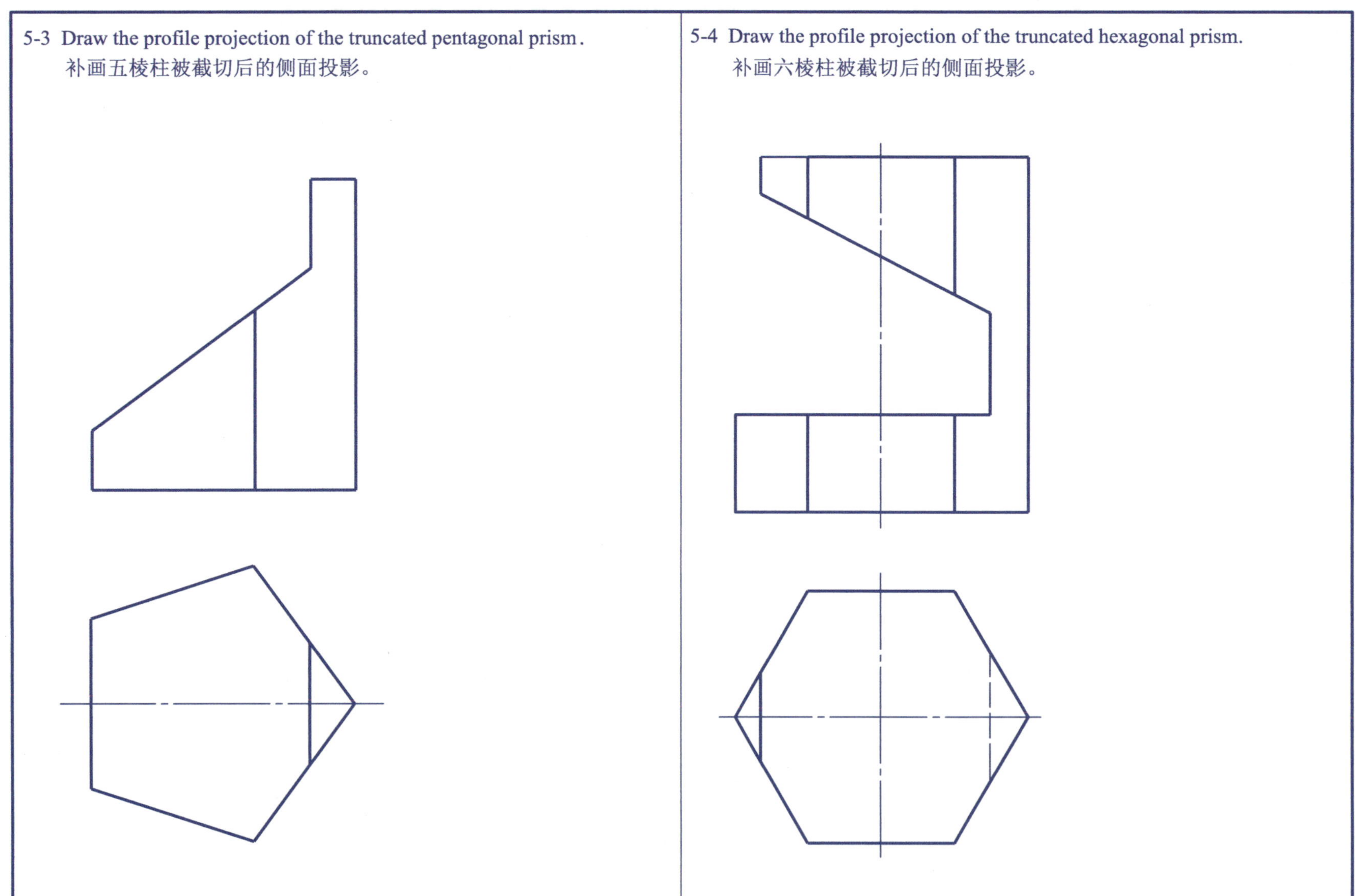

Class 班级： Name 姓名： ID Number 学号：

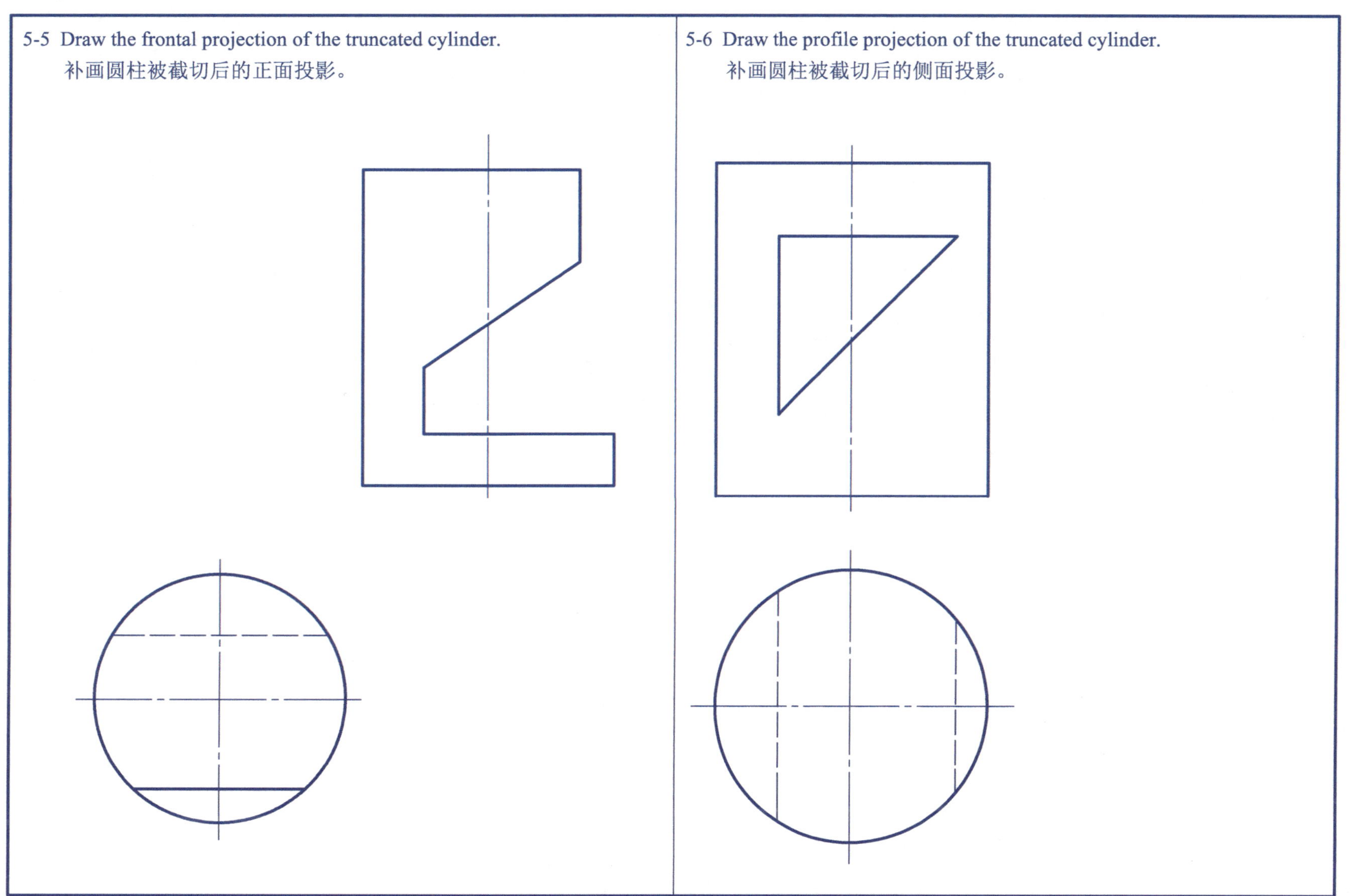

5-5 Draw the frontal projection of the truncated cylinder.
补画圆柱被截切后的正面投影。

5-6 Draw the profile projection of the truncated cylinder.
补画圆柱被截切后的侧面投影。

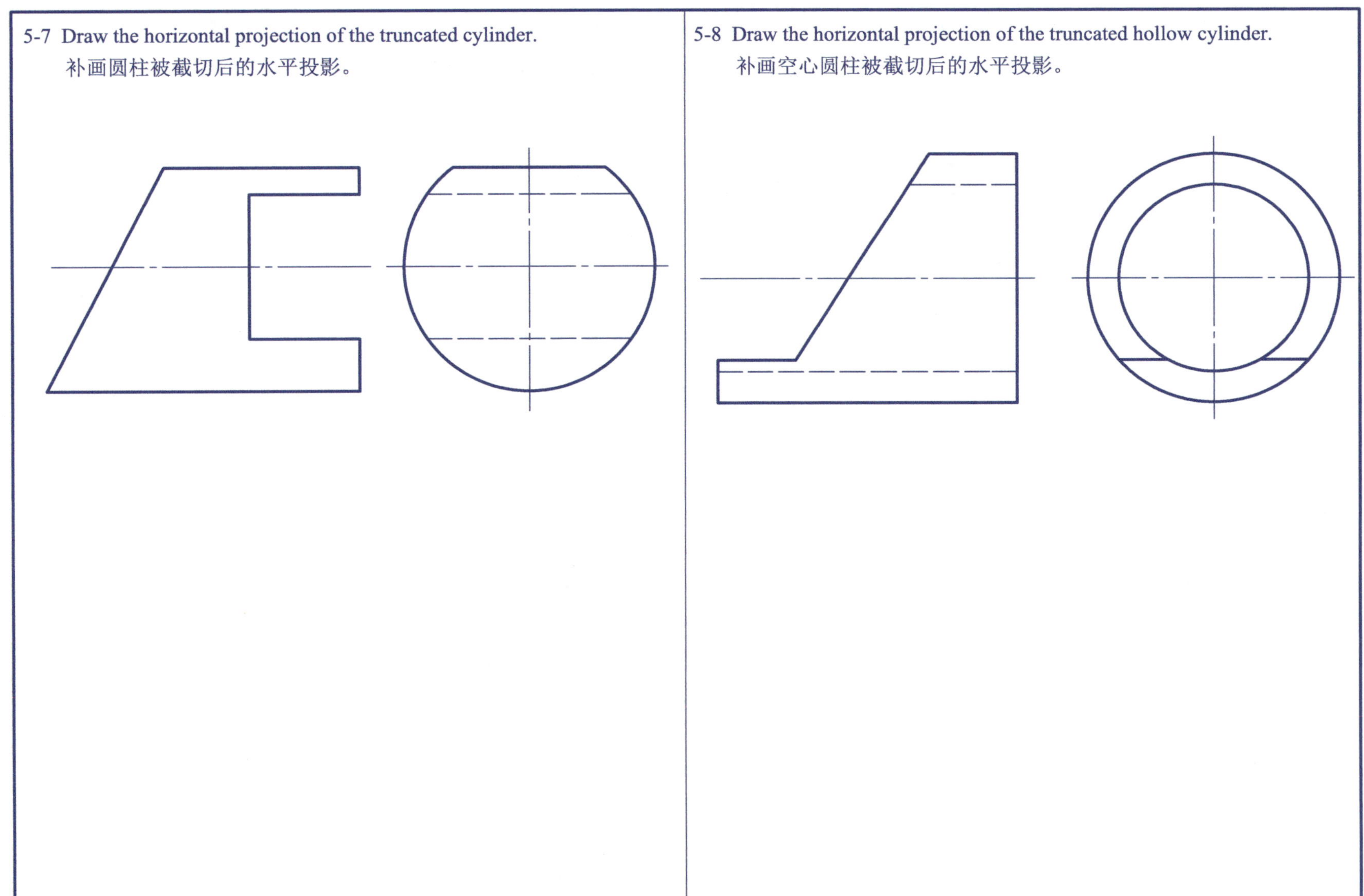

5-7 Draw the horizontal projection of the truncated cylinder.

补画圆柱被截切后的水平投影。

5-8 Draw the horizontal projection of the truncated hollow cylinder.

补画空心圆柱被截切后的水平投影。

5-9 Draw the profile projection of the truncated hollow cylinder.
补画空心圆柱被截切后的侧面投影。

5-10 Draw the horizontal projection and profile projection of the truncated circular cone.
补画圆锥被截切后的水平投影和侧面投影。

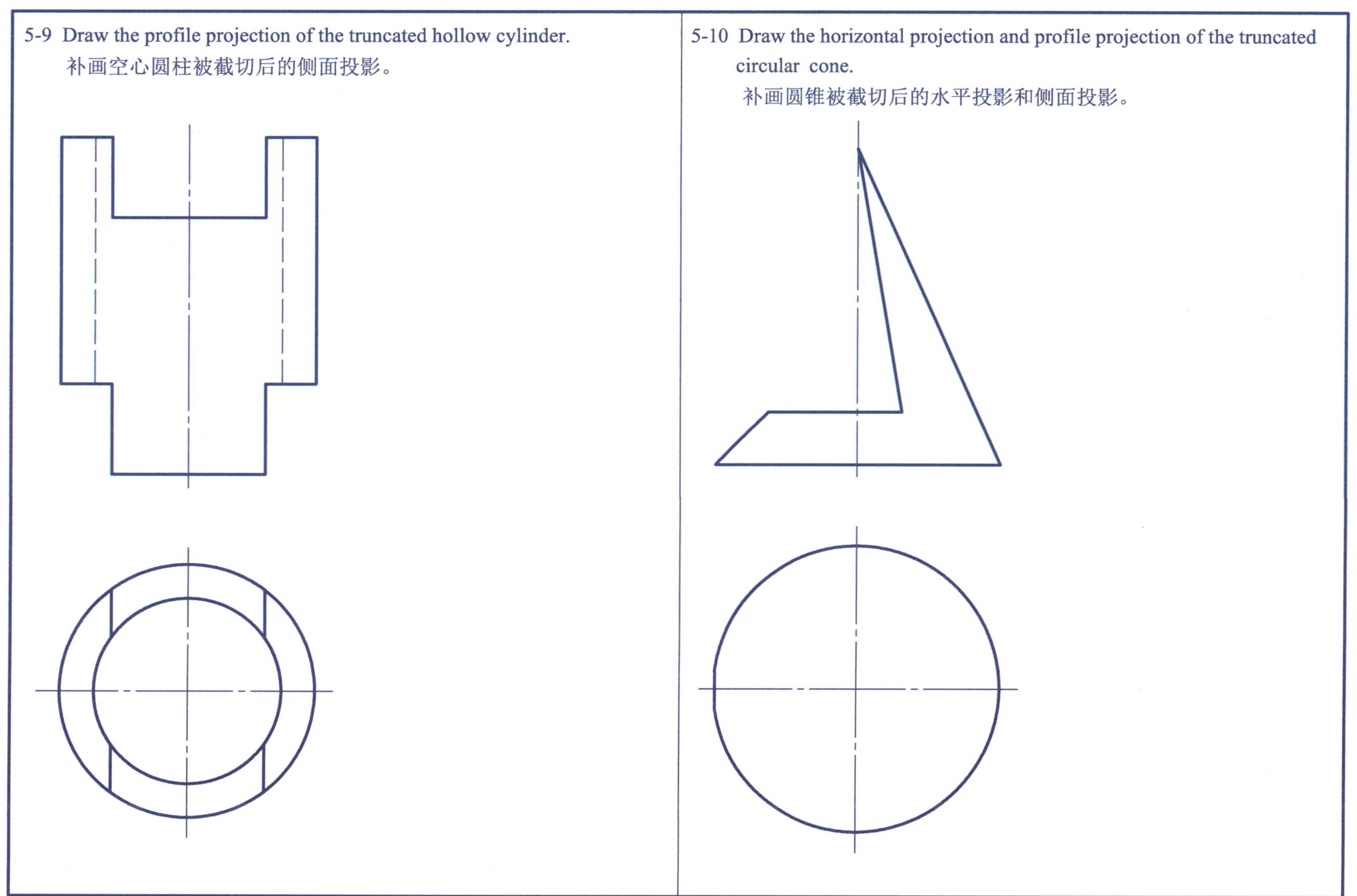

Class 班级：　　Name 姓名：　　ID Number 学号：

5-11 Draw the horizontal projection and profile projection of the truncated circular cone.

补画圆锥被截切后的水平投影和侧面投影。

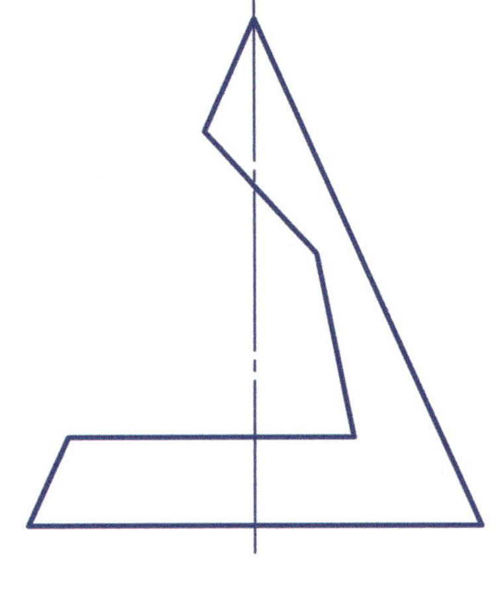

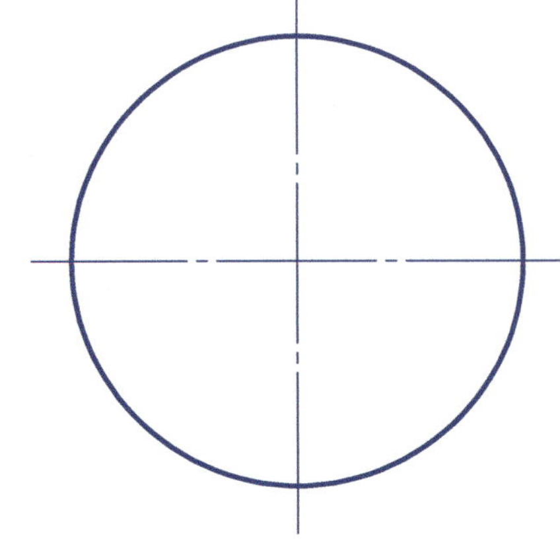

5-12 Draw the horizontal projection and profile projection of the truncated circular cone.

补画圆锥被截切后的水平投影和侧面投影。

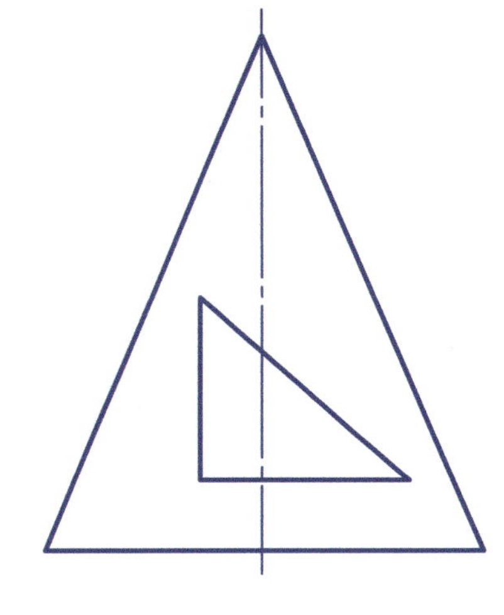

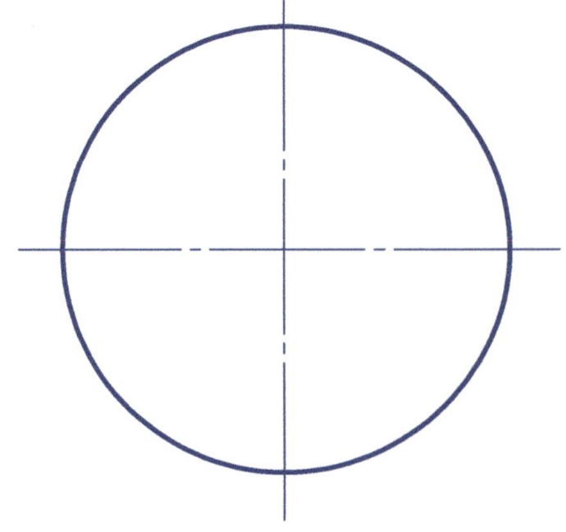

Class 班级：　Name 姓名：　ID Number 学号：

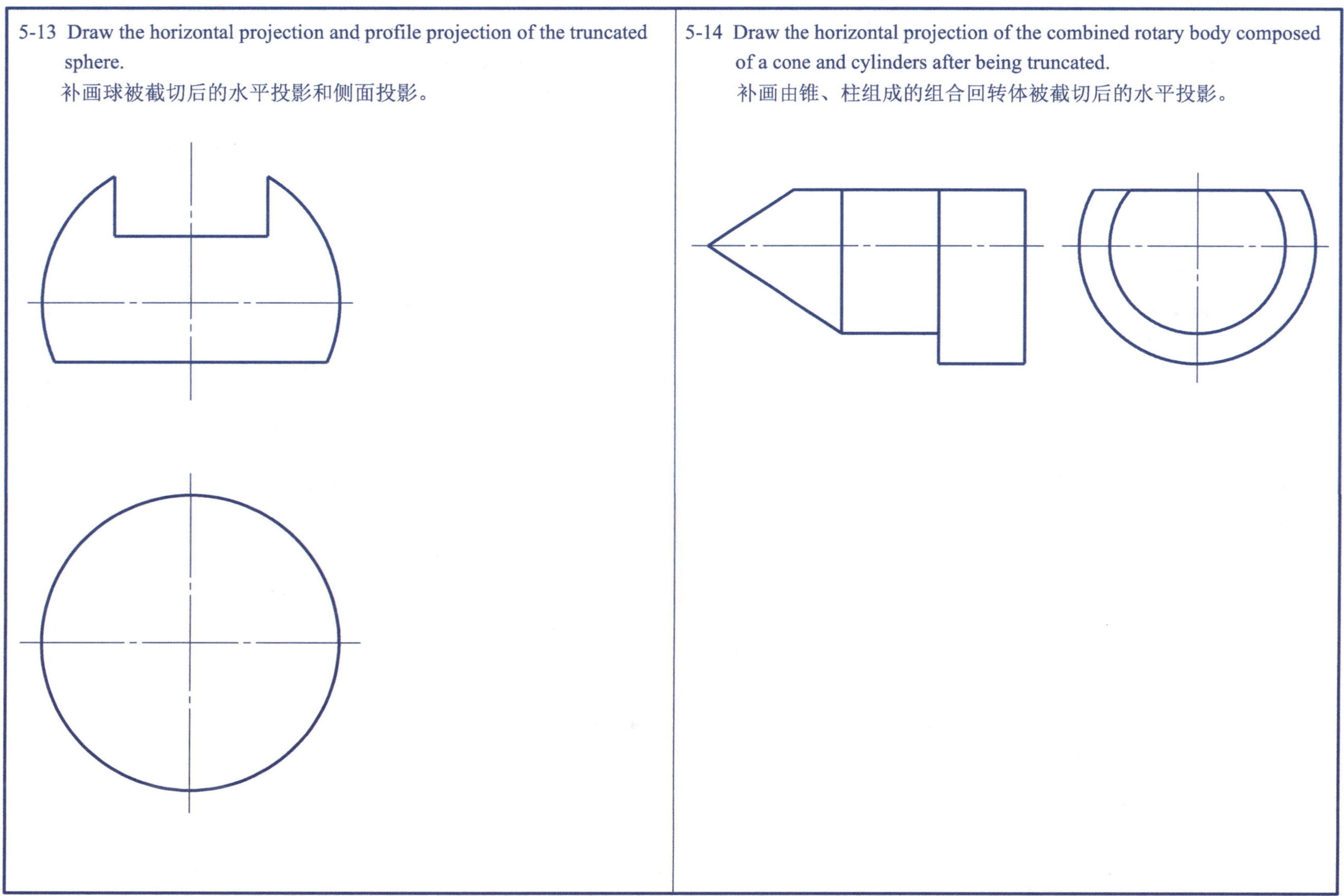

5-13 Draw the horizontal projection and profile projection of the truncated sphere.

补画球被截切后的水平投影和侧面投影。

5-14 Draw the horizontal projection of the combined rotary body composed of a cone and cylinders after being truncated.

补画由锥、柱组成的组合回转体被截切后的水平投影。

5-15 Draw the horizontal projection of the combined rotary body composed of a sphere and a cylinder after being truncated.

补画由球、柱组成的组合回转体被截切后的水平投影。

5-16 Draw the profile projection of the combined rotary body composed of a cylinder and a sphere after being truncated.

补画由柱、球组成的组合回转体被截切后的侧面投影。

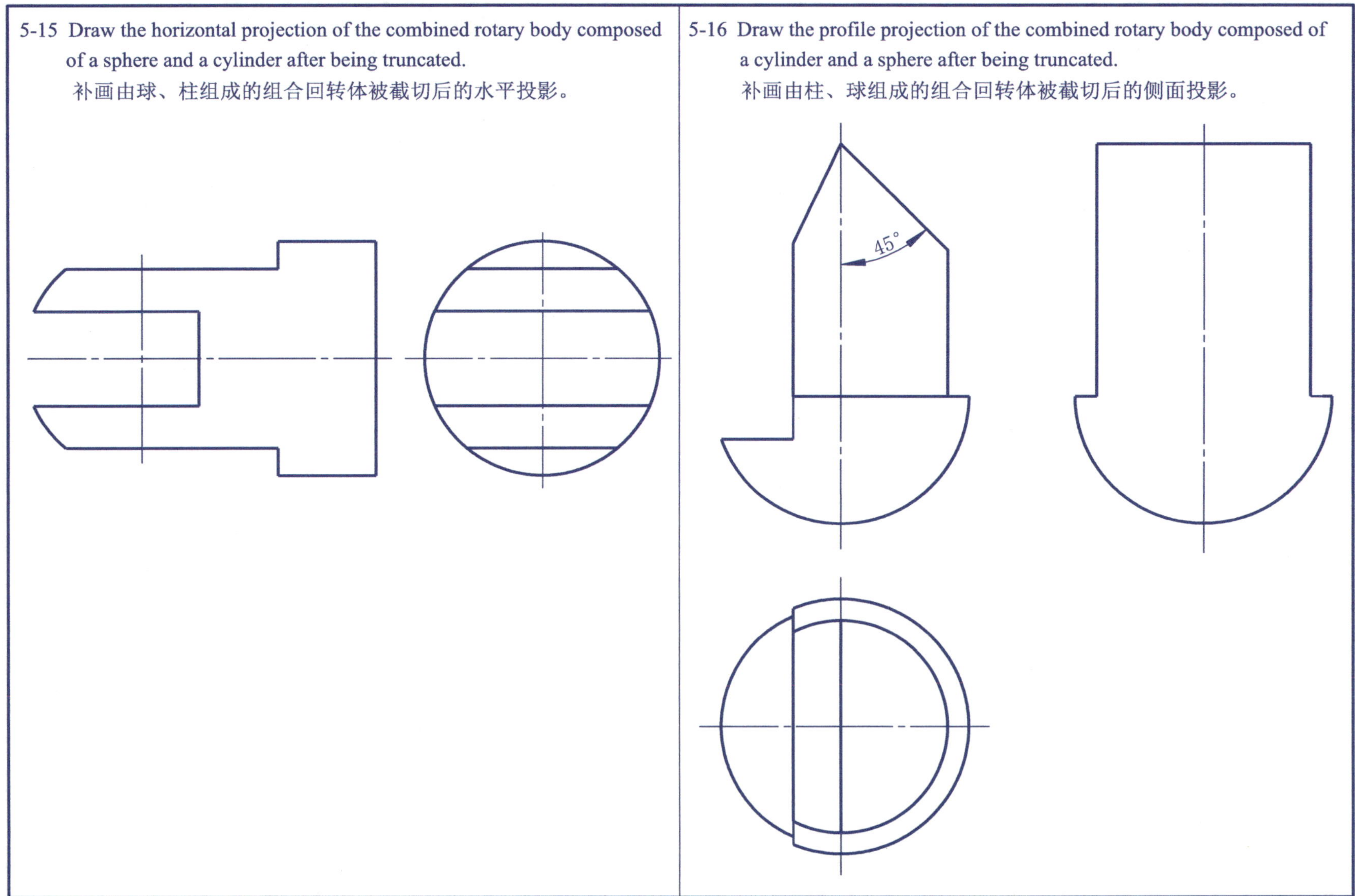

Class 班级： Name 姓名： ID Number 学号：

6-1 Complete the missing intersecting lines in the projection.
补画立体投影图中缺少的相贯线。

6-2 Complete the missing intersecting lines in the projection.
补画立体投影图中缺少的相贯线。

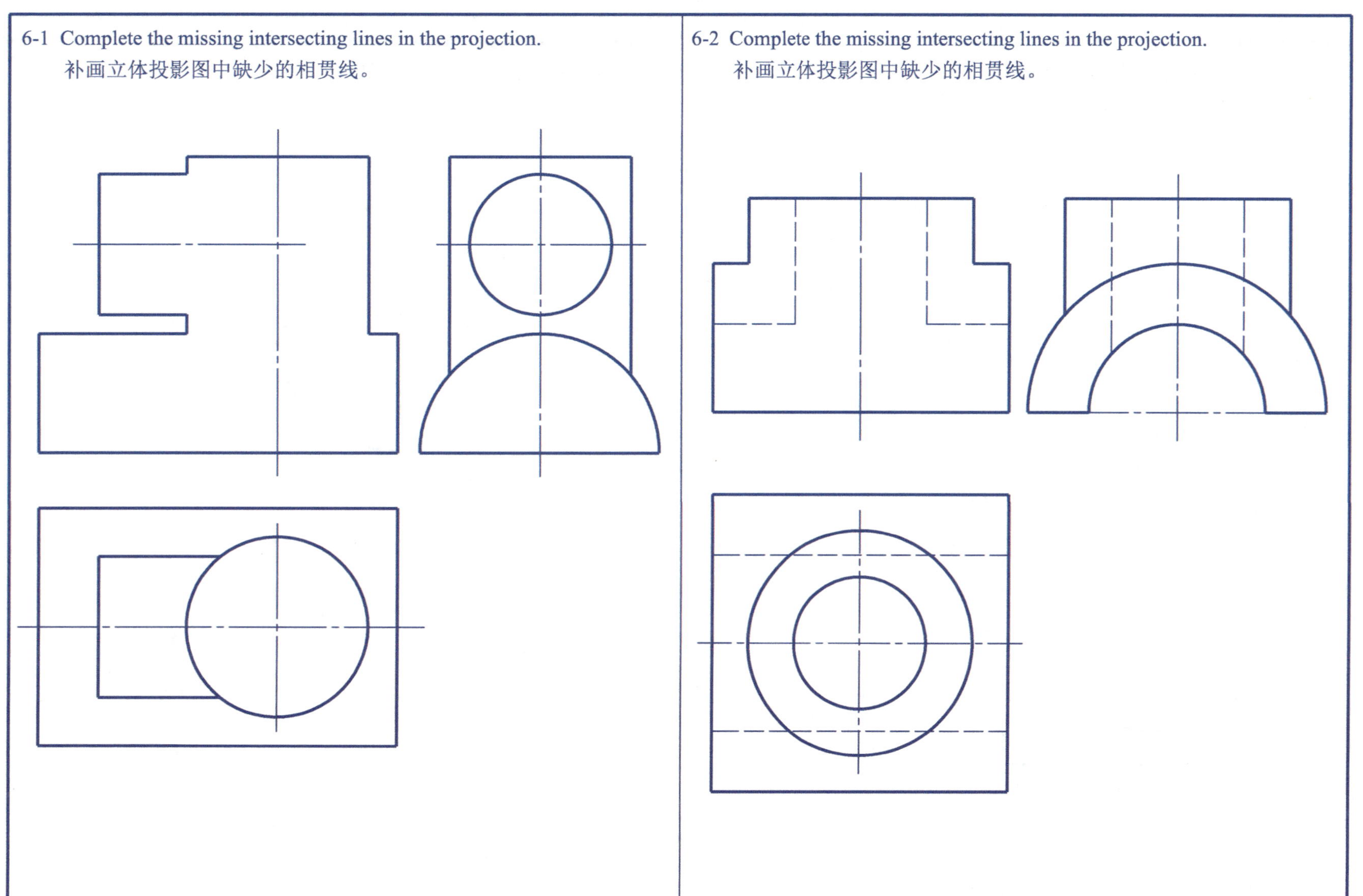

6-3 Complete the missing intersecting lines in the projection.
补画立体投影图中缺少的相贯线。

6-4 Draw the solid intersecting lines and the missing outlines.
作出立体的相贯线，并补齐缺少的轮廓线。

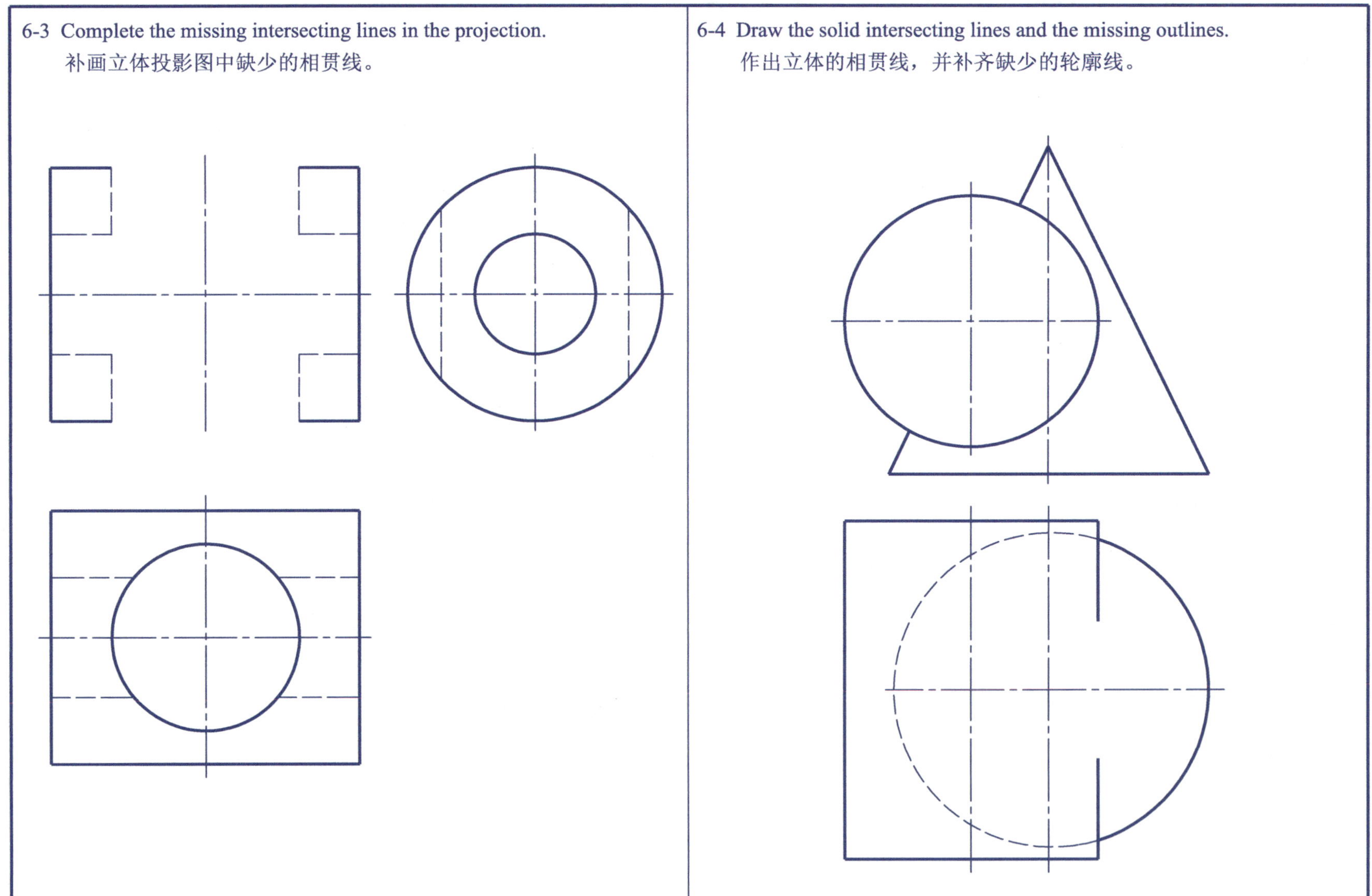

6-5 Draw the solid intersecting lines and the missing outlines.
作出立体的相贯线，并补齐缺少的轮廓线。

6-6 Draw the solid intersecting lines and the missing outlines.
作出立体的相贯线，并补齐缺少的轮廓线。

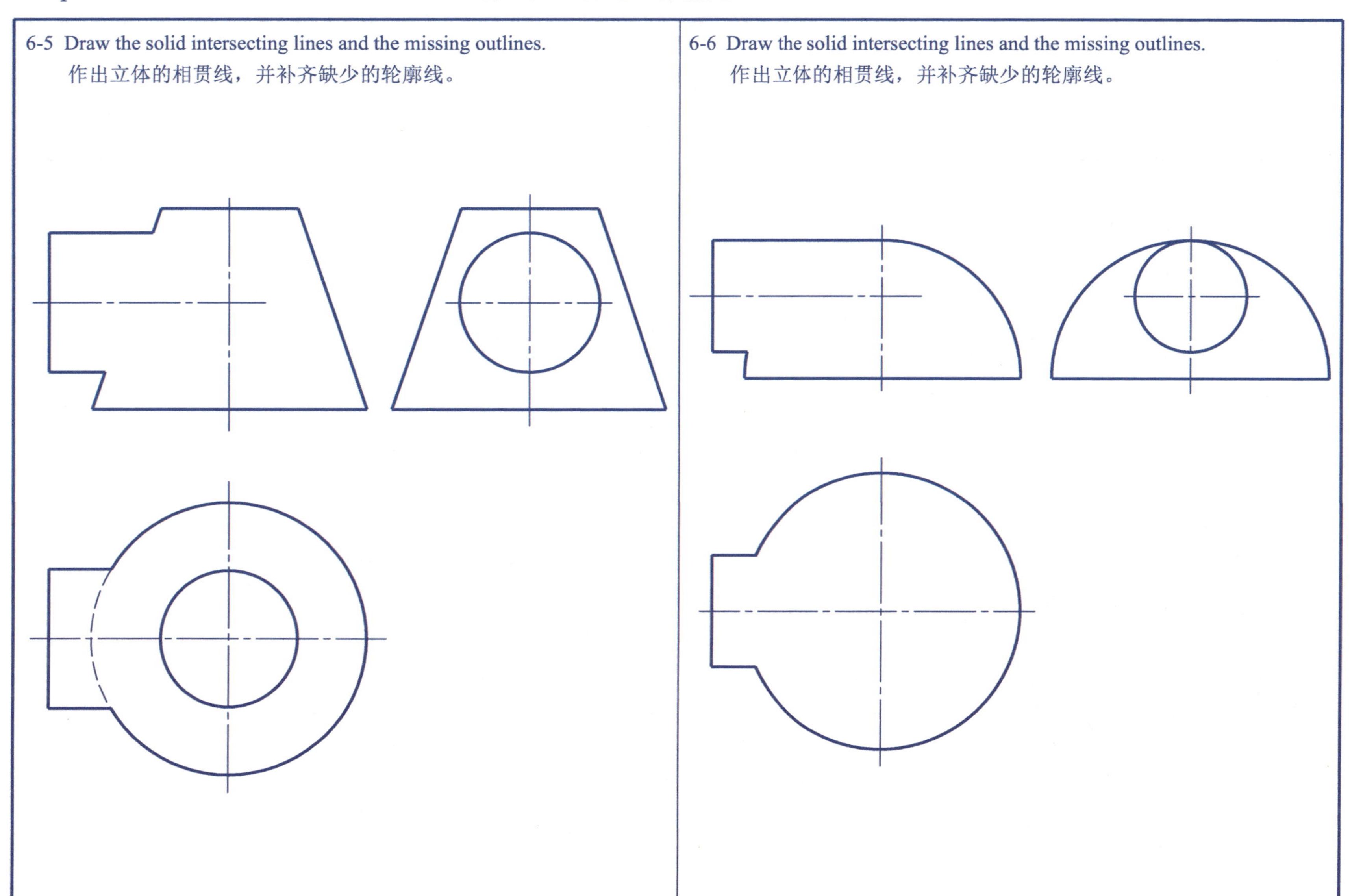

6-7 Draw the profile projection of the solid intersecting lines.
画出立体相贯线的侧面投影。

6-8 Draw the projection of the solid intersecting lines.
画出立体相贯线的投影。

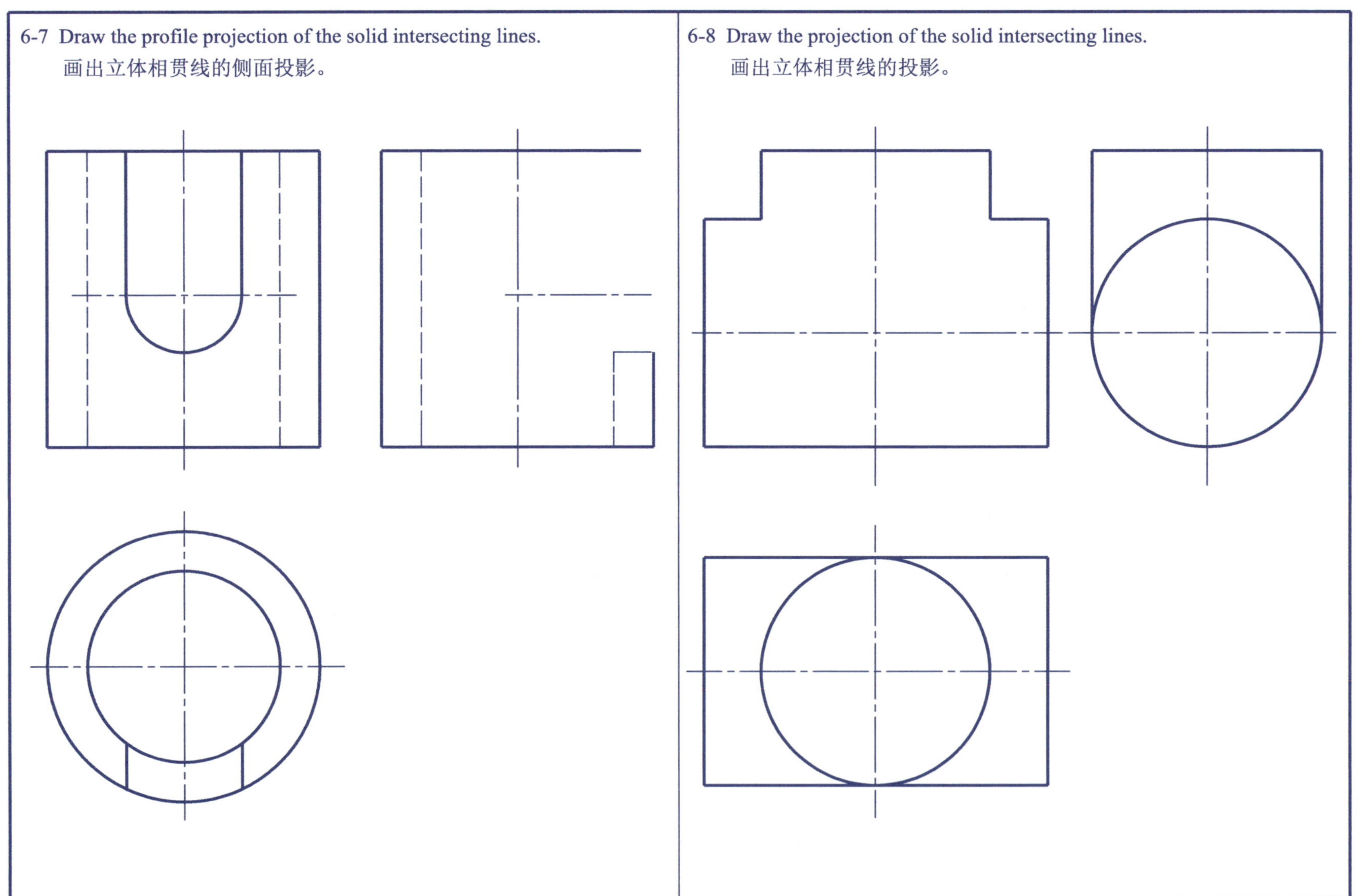

Class 班级： Name 姓名： ID Number 学号：

6-9 Complete the missing intersecting lines in the projection.
补画立体投影图中缺少的相贯线。

6-10 Draw the projection of the solid intersecting lines.
画出立体相贯线的投影。

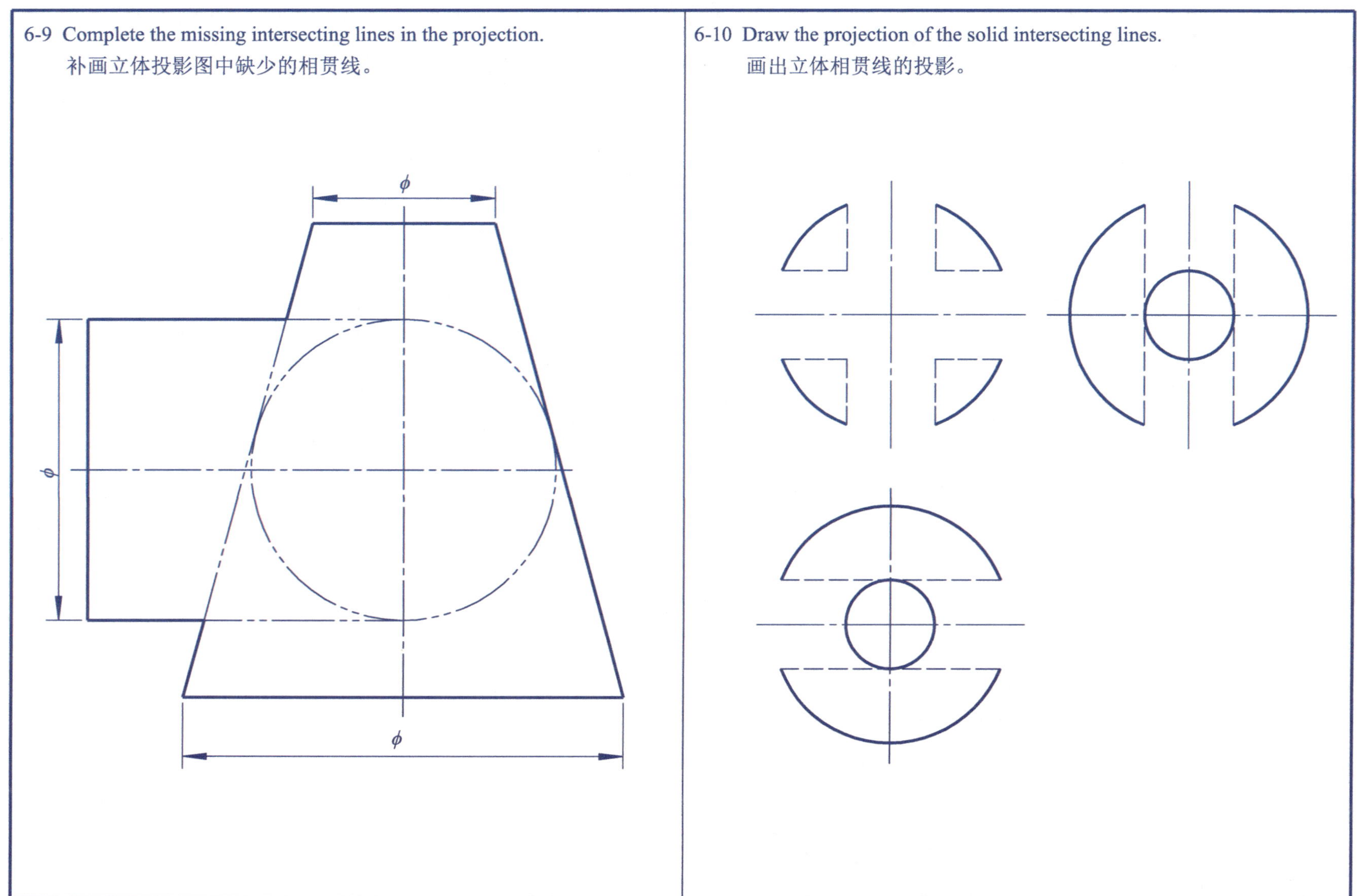

Class 班级： Name 姓名： ID Number 学号：

6-11 Complete the missing intersecting lines in the projection.
补画立体投影图中缺少的相贯线。

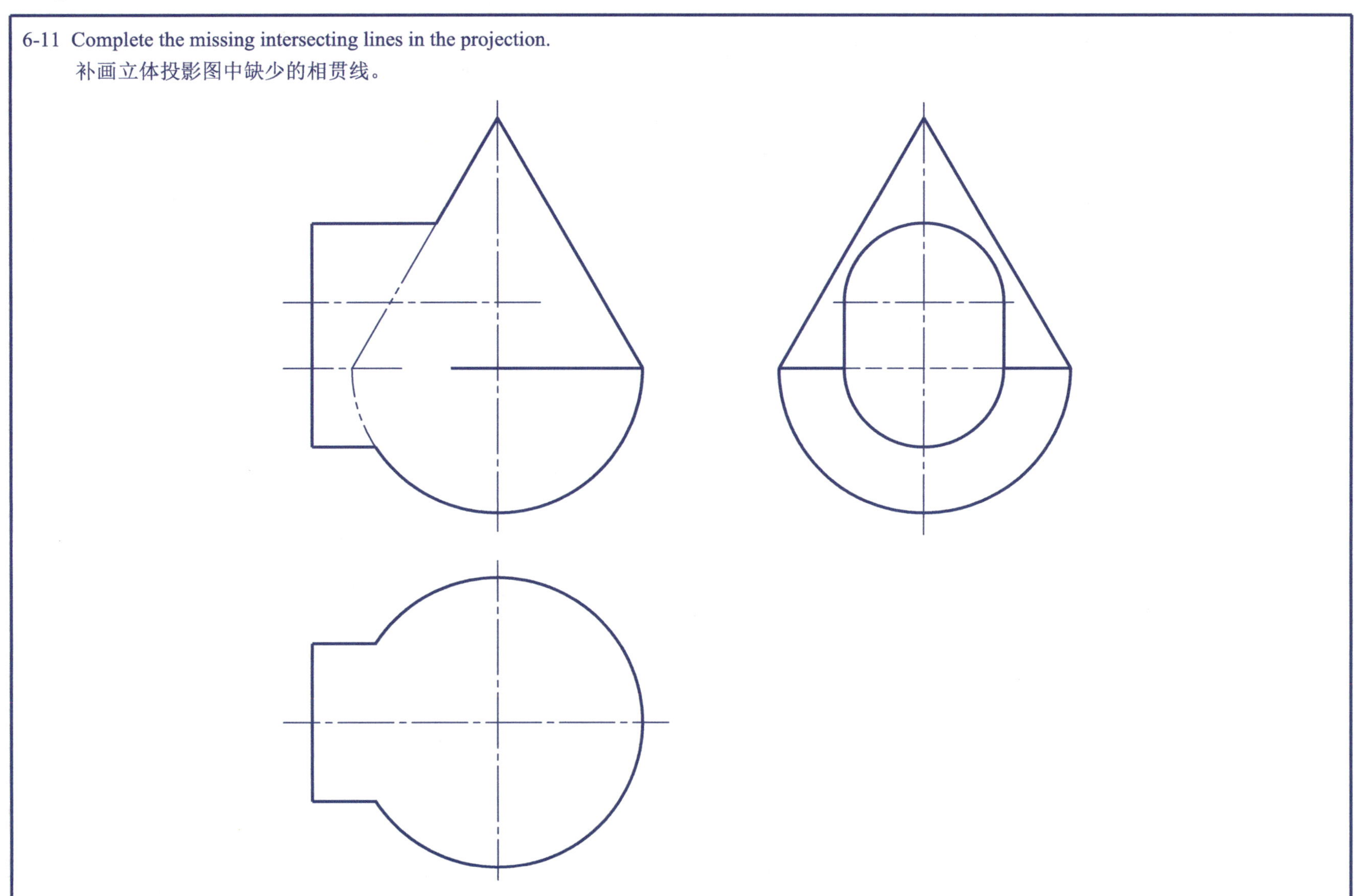

Class 班级：   Name 姓名：   ID Number 学号：

6-12 Complete the missing intersecting lines in the projection.
补画立体投影图中缺少的相贯线。

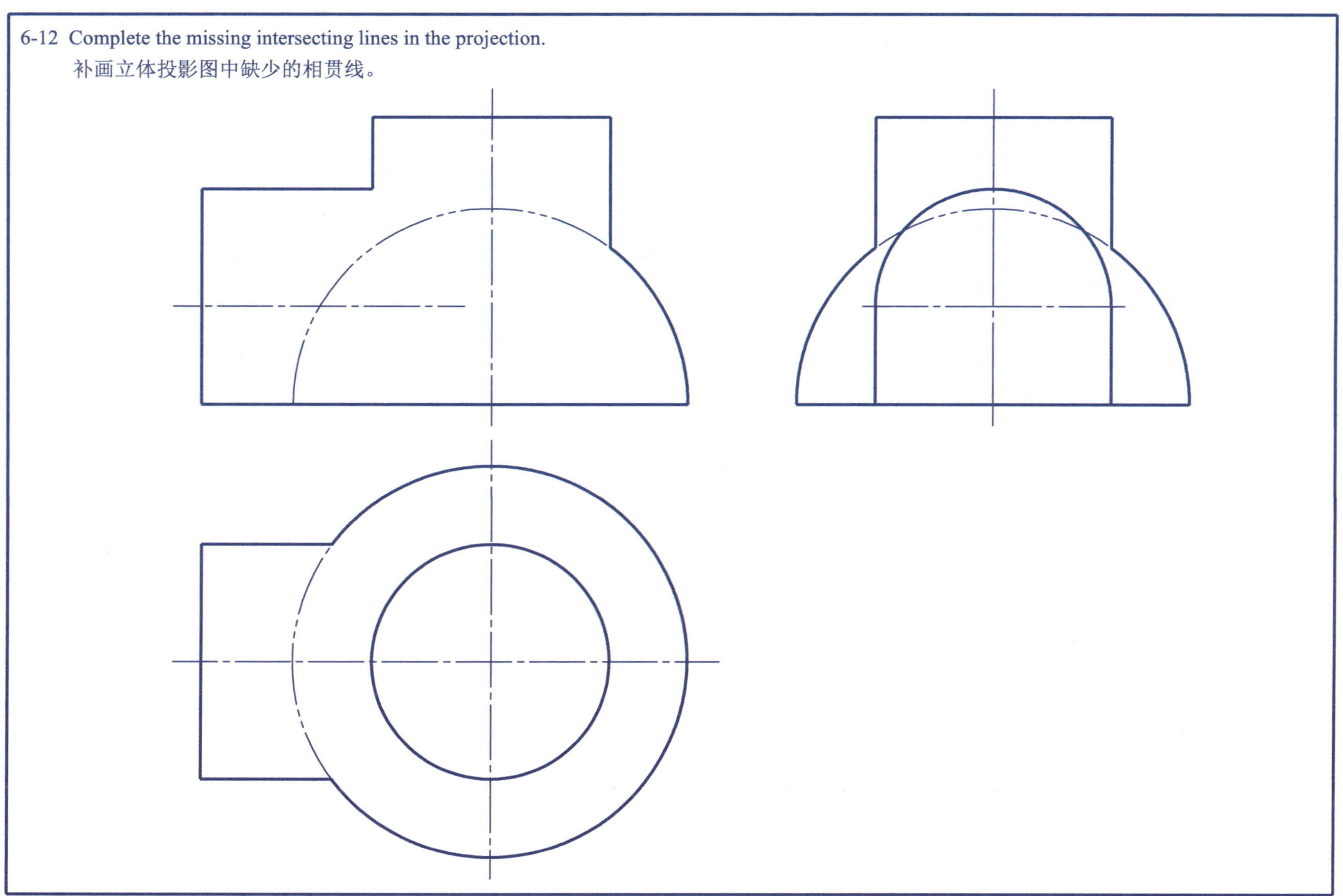

Class 班级： Name 姓名： ID Number 学号：

6-13 Complete the missing intersecting lines in the projection.
补画立体投影图中缺少的相贯线。

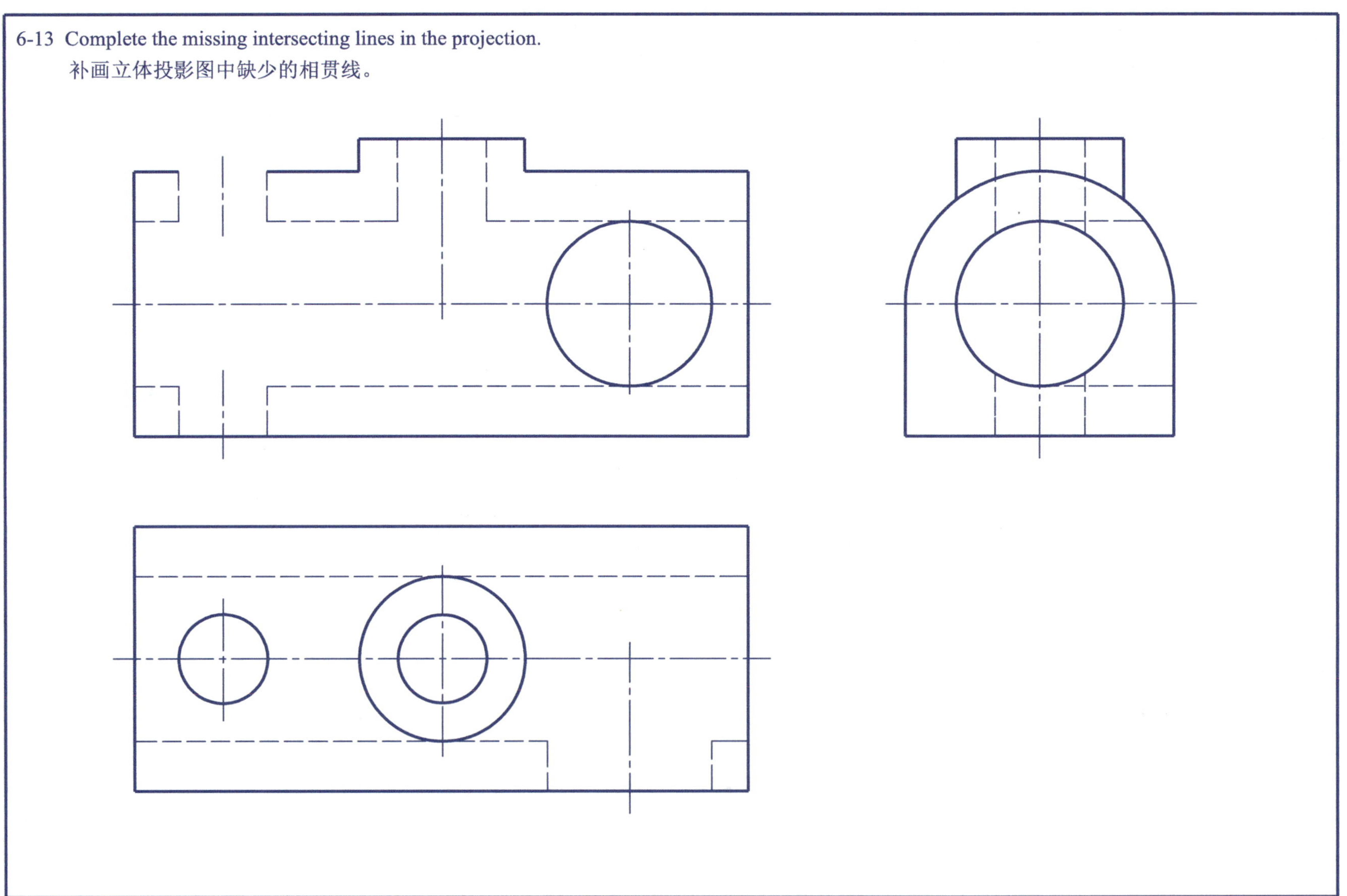

 Class 班级： Name 姓名： ID Number 学号：

6-14 Complete the missing intersecting lines in the projection.
补画立体投影图中缺少的相贯线。

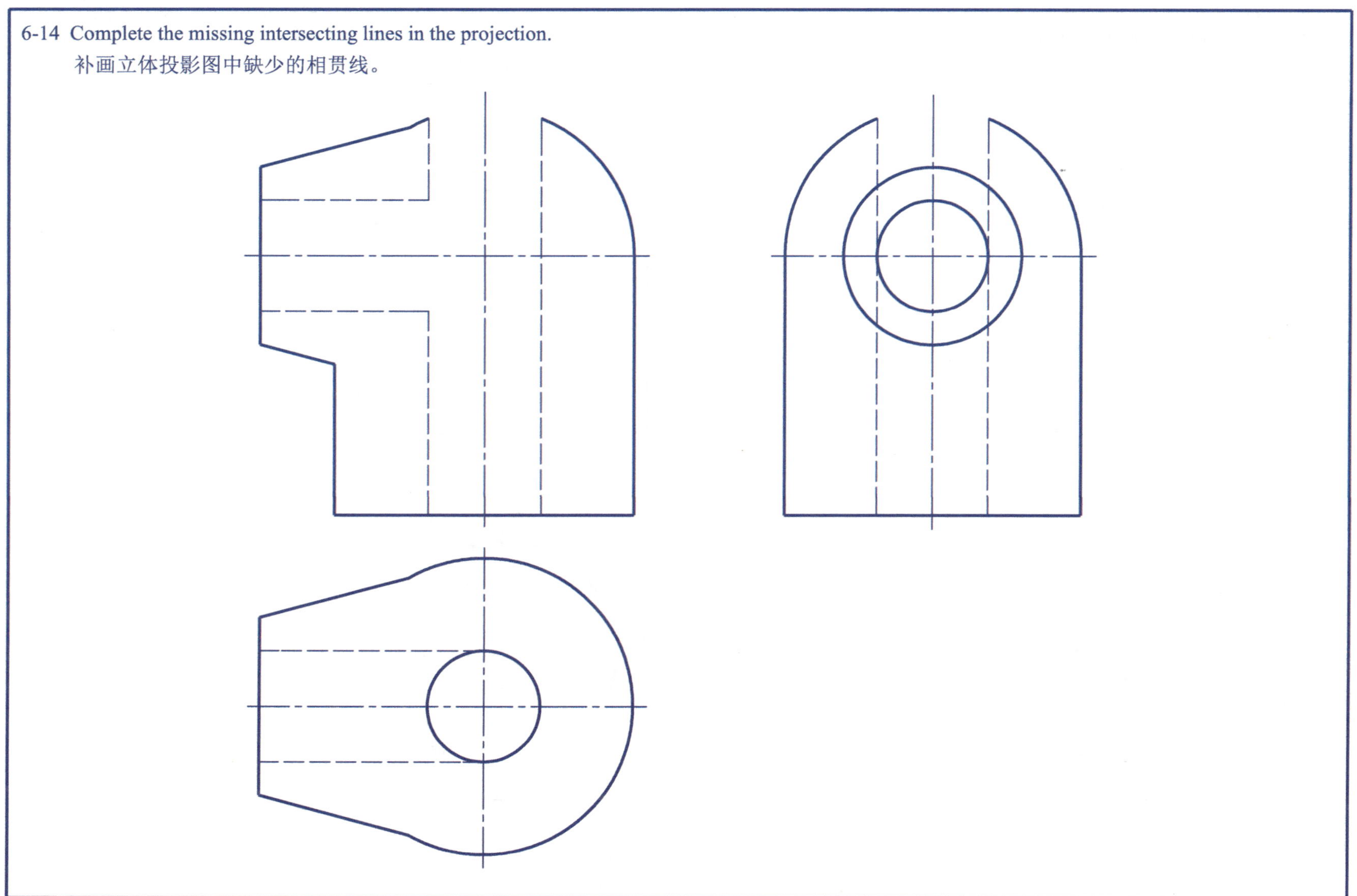

*6-15 Draw the profile projection of the solid.
画出立体的侧面投影。

*6-16 Draw the profile projection of the solid.
画出立体的侧面投影。

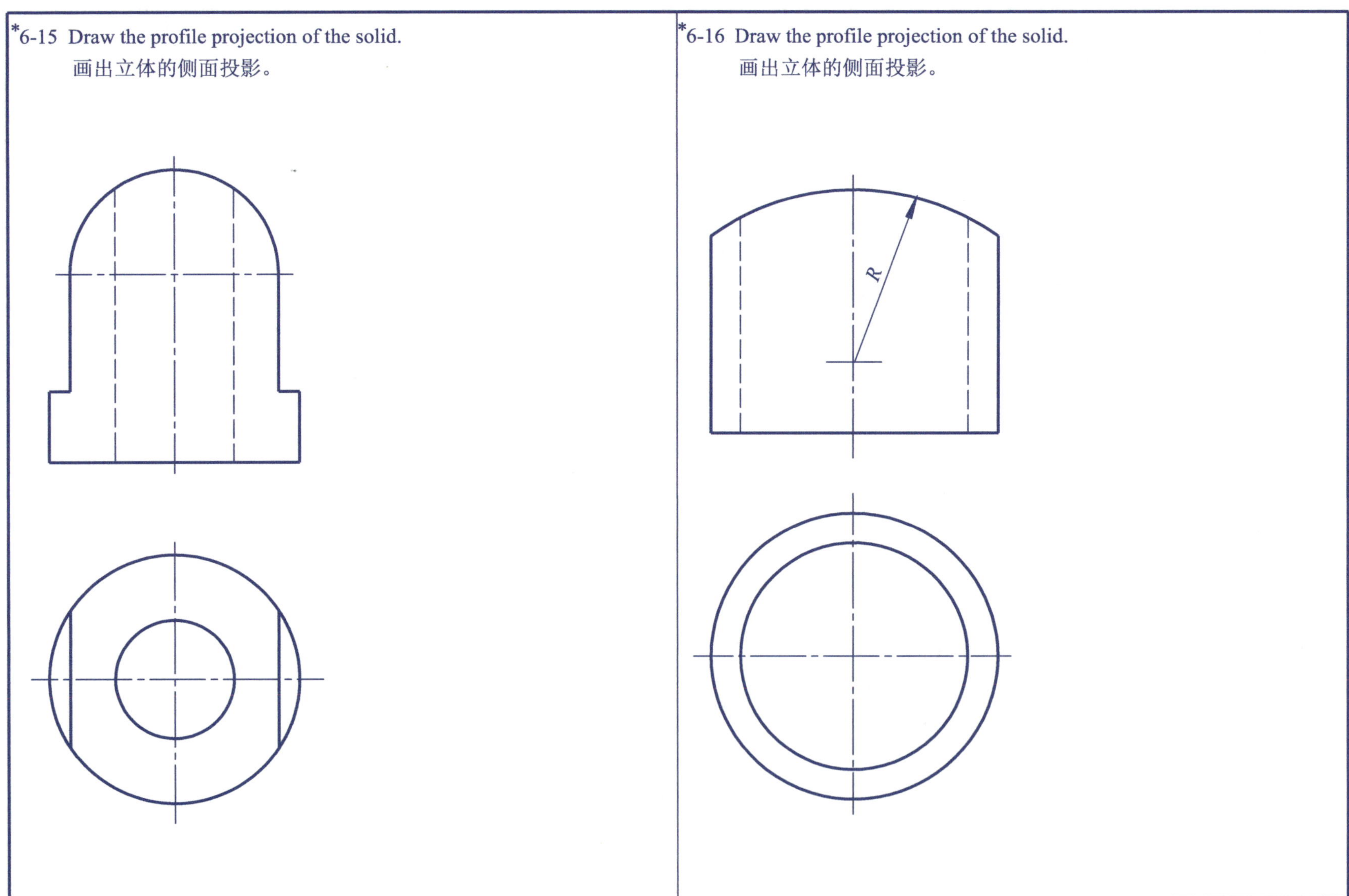

Class 班级： Name 姓名： ID Number 学号：

7-1 Draw the three views of the illustrated composite solid (scale 1 ∶ 1).
画出图示组合体的三视图（比例 1∶1）。

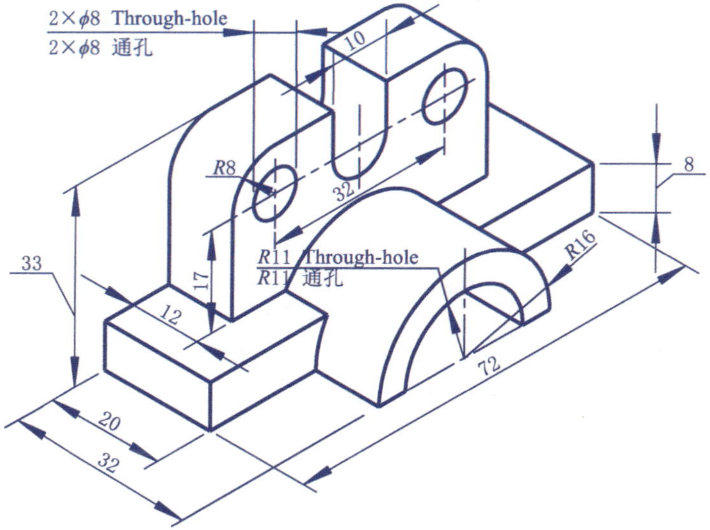

2×ϕ8 Through-hole
2×ϕ8 通孔
10
R8
32
8
33
17
12
R11 Through-hole
R11 通孔
R16
72
20
32

7-2 Draw the three views of the illustrated composite solid (scale 1：1).
画出图示组合体的三视图（比例 1：1）。

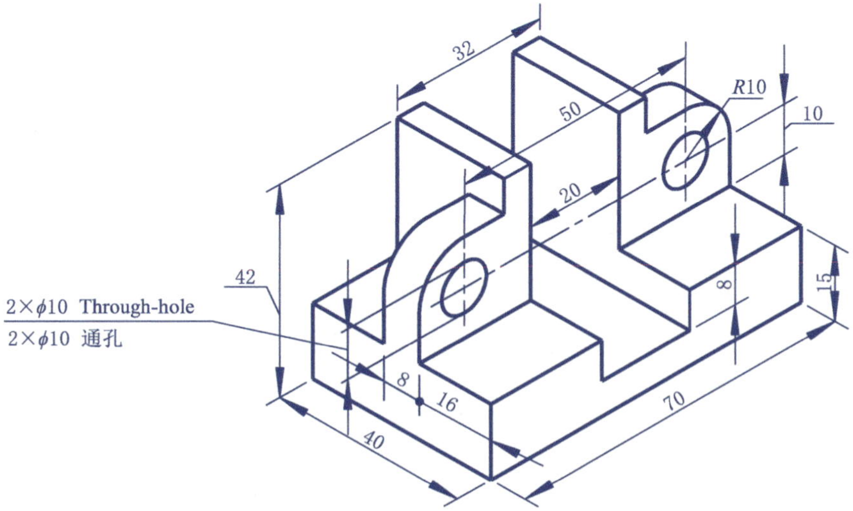

2×ϕ10 Through-hole
2×ϕ10 通孔

7-3 Draw the three views of the illustrated composite solid (scale 1∶1).
画出图示组合体的三视图（比例 1∶1）。

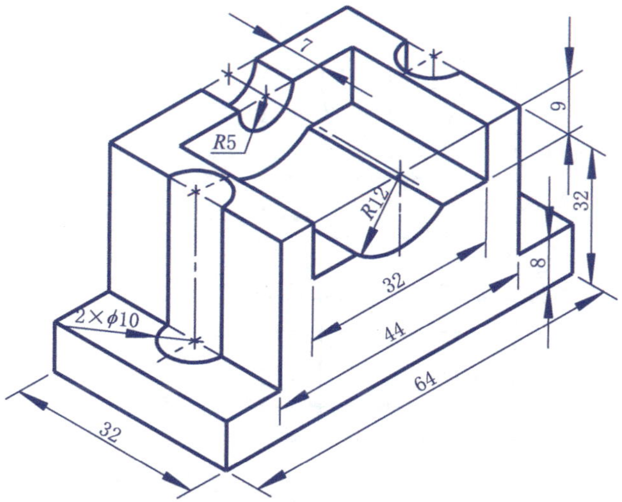

7
R5
R12
2×φ10
9
32
8
32
44
64
32

7-4 Draw the third view based on the two views of the composite solid.
根据组合体的两个视图，补画第三个视图。

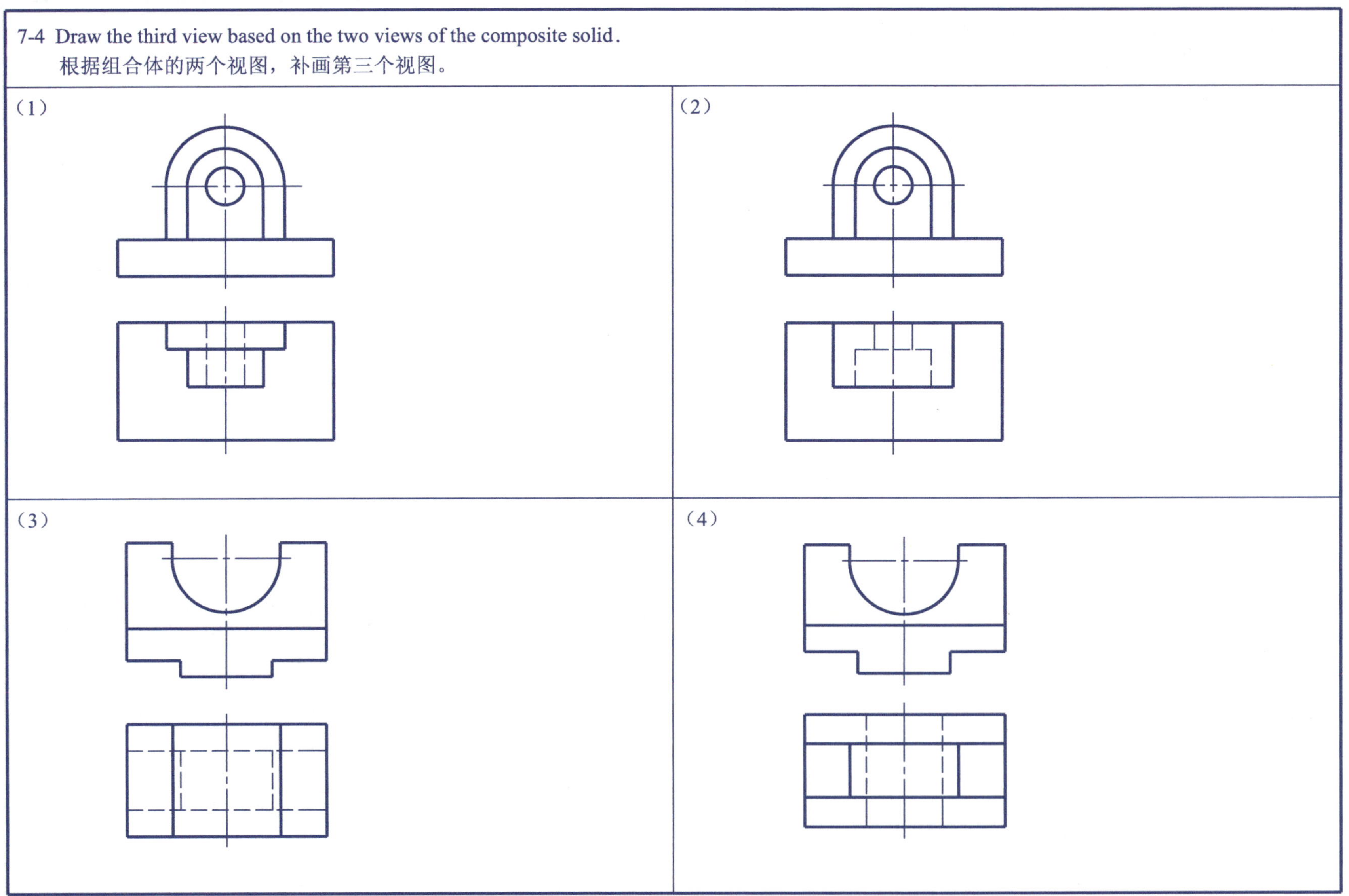

7-5 Complete the missing lines in the front view based on the top view of the composite solids.
根据组合体的俯视图，补全主视图中缺少的图线。

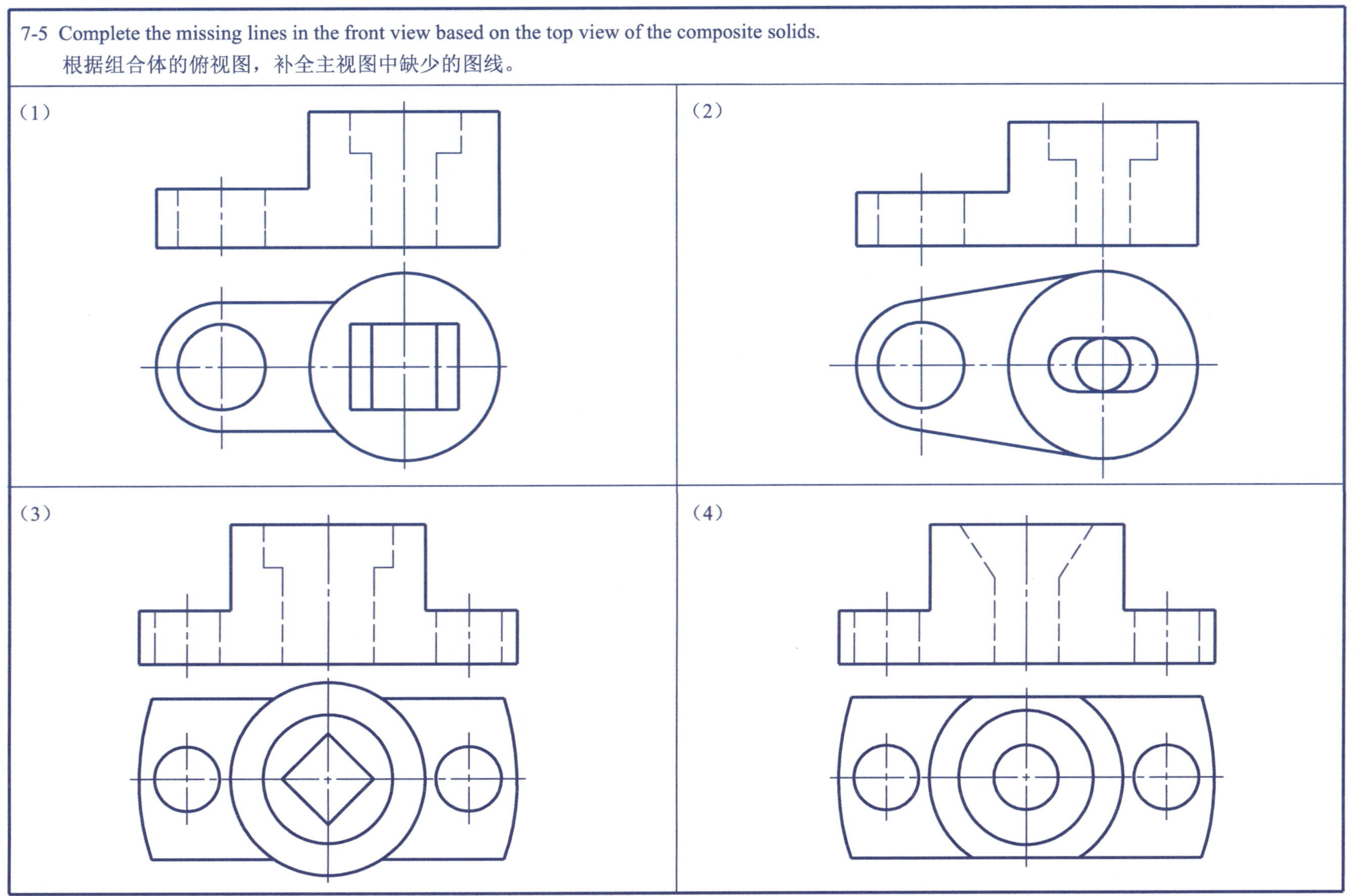

Class 班级：　　Name 姓名：　　ID Number 学号：

7-6 Based on the two views of the composite solid, draw the third view.
根据组合体的两个视图，补画第三个视图。

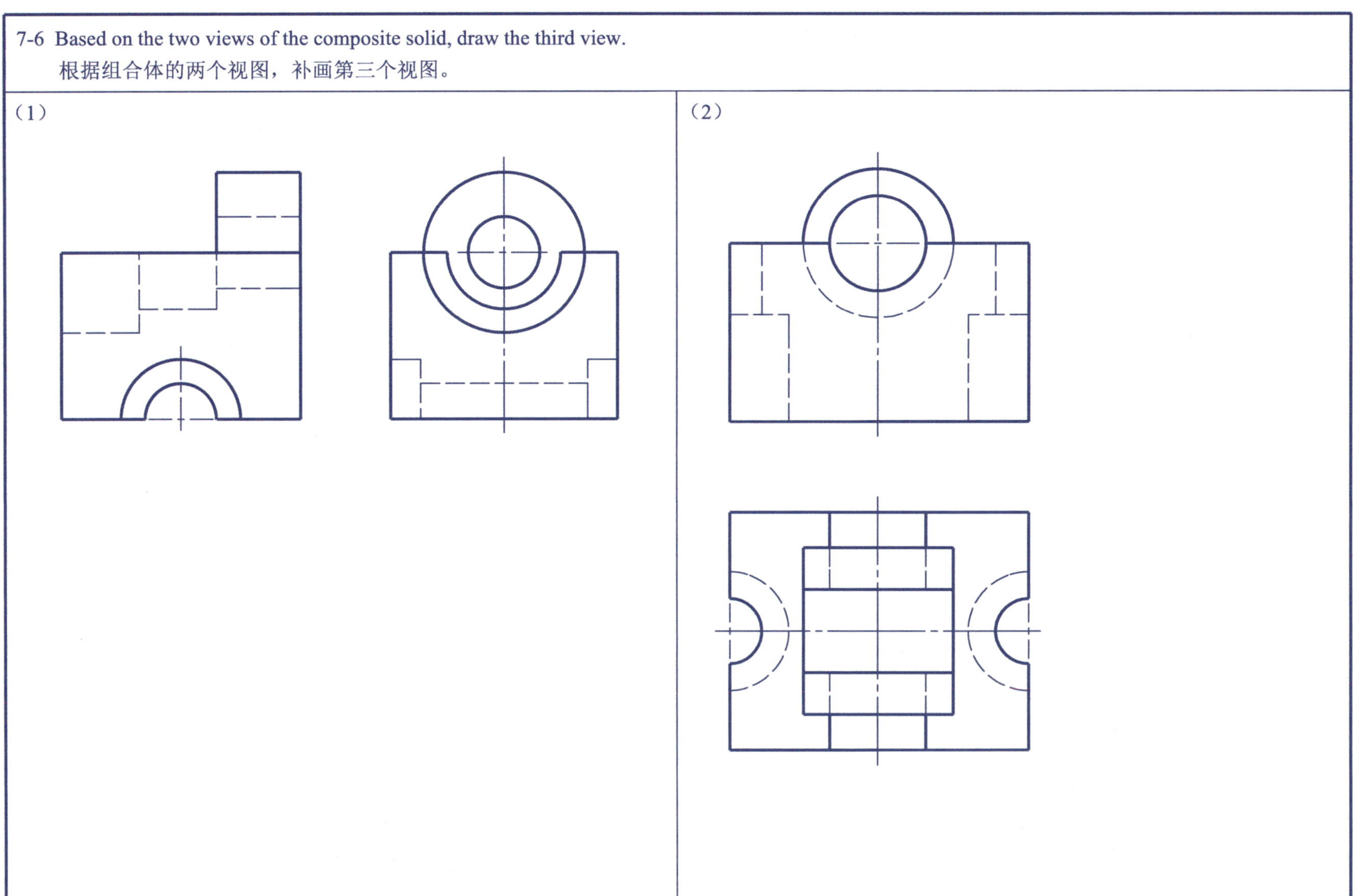

 Class 班级： Name 姓名： ID Number 学号：

7-7 Based on the two views of the composite solid, draw the third view.
根据组合体的两个视图，补画第三个视图。

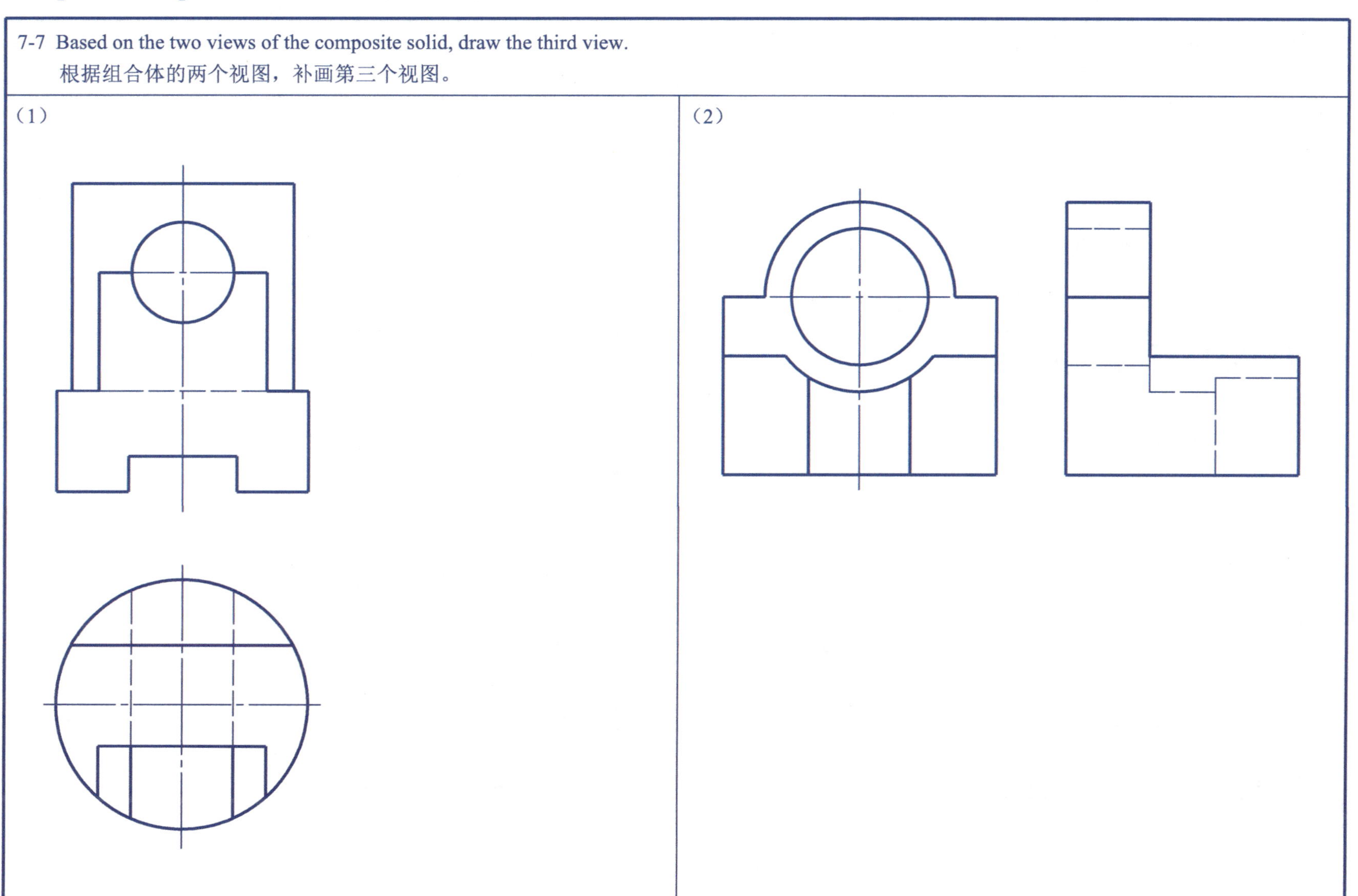

7-8 Based on the two views of the composite solid, draw the third view.
根据组合体的两个视图，补画第三个视图。

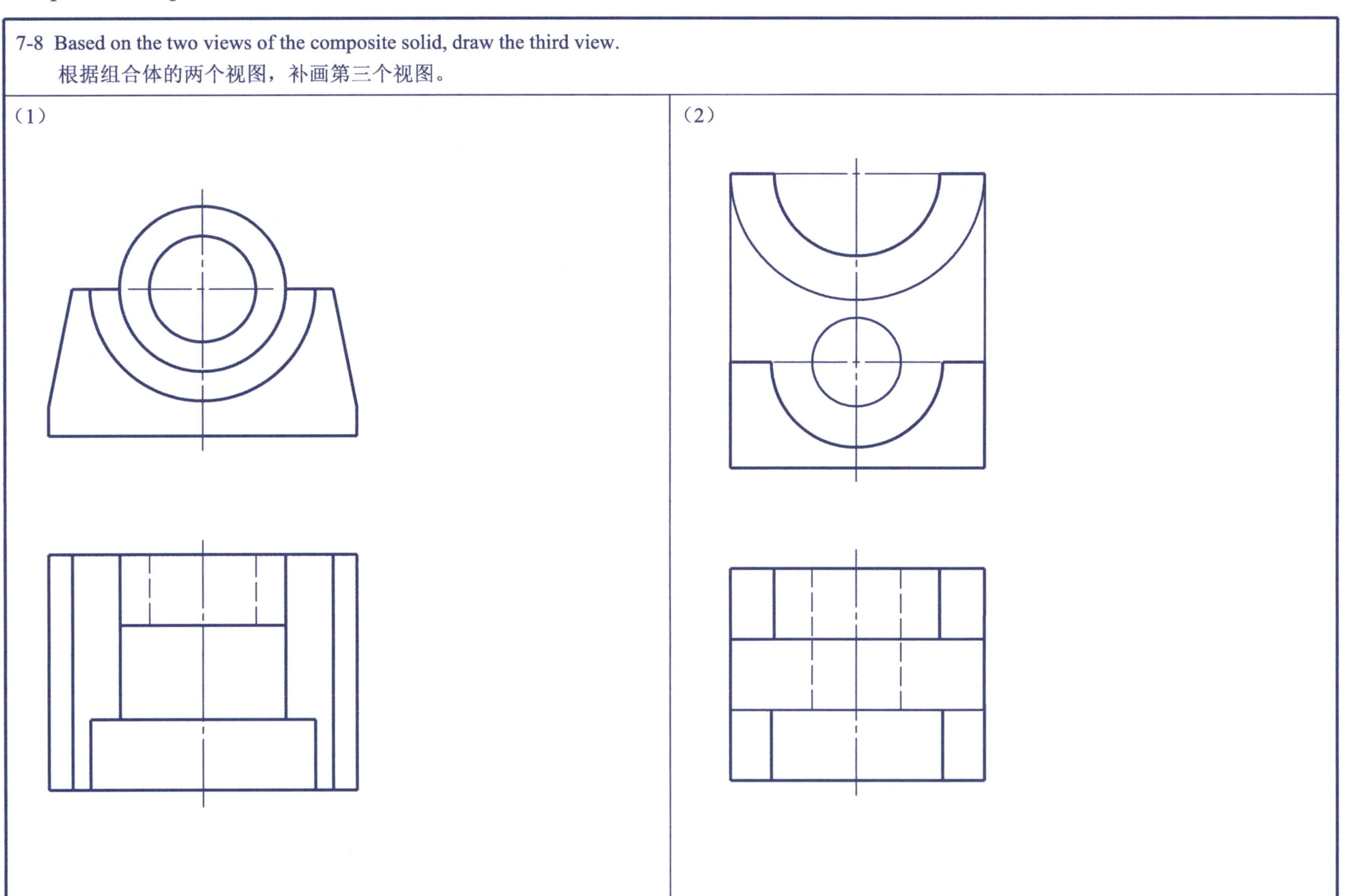

7-9 Based on the two views of the composite solid, draw the third view.
根据组合体的两个视图，补画第三个视图。

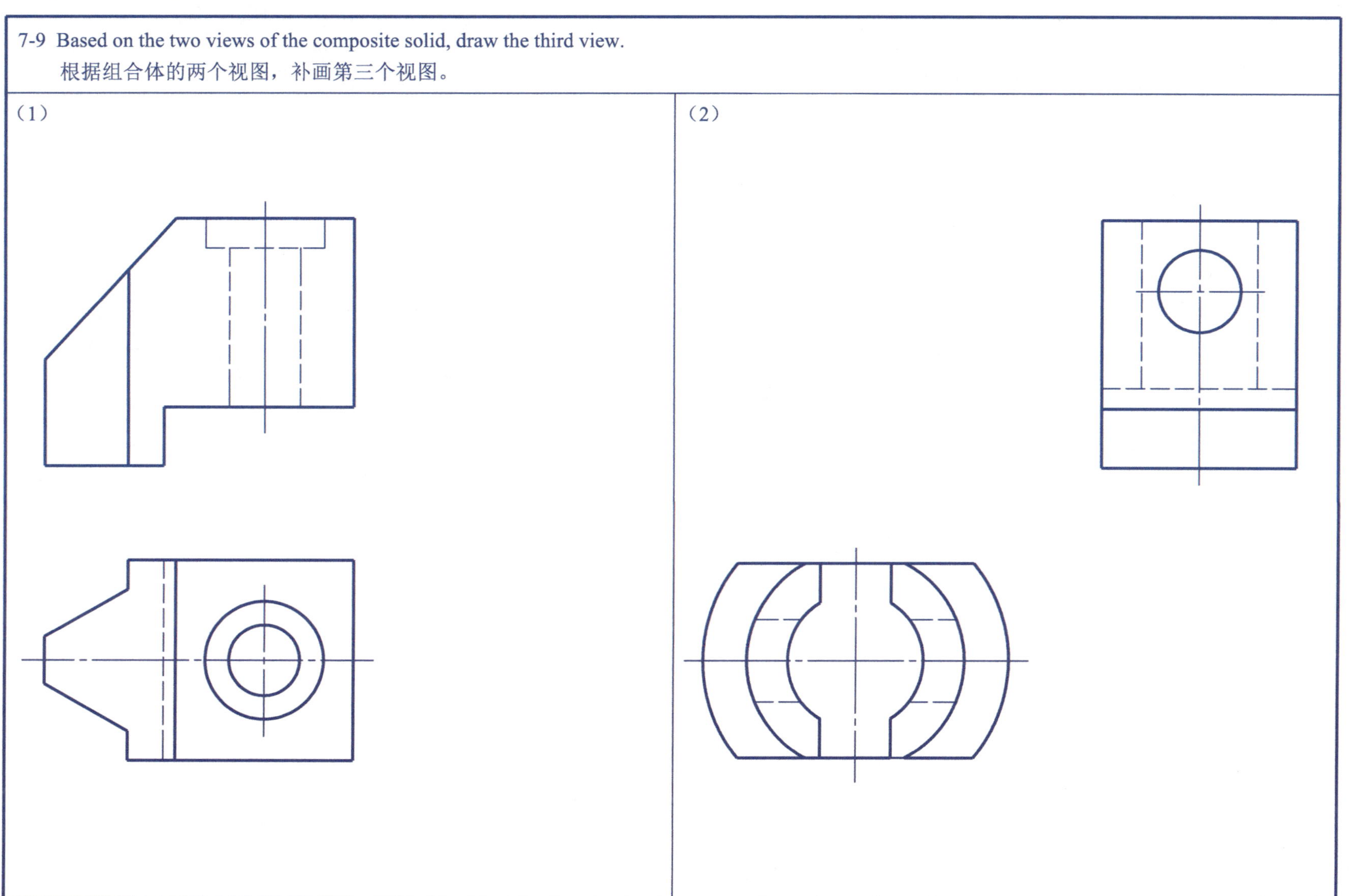

7-10 Based on the two views of the composite solid, draw the third view.
根据组合体的两个视图，补画第三个视图。

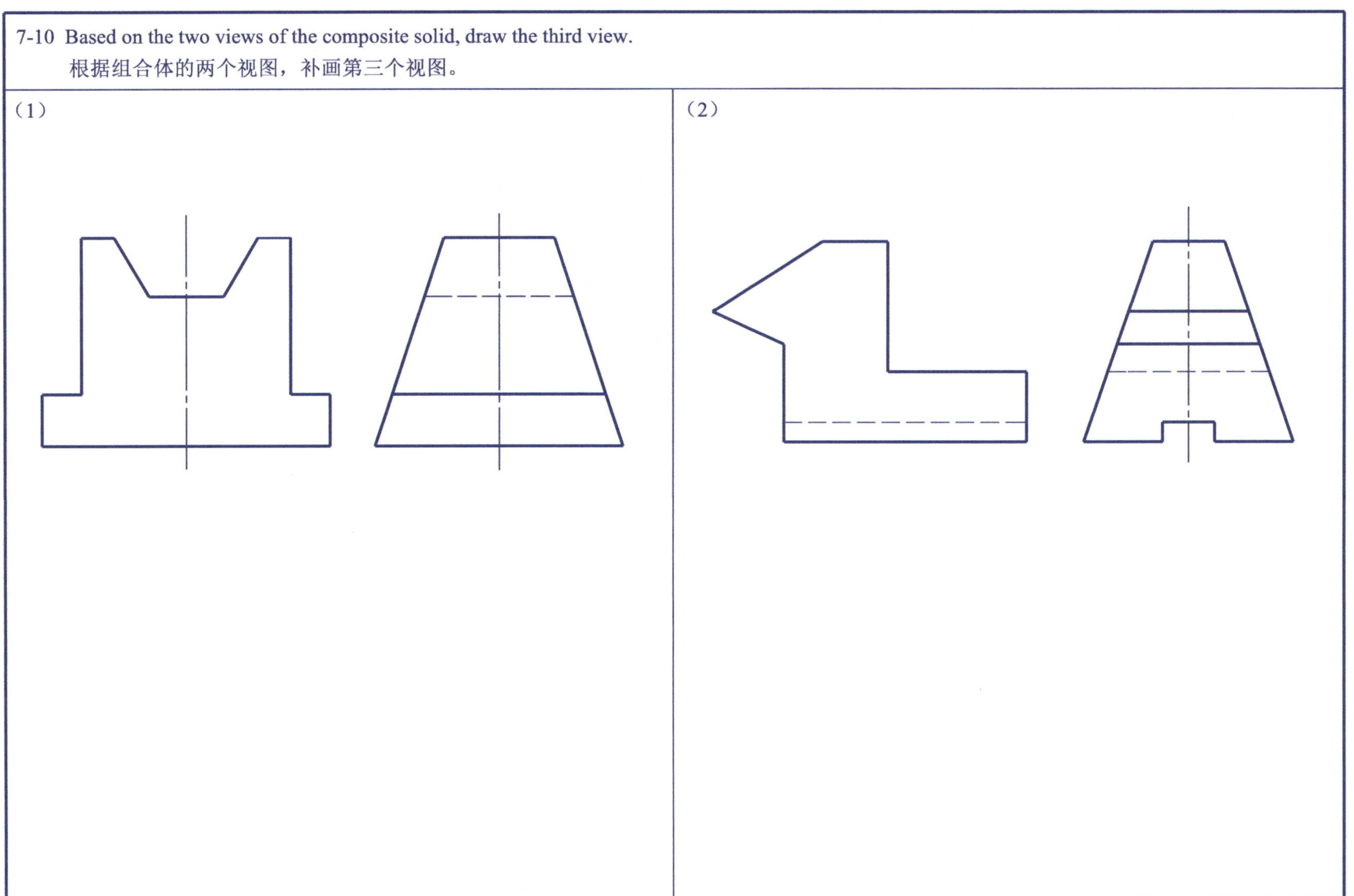

7-11 Based on the two views of the composite solid, draw the third view.
根据组合体的两个视图，补画第三个视图。

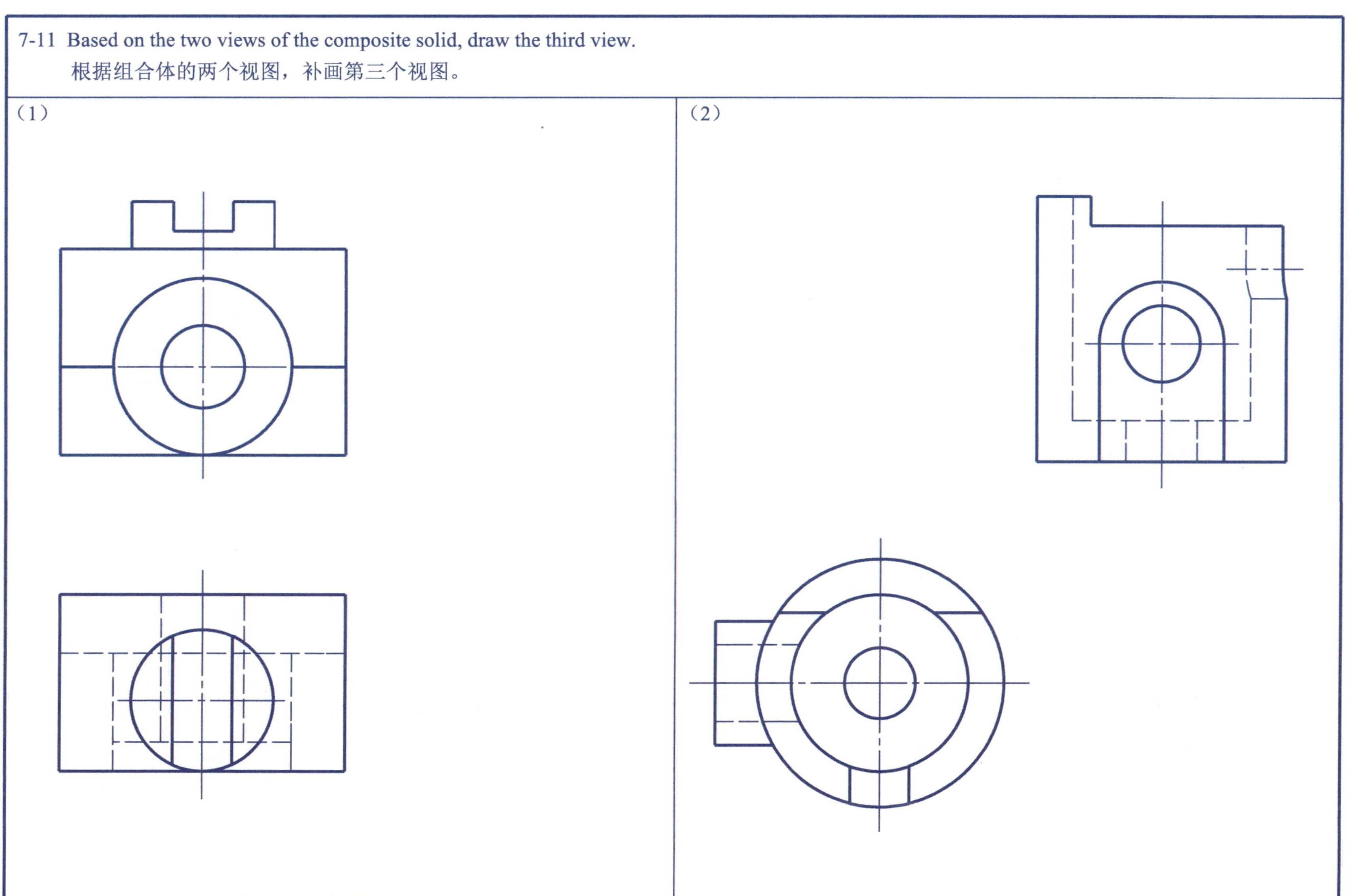

7-12 Mark the dimensions of the composite solid (measured 1∶1 on the diagram and rounded).

标注组合体的尺寸（在图上1∶1量取，并取整）。

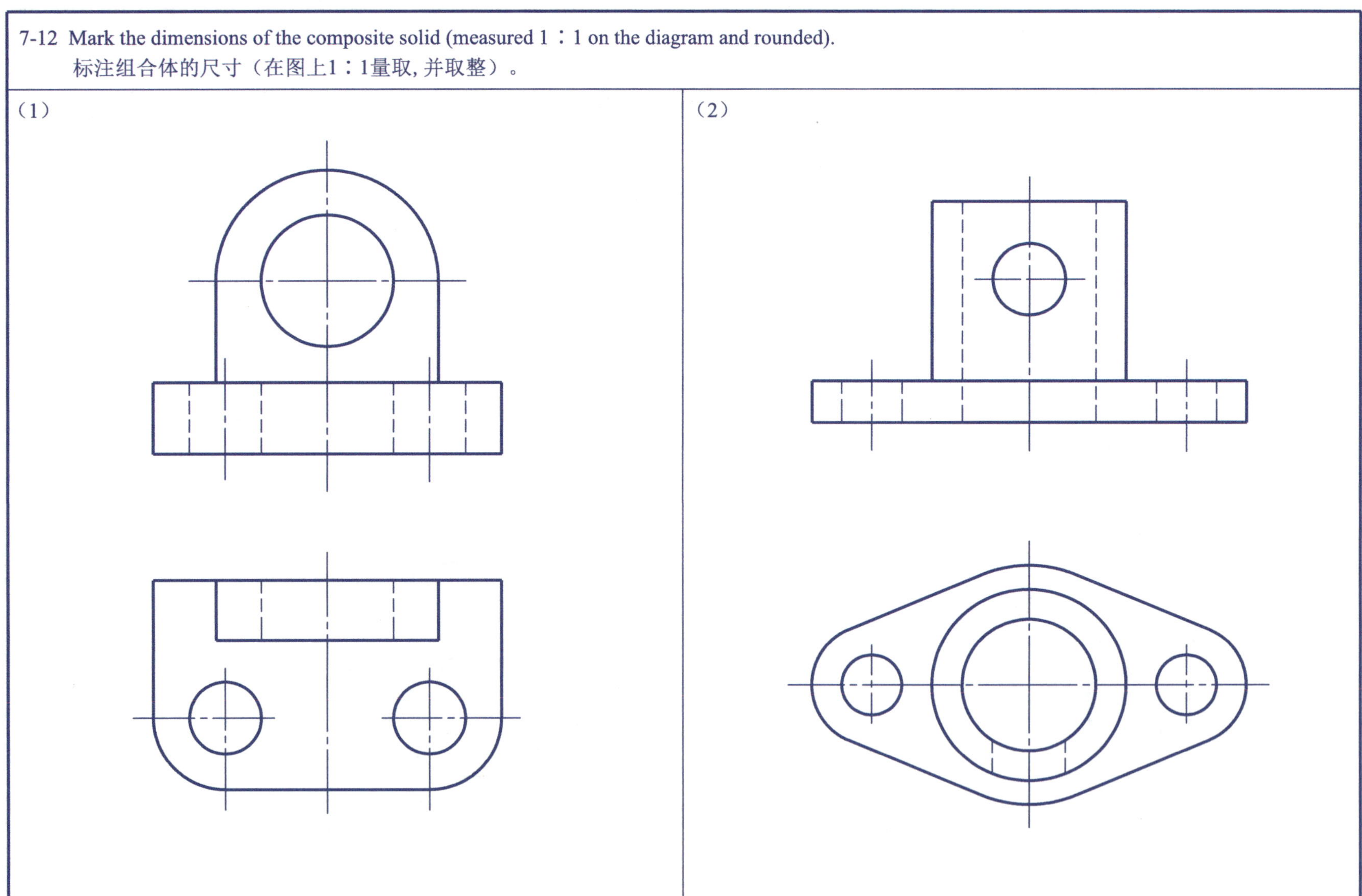

7-13 Mark the dimensions of the composite solid (measured 1：1 on the diagram and rounded).
标注组合体的尺寸（在图上1：1量取，并取整）。

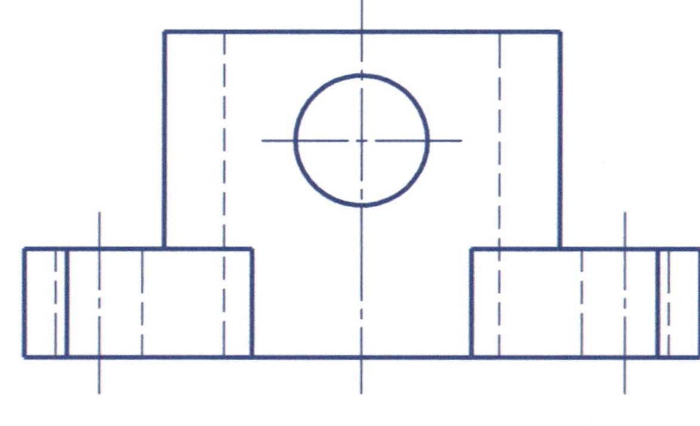

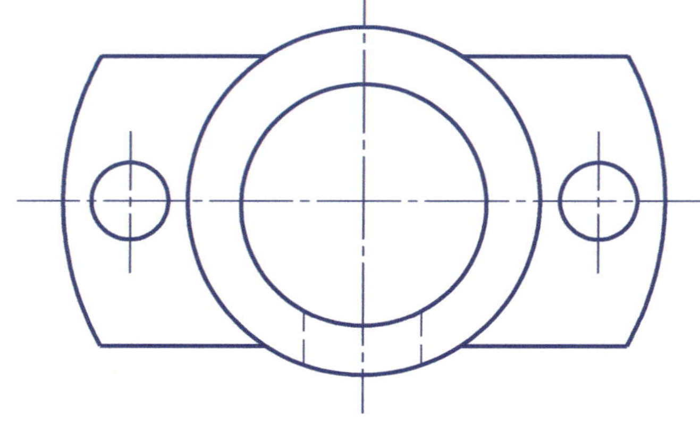

7-14 Mark the dimensions of the composite solid (measured 1：1 on the diagram and rounded).
标注组合体的尺寸（在图上1：1量取，并取整）。

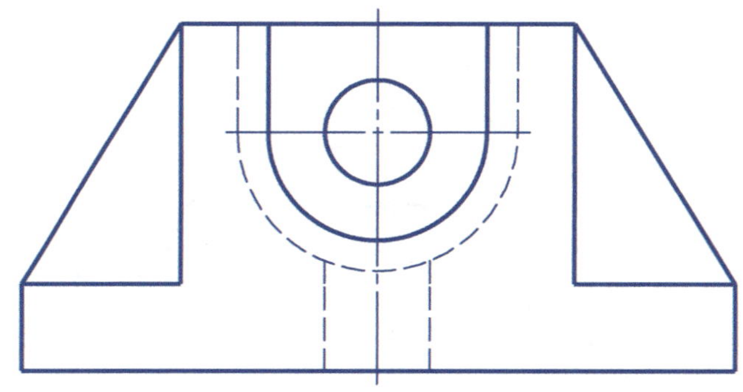

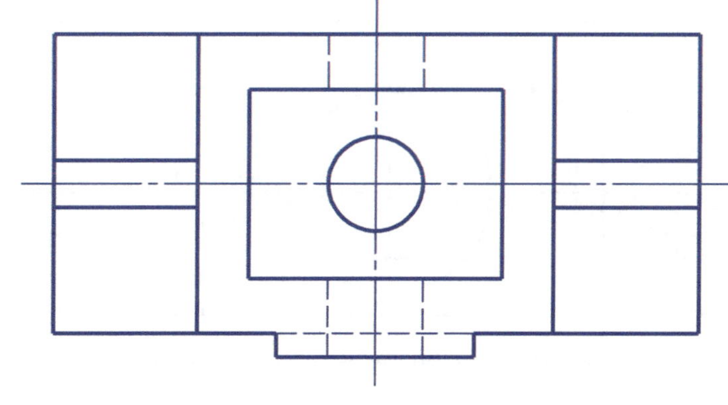

Class 班级： Name 姓名： ID Number 学号：

7-15 Draw the third view based on the two views of the composite solid, and mark the dimensions of the composite solid (measured 1 ∶1 from the diagram and rounded). Draw it on the A3 drawing with a scale of 2 ∶1.

根据组合体的两个视图，补画第三个视图，并标注组合体的尺寸（从图上1∶1量取，并取整）；用2∶1的比例画到A3图纸上。

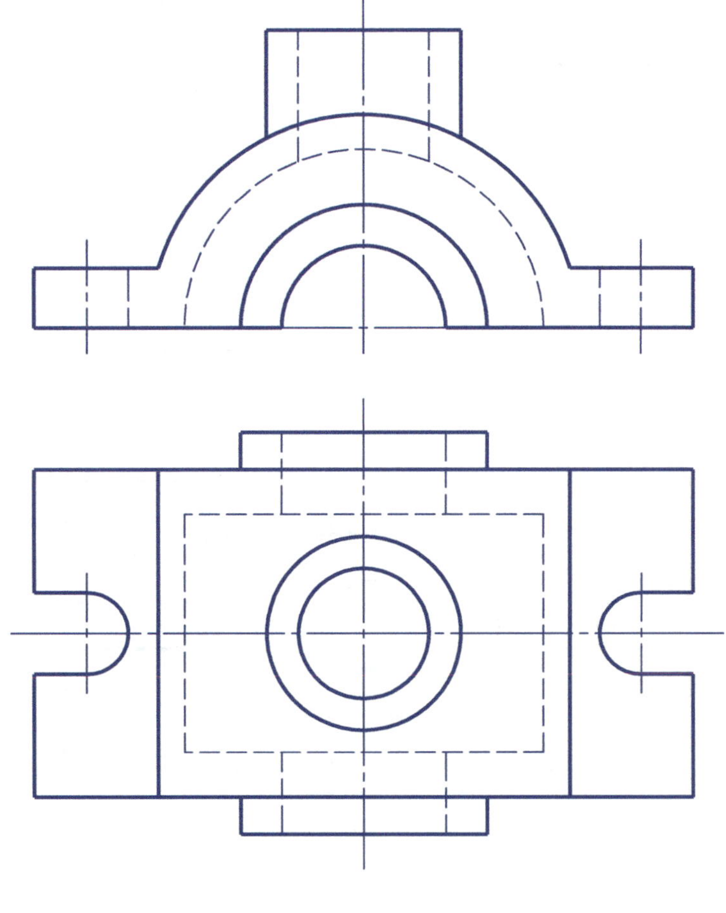

　　Class 班级：　　Name 姓名：　　ID Number 学号：

7-16 Based on the two views of the composite solid, draw the third view.
根据组合体的两个视图，补画第三个视图。

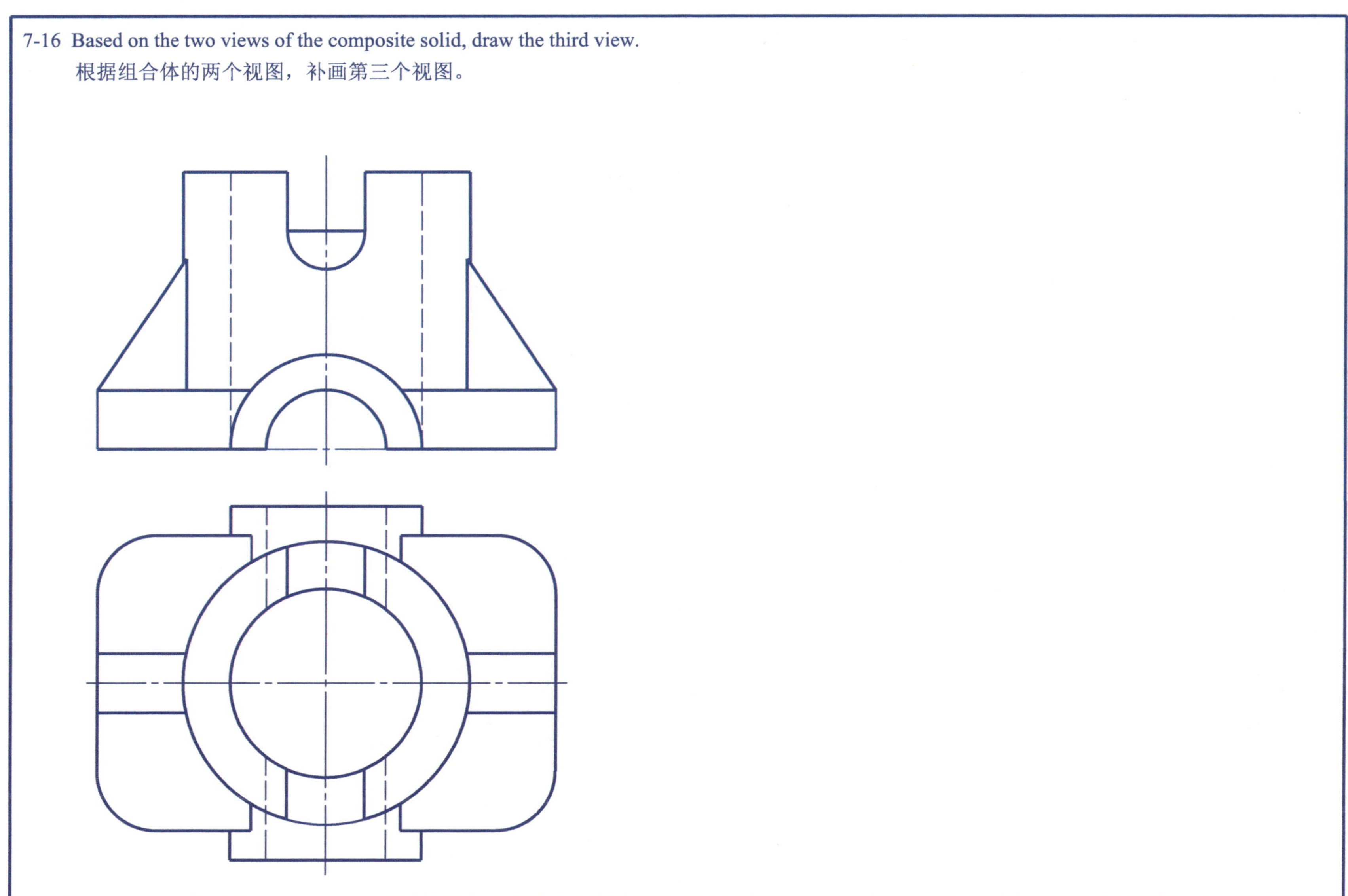

Class 班级：　　Name 姓名：　　ID Number 学号：

7-17 Based on the two views of the composite solid, draw the third view.
根据组合体的两个视图，补画第三个视图。

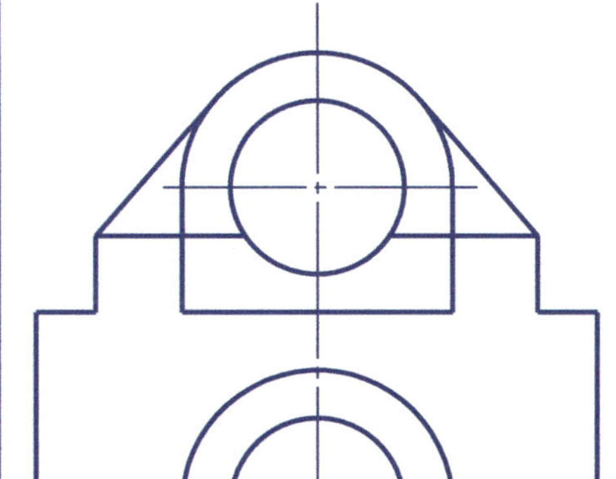

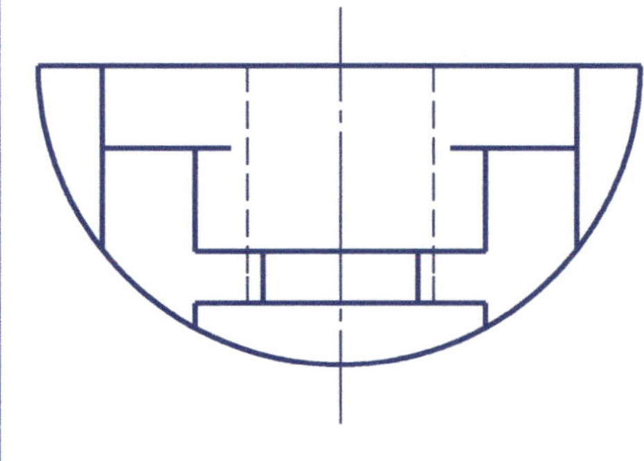

7-18 Based on the two views of the composite solid, draw the third view.
根据组合体的两个视图，补画第三个视图。

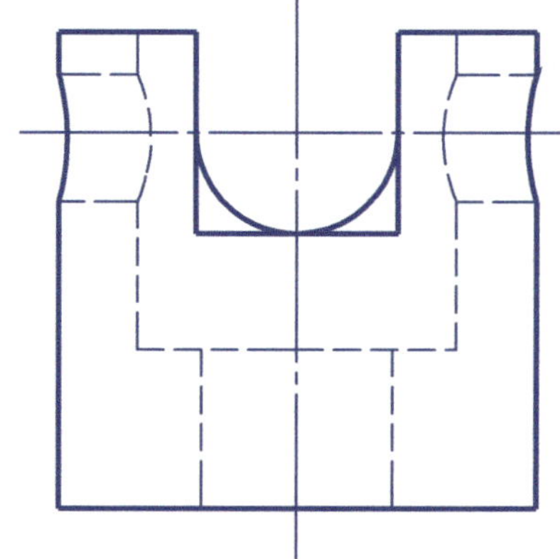

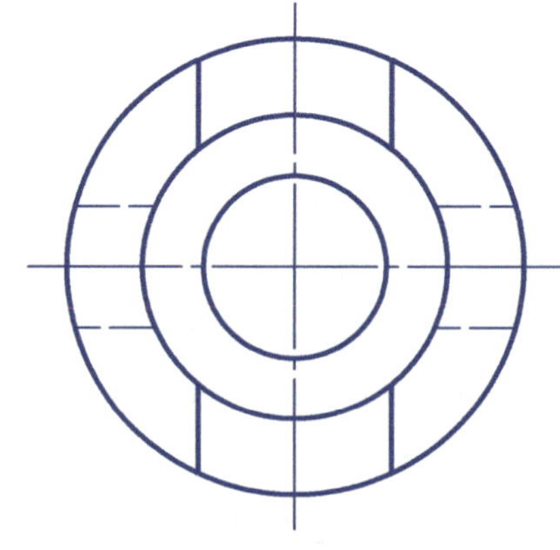

Class 班级：　　Name 姓名：　　ID Number 学号：

7-19 Based on the two views of the composite solid, draw the third view.
根据组合体的两个视图，补画第三个视图。

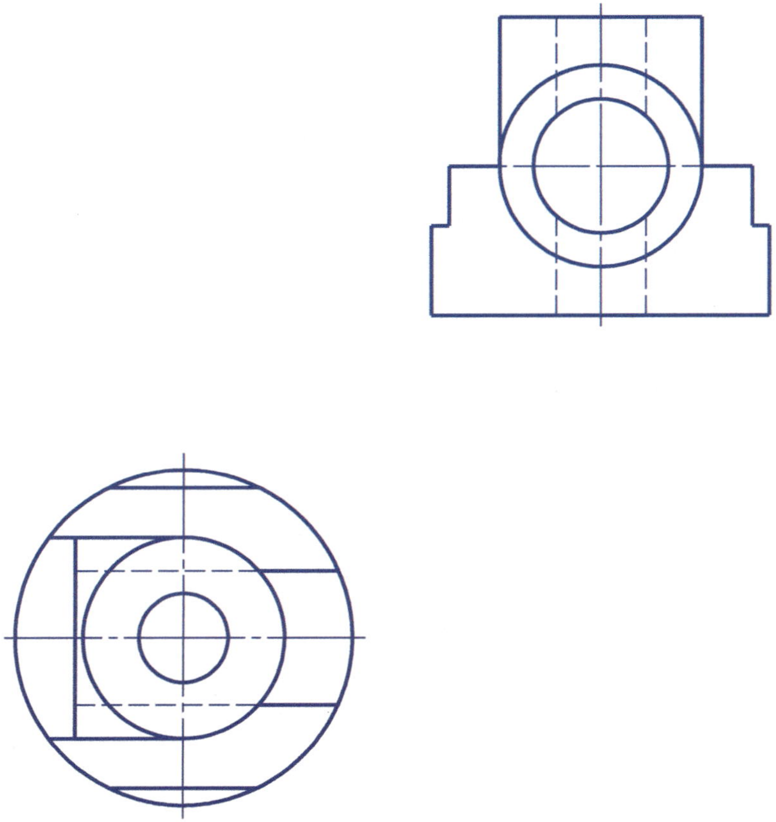

7-20 Based on the two views of the composite solid, draw the third view.
根据组合体的两个视图，补画第三个视图。

Class 班级： Name 姓名： ID Number 学号：

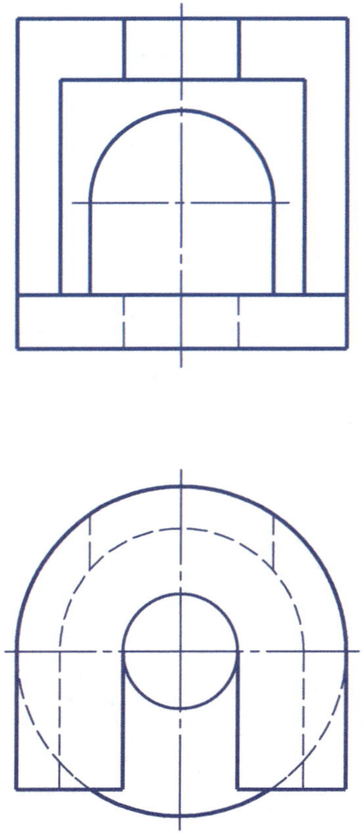

7-21 Based on the two views of the composite solid, draw the third view.
根据组合体的两个视图，补画第三个视图。

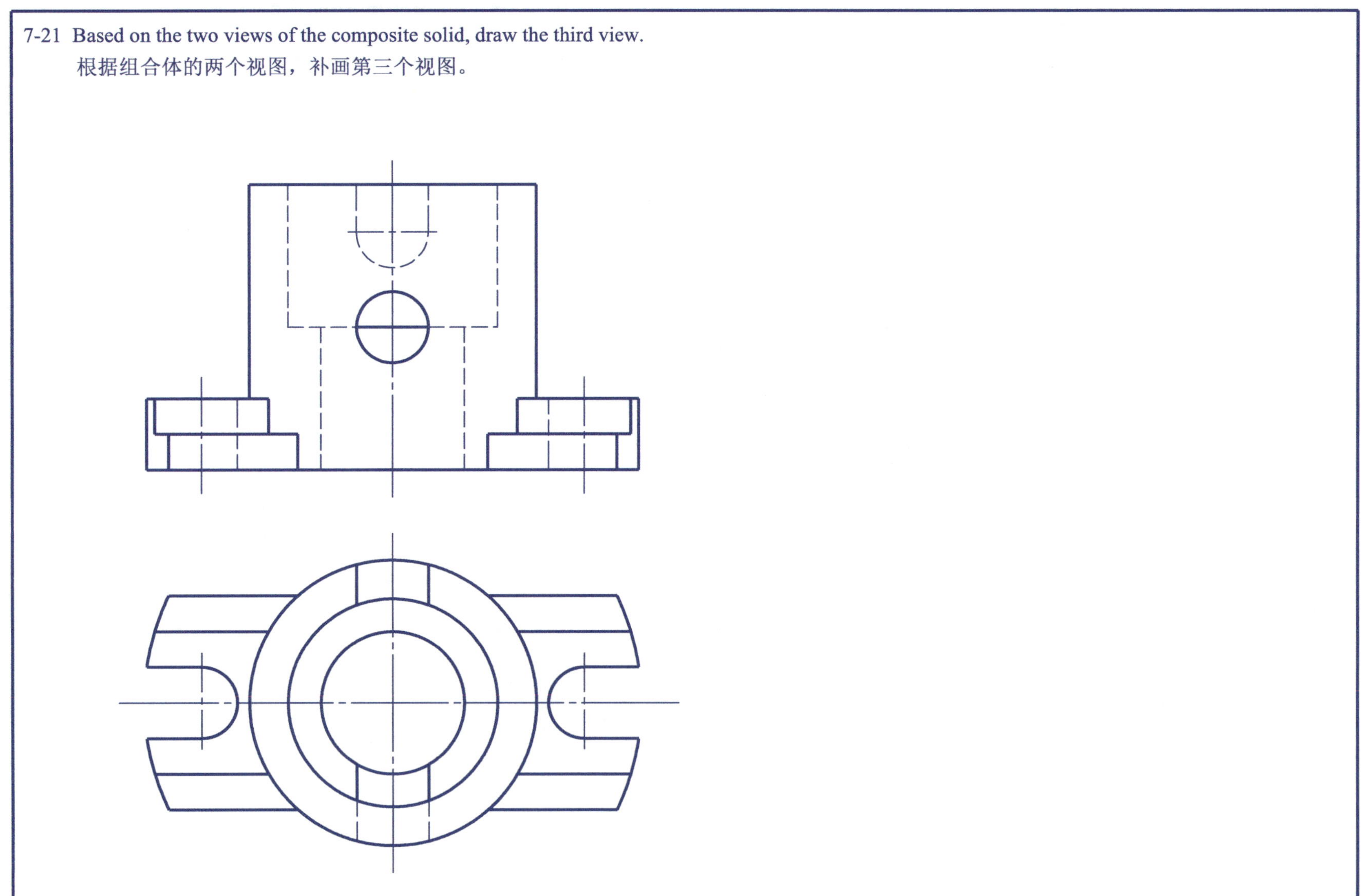

 Class 班级： Name 姓名： ID Number 学号：

7-22 Based on the front view given, conceive a composite solid of different shapes and draw the other two views.
根据所给的主视图，构思不同形状的组合体，并画出该组合体的另外两个视图。

The given front view
题给的主视图

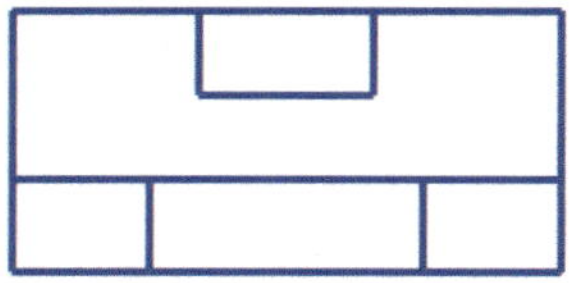

The front view given can not uniquely determine the shape of the composite solid. Each line box in the front view can represent a surface, whose position can be front or back, be convex or concave, and be flat or curved, resulting in a variety of solutions. The right axonometric projection is one of the solutions.

题中所给的主视图并不能唯一地确定组合体形状，因为主视图中的线框可表示一个面，且每面位置可前、可后，形状可凸、可凹、可平、可曲，从而产生多种解。右侧轴测图即为其中一解。

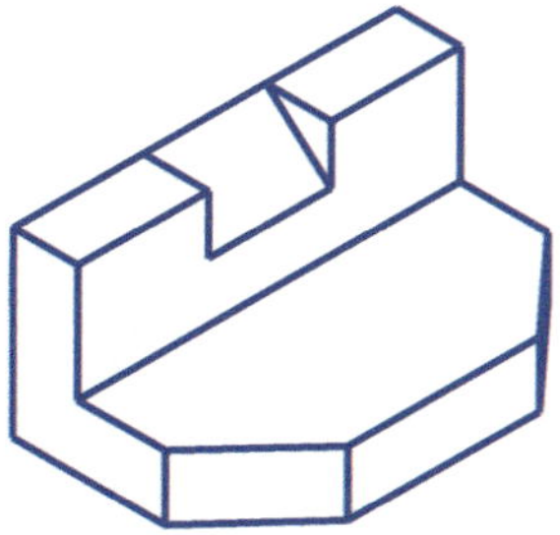

Solution①:
解①：

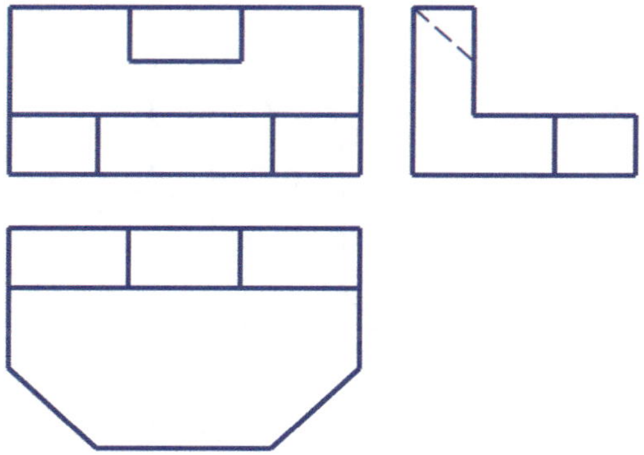

Solution②:
解②：

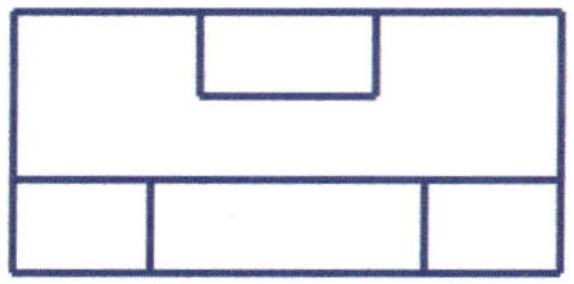

Solution③:
解③：

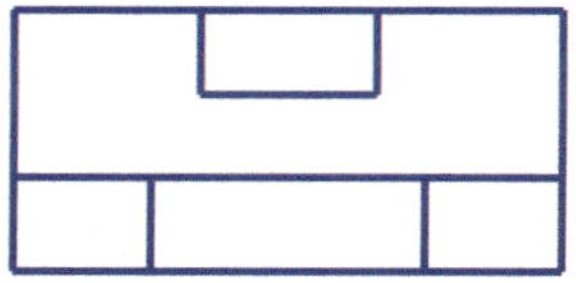

Class 班级： Name 姓名： ID Number 学号：

8-1 Based on the axonometric projection given, the front view, the top view and the left view, draw its other three basic views (Draw according to the view expansion drawing method specified in the national standard).

根据机件的轴测图及其主、俯、左三视图，补画其他三个基本视图（按照国家标准规定的视图展开画法绘制）。

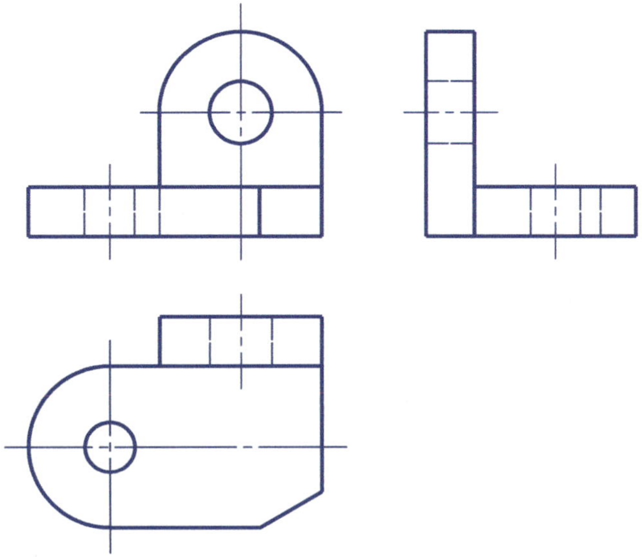

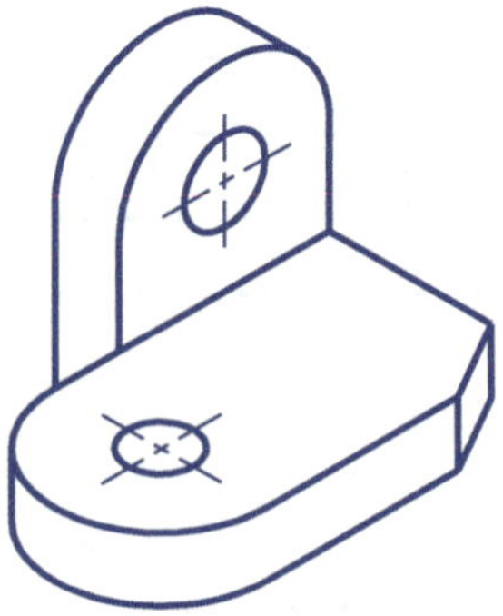

8-2 Based on the axonometric projection and two views given, draw the partial view from direction *A*.
根据所给的两个视图，参照轴测图画出*A*向局部视图。

8-3 Based on the axonometric projection and the front view given, draw the oblique view from direction *A*.
根据轴测图及主视图，画出*A*向斜视图。

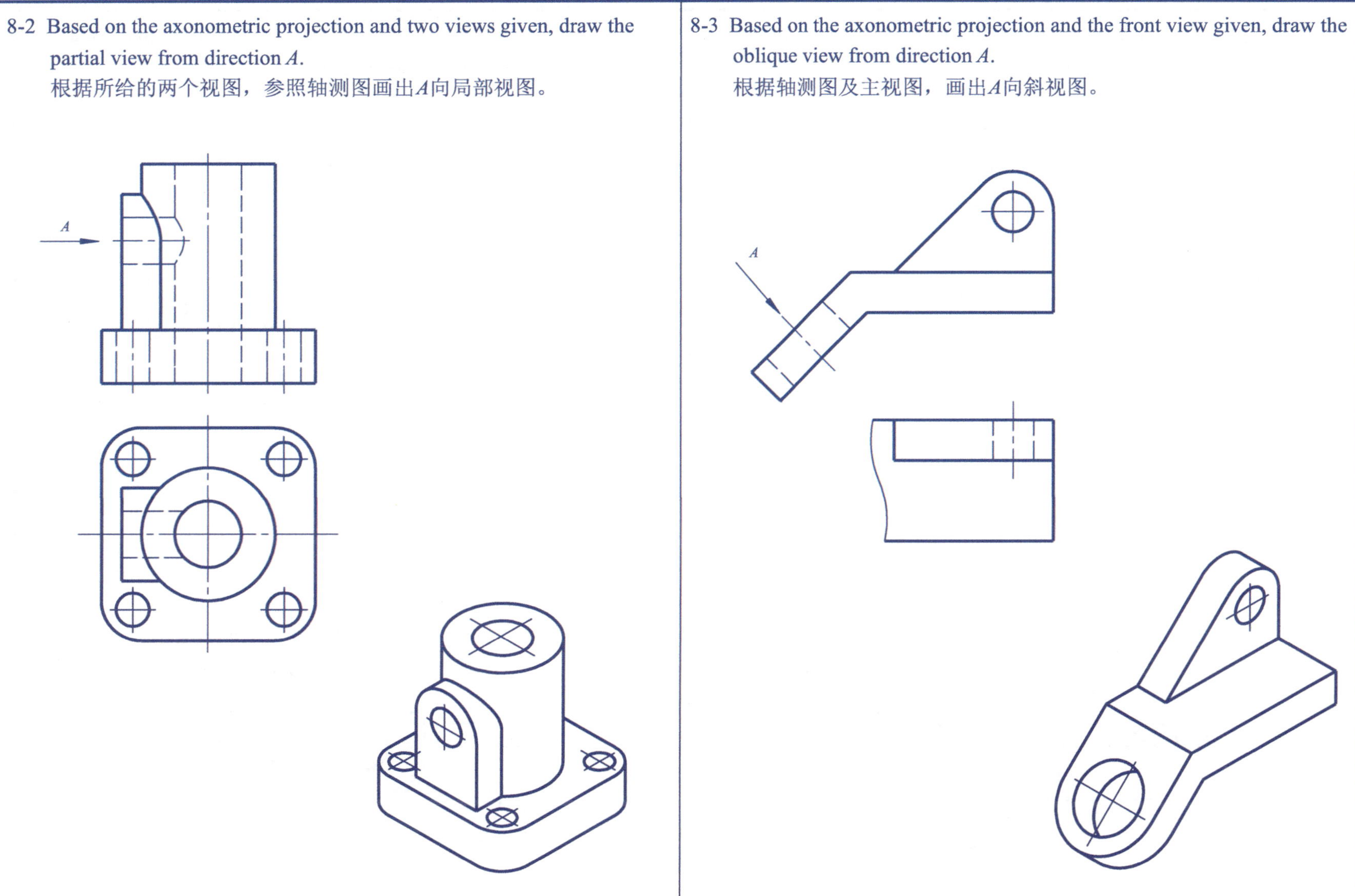

8-4 In the space given, change the front view into the full sectional view.
在中间空白处将主视图改画成全剖视图。

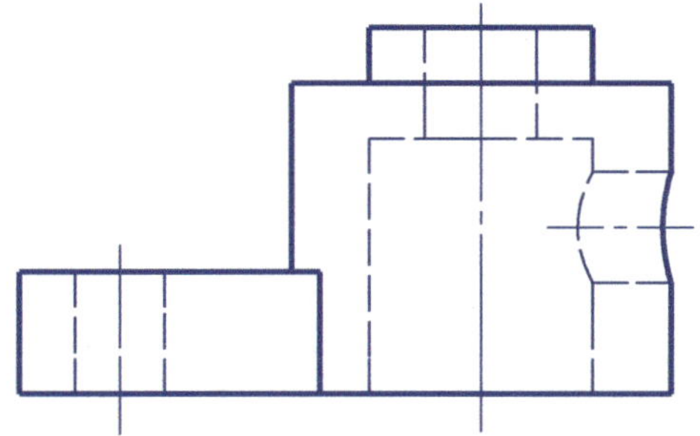

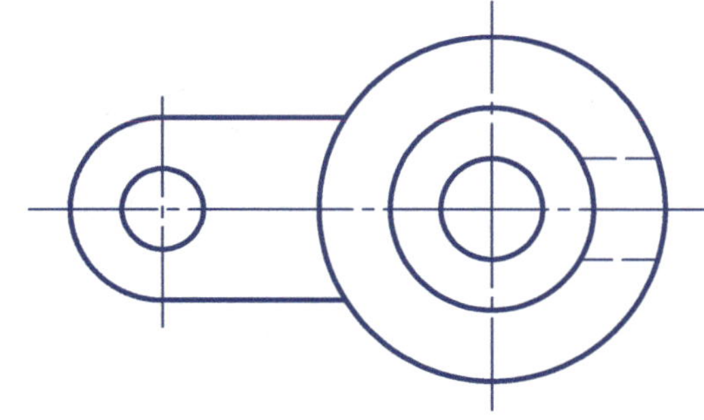

8-5 In the space given, change the front view into the full sectional view.
在中间空白处将主视图改画成全剖视图。

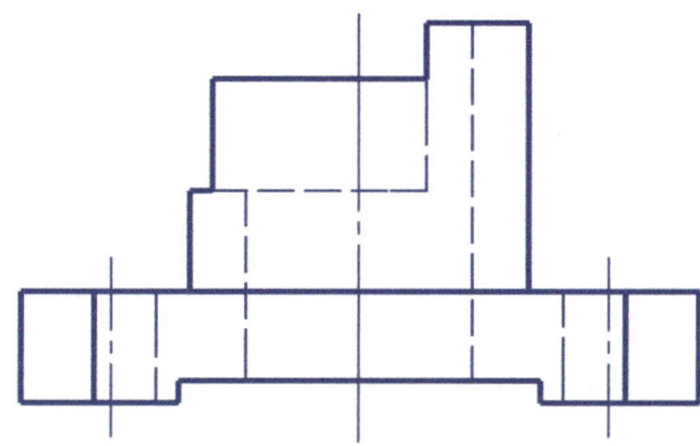

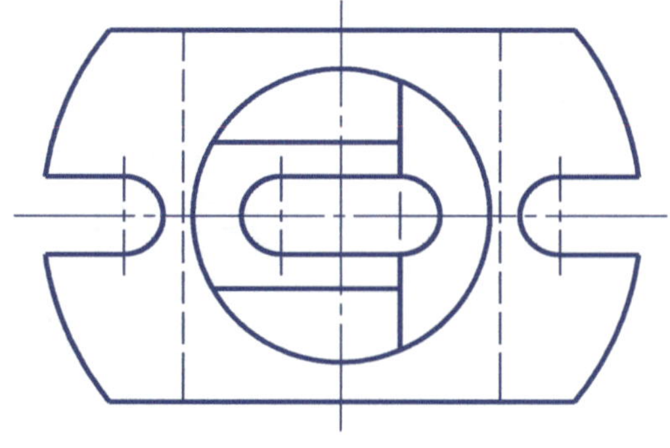

8-6 Draw the missing lines in the sectional views given below.
补画剖视图中缺少的图线。

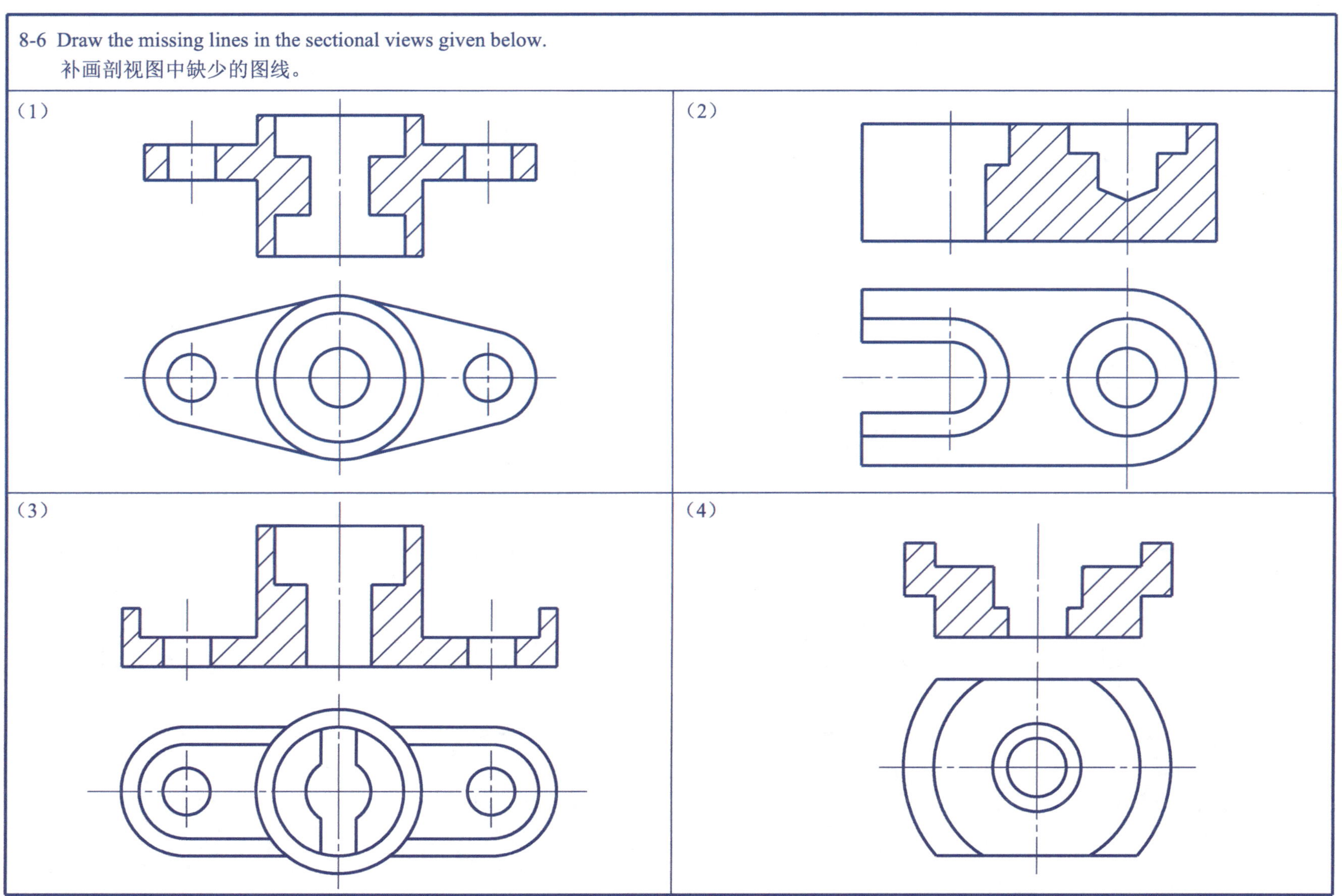

Class 班级： Name 姓名： ID Number 学号：

8-7 Draw the missing lines in the sectional views given below.
补画剖视图中缺少的图线。

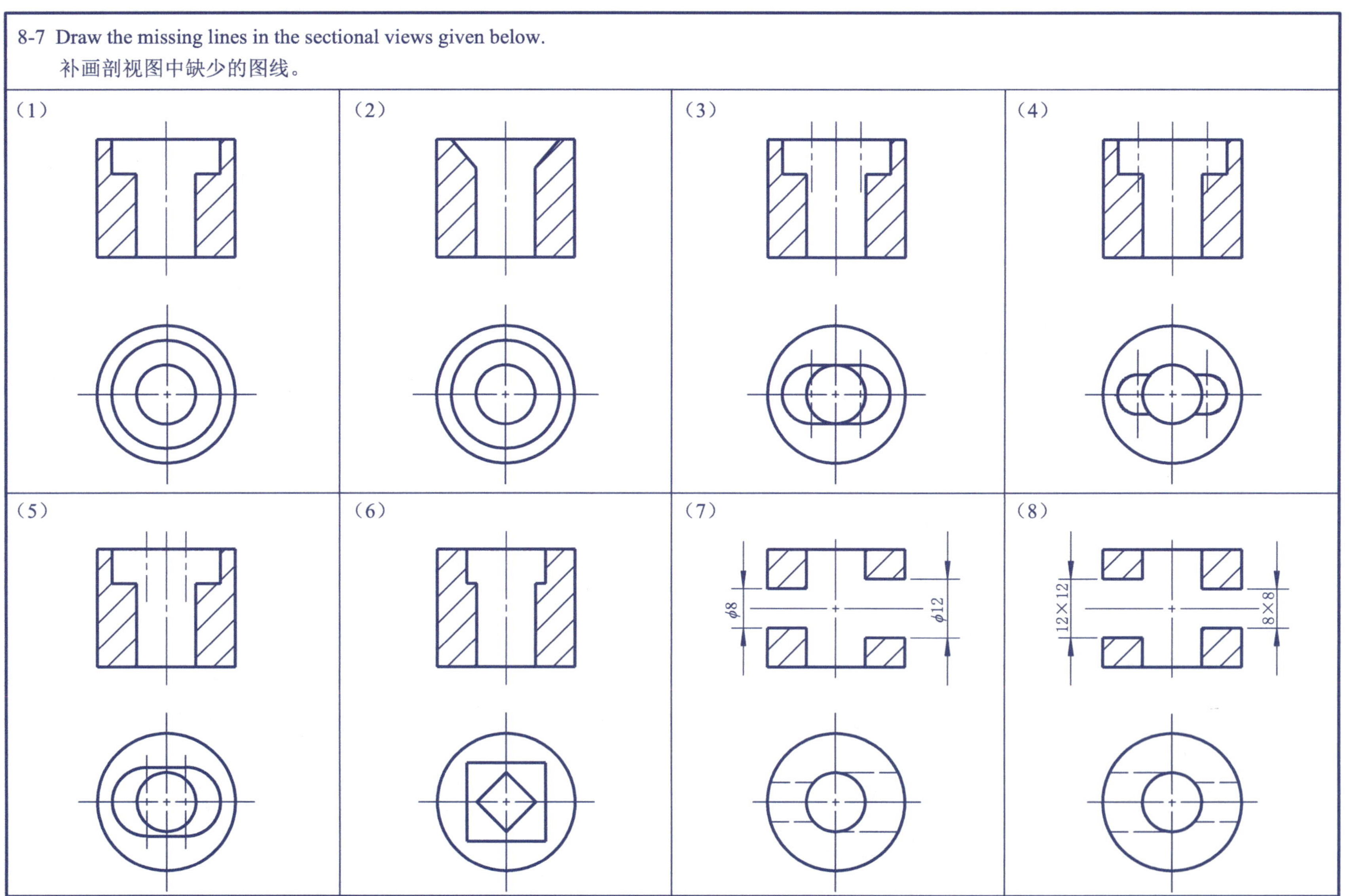

Class 班级： Name 姓名： ID Number 学号：

8-8 Draw the missing lines in the front view and left view given below.
补全主视图和左视图中缺少的图线。

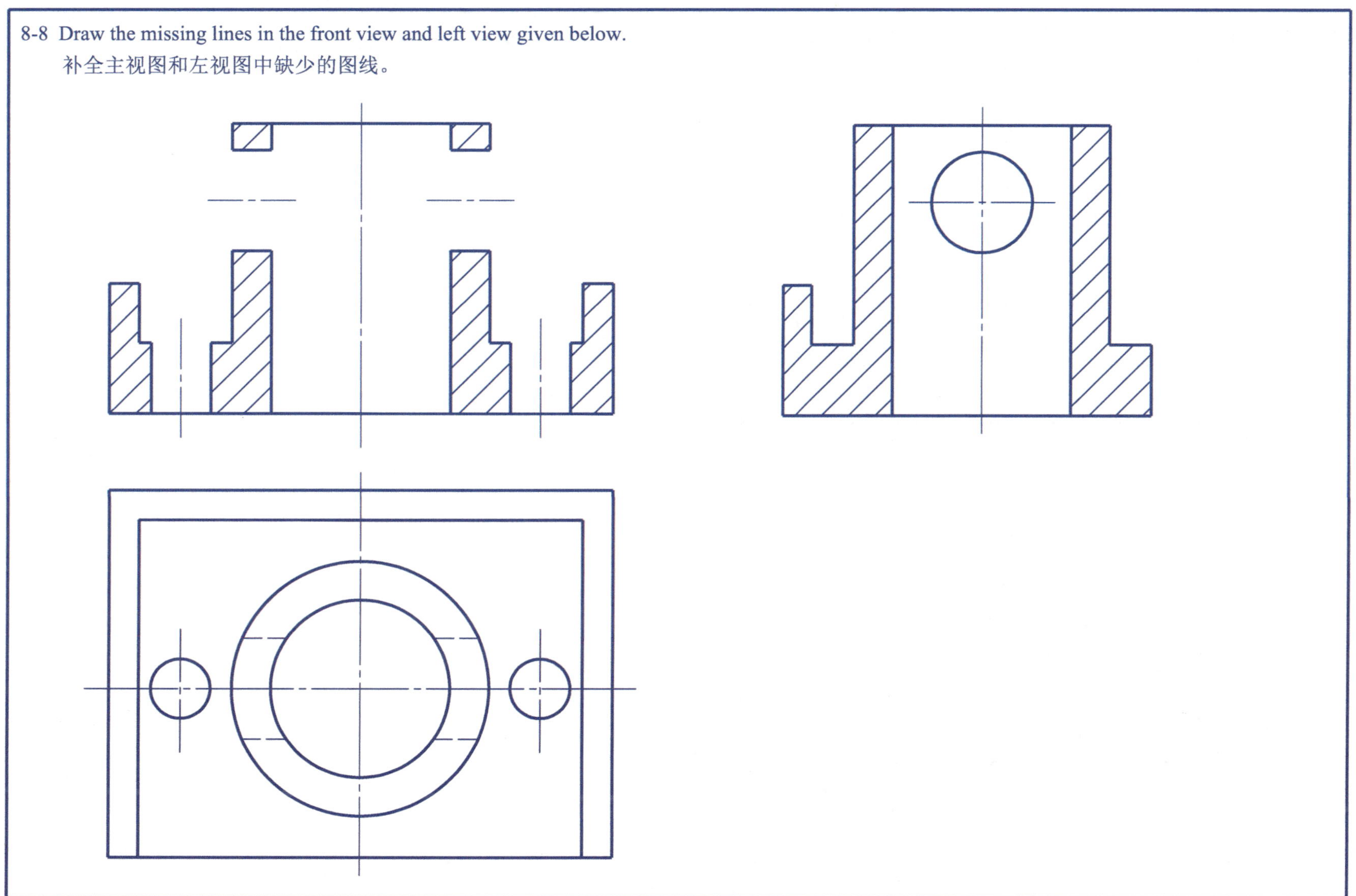

Class 班级： Name 姓名： ID Number 学号：

8-9 Draw the missing lines in the half-sectional front view, and draw the left view as a full sectional view.
补画半剖的主视图中缺少的图线，并补画全剖的左视图。

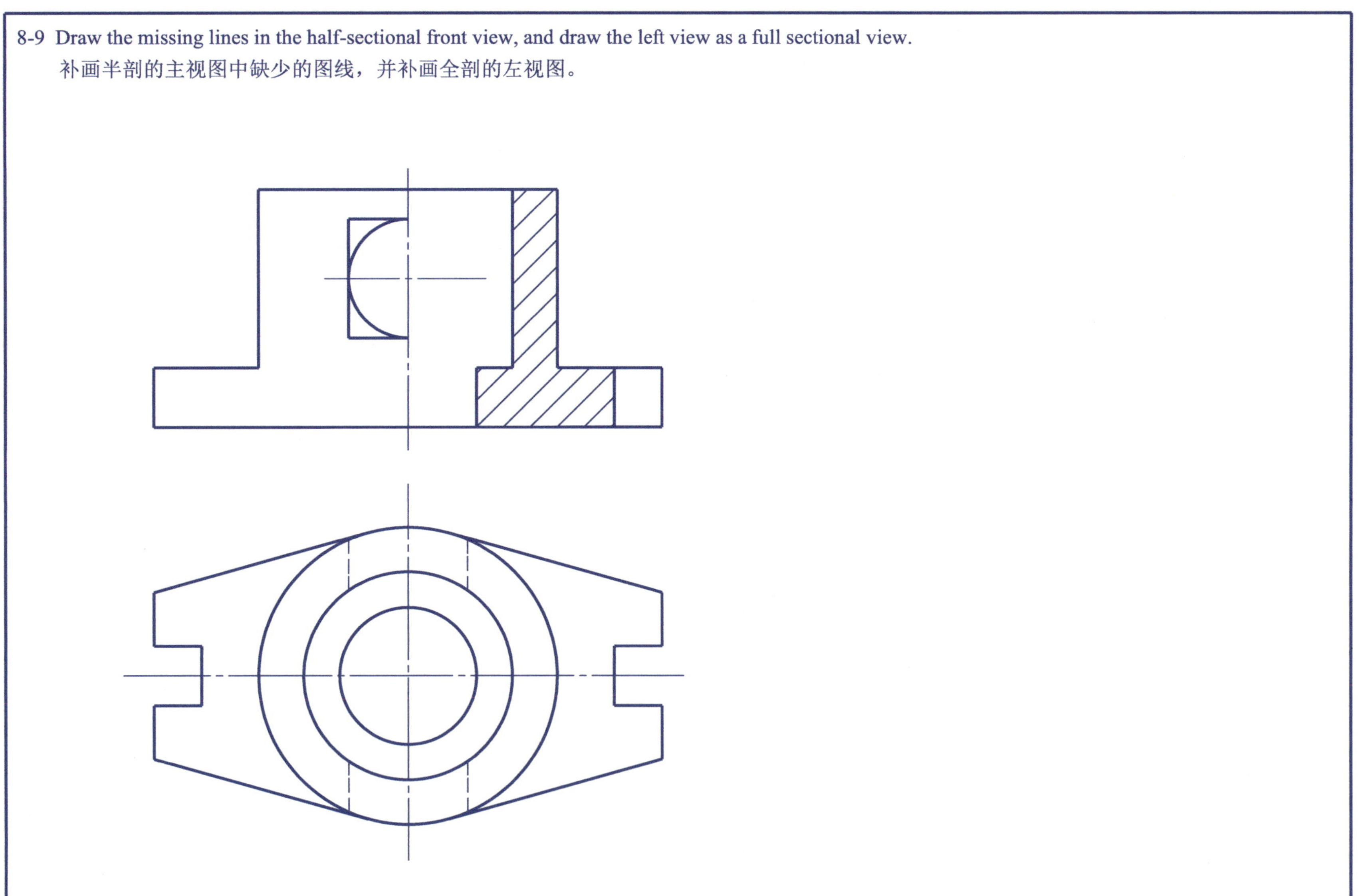

8-10 Draw the missing lines in the half-sectional view given below.
补画半剖视图中所漏画的图线。

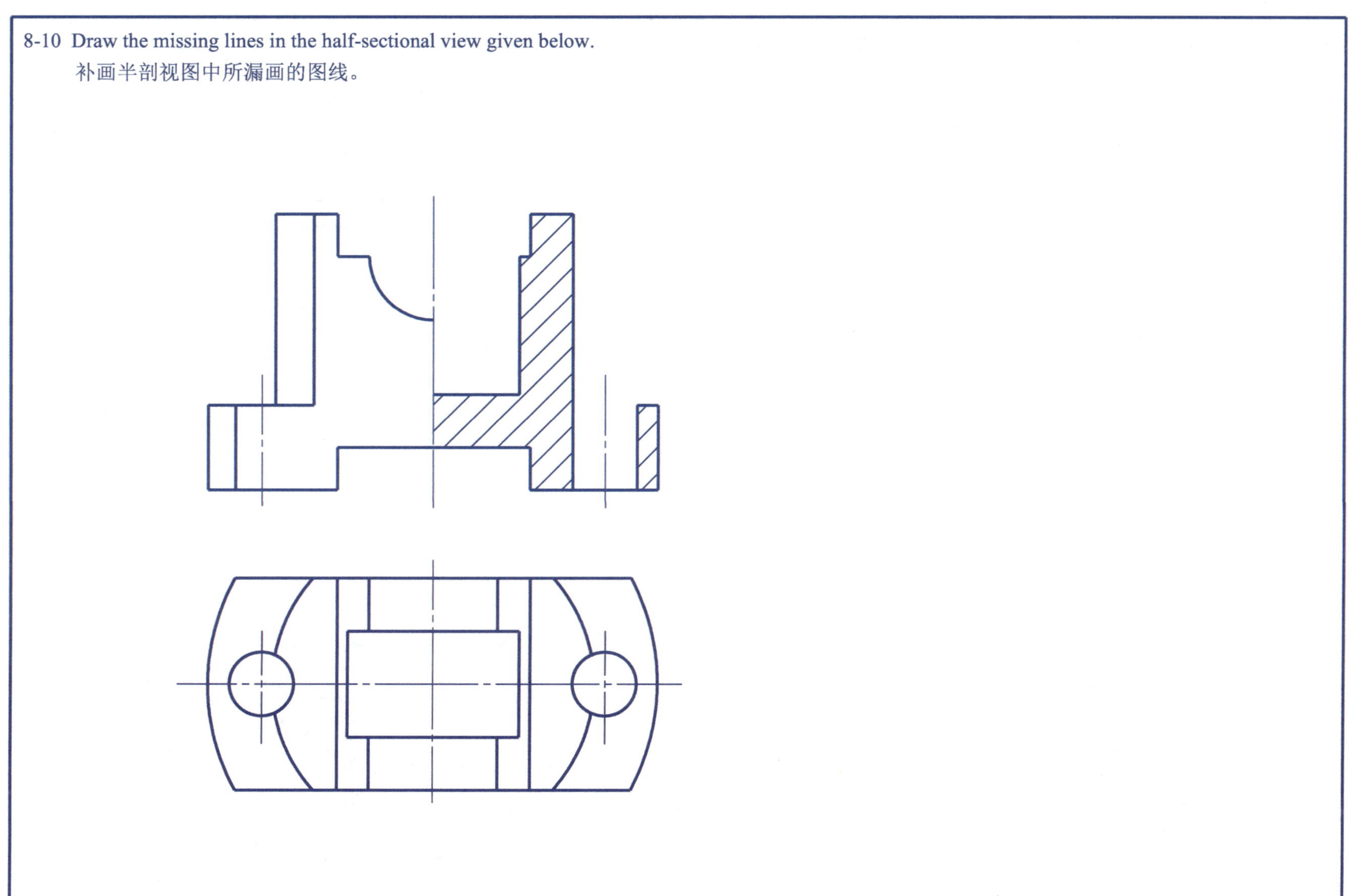

8-11 In the space given, change the front view into the half-sectional view.
在中间空白处将机件的主视图画成半剖视图。

8-12 In the space given, change the front view into the half-sectional view.
在中间空白处将机件的主视图画成半剖视图。

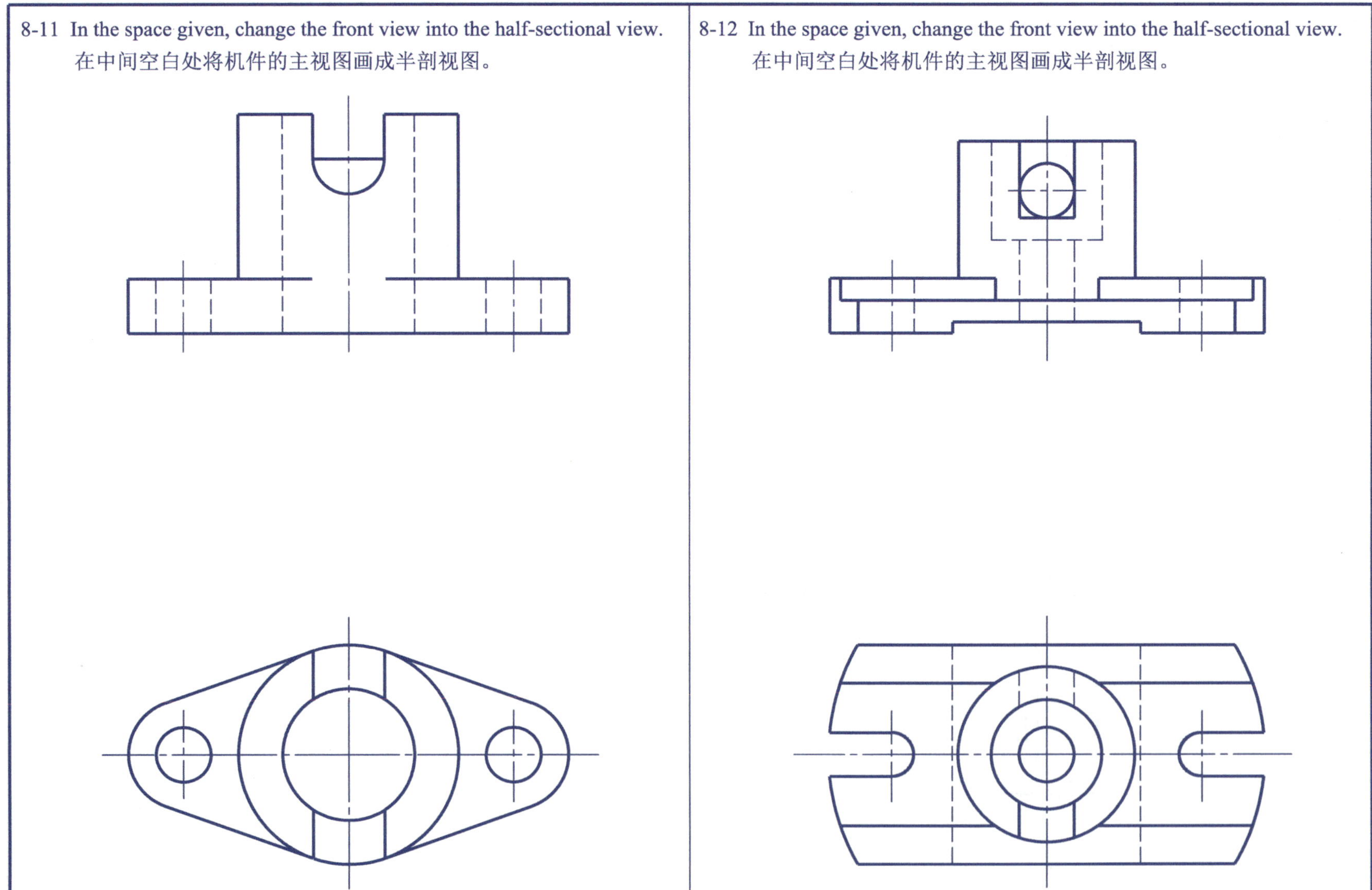

8-13 Based on the two views given, draw the half-sectional views of the front view and left view.

根据给出的两个视图，画出主视图和左视图的半剖视图。

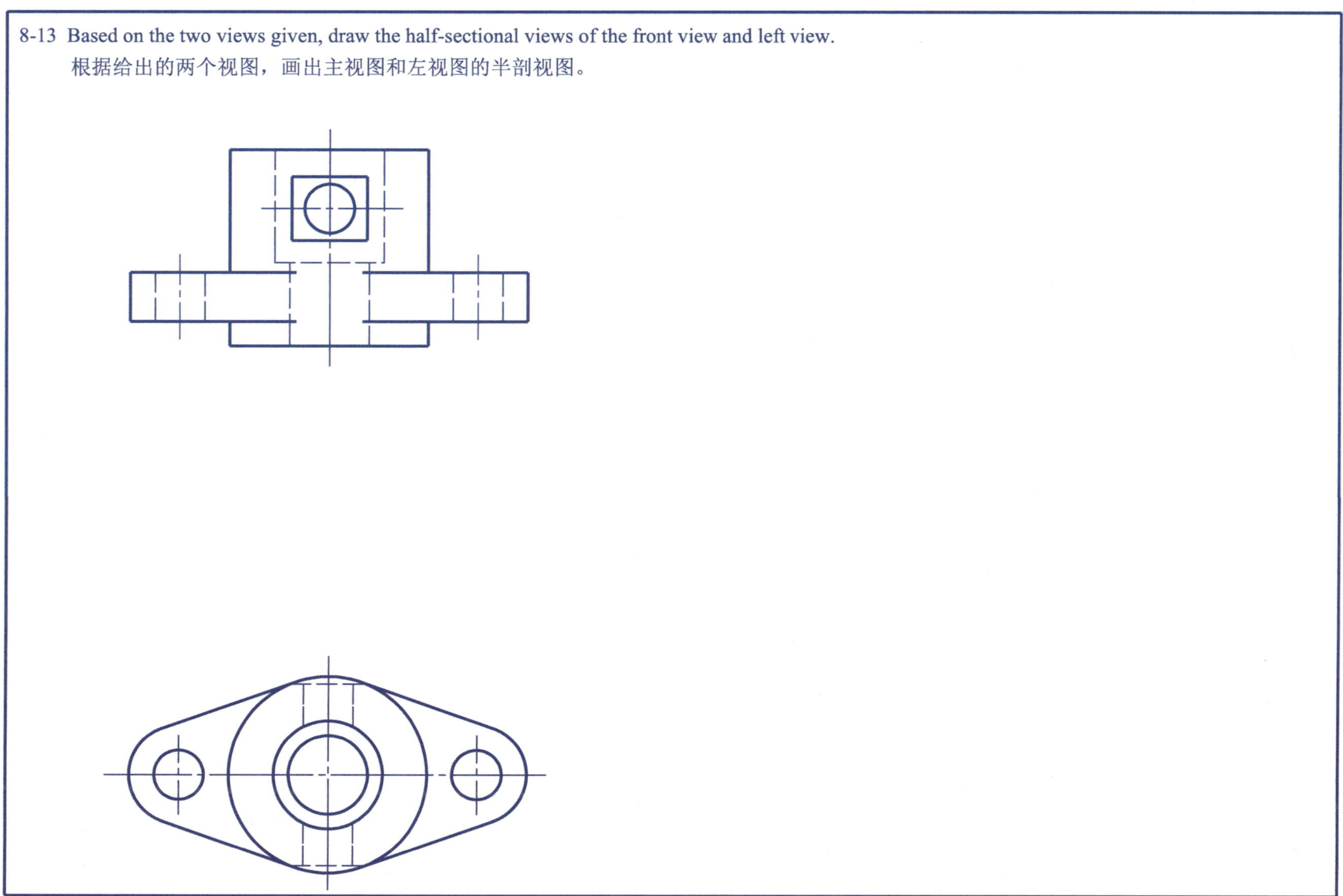

8-14 In the right space given, change the front view and top view into the local sectional views.
在右侧的指定位置将主、俯视图画成局部剖视图。

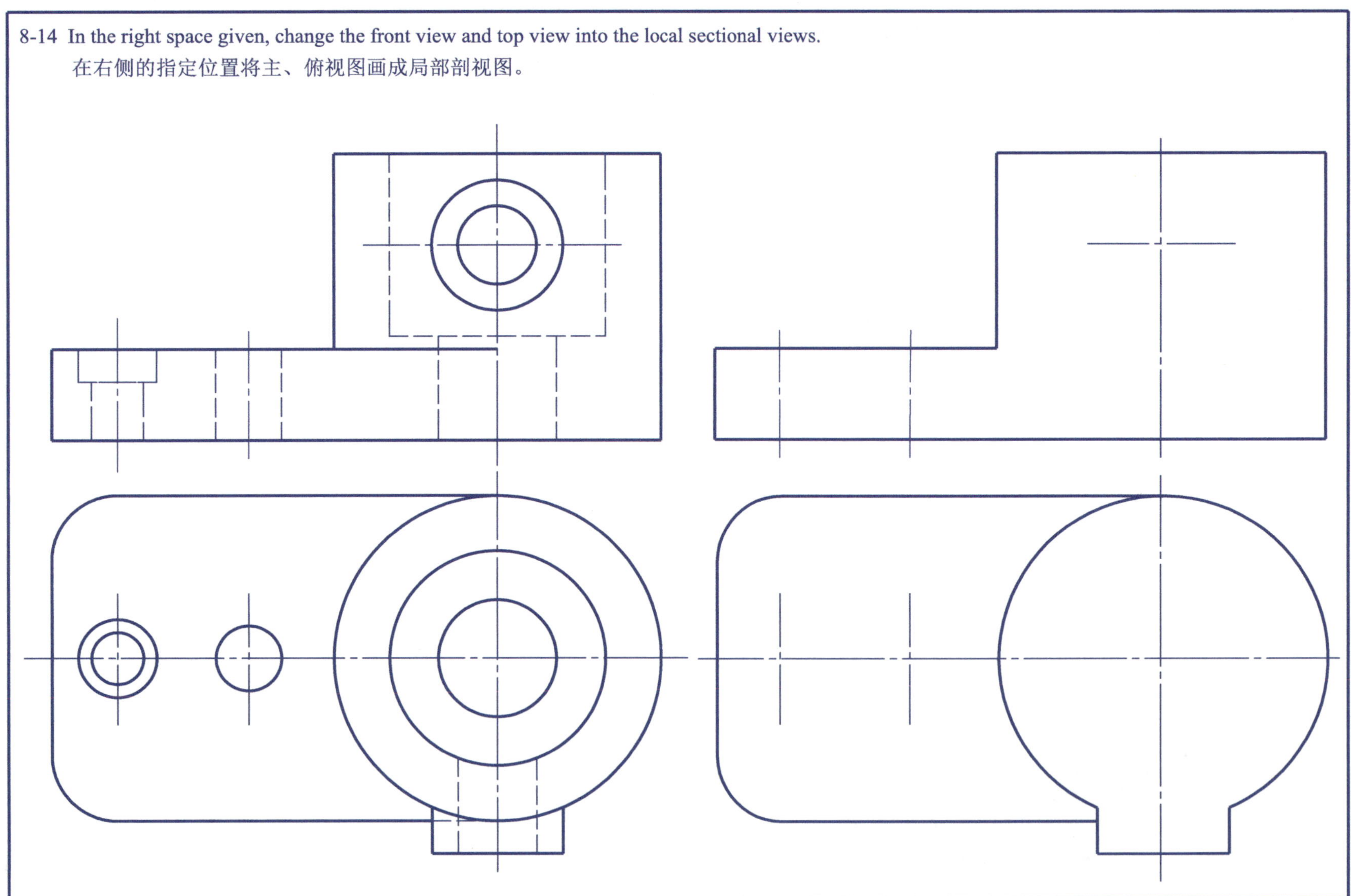

 Class 班级： Name 姓名： ID Number 学号：

8-15 Draw the two views given as local sectional views (mark × for unnecessary lines).

将给出的两个视图画成局部剖视图（不要的线打×）。

8-16 Draw the full sectional view according to the cutting position *A*-*A* as shown.

按给出的剖切位置，画出*A*-*A*全剖视图。

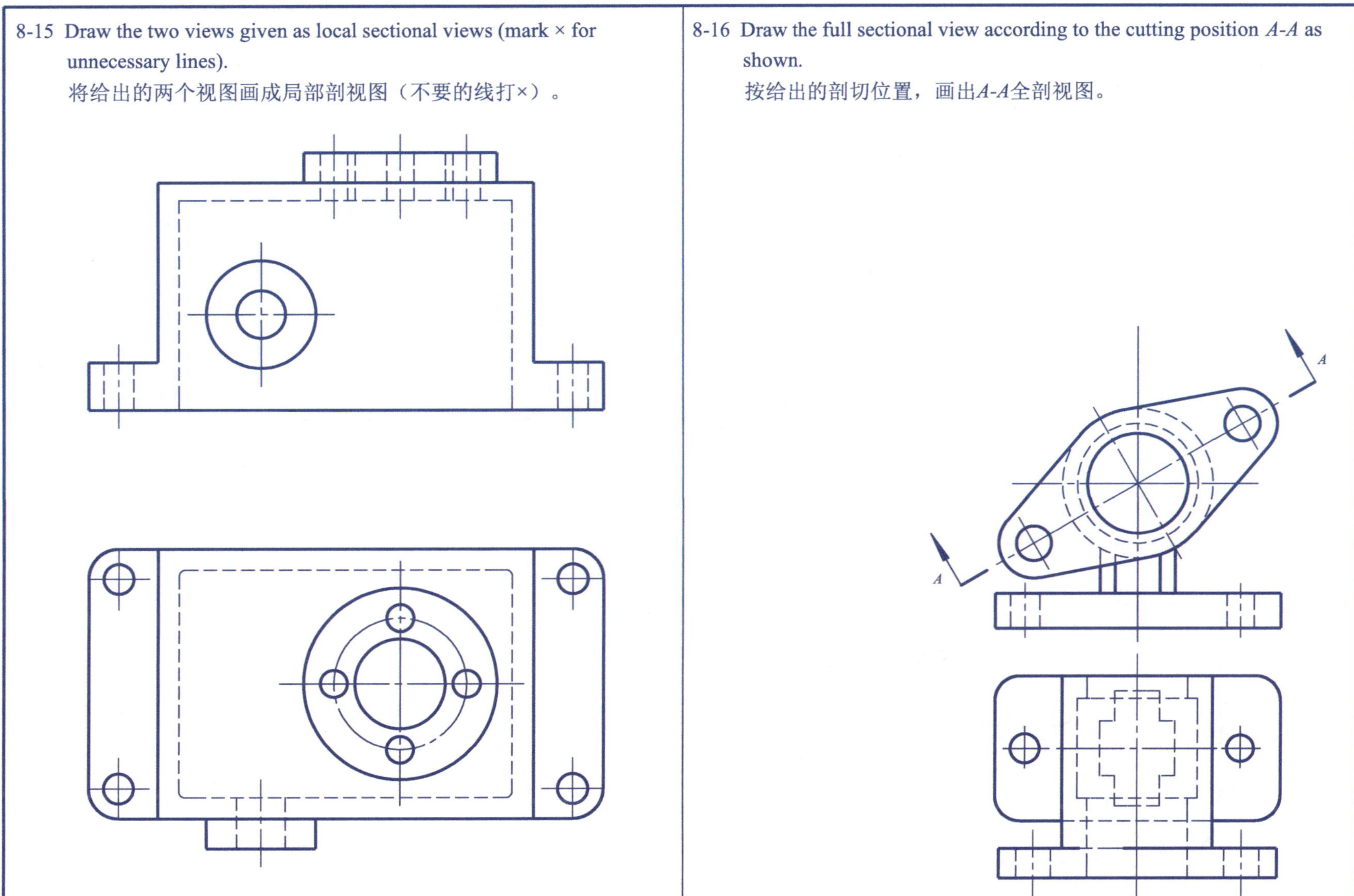

Class 班级：　　Name 姓名：　　ID Number 学号：

8-17 In the space given, draw the full sectional view according to the cutting position *A-A* as shown.

在中间空白处按给出的剖切位置，画出*A-A*全剖视图。

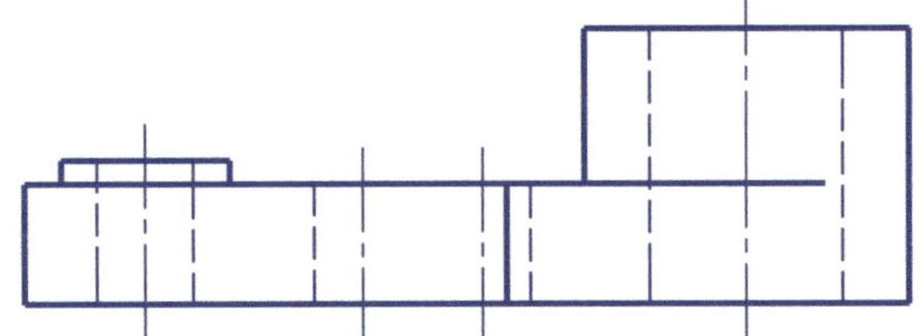

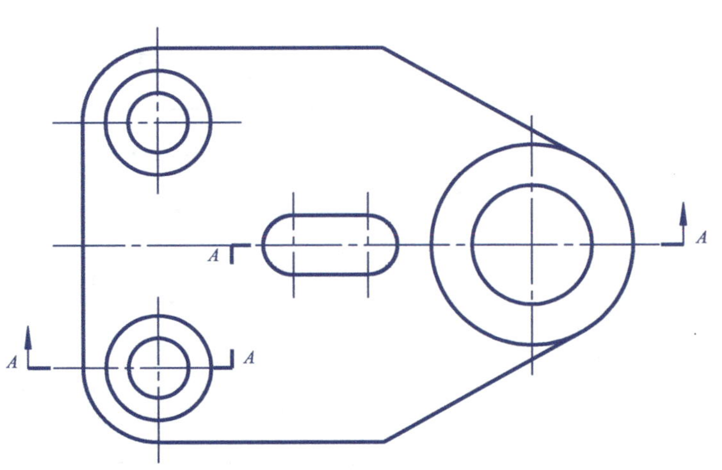

8-18 In the space given, change the top view into the full sectional view according to the cutting position *A-A* as shown.

在适当位置将俯视图画成*A-A*全剖视图。

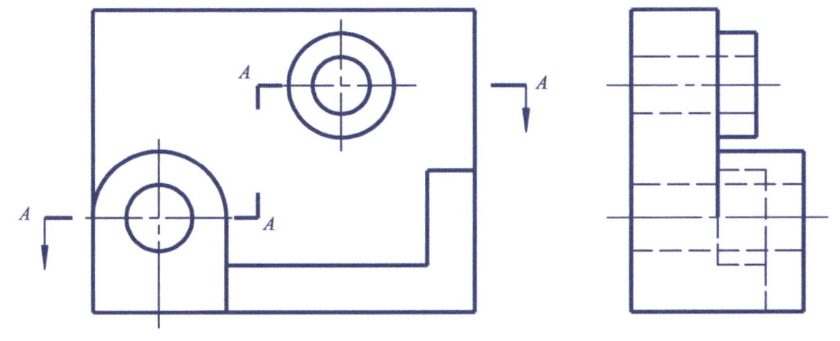

Class 班级：　　Name 姓名：　　ID Number 学号：

8-19 In the space given, draw the full sectional view according to the cutting position *A*-*A* as shown.
在中间空白处按给出的剖切位置，画出*A*-*A*全剖视图。

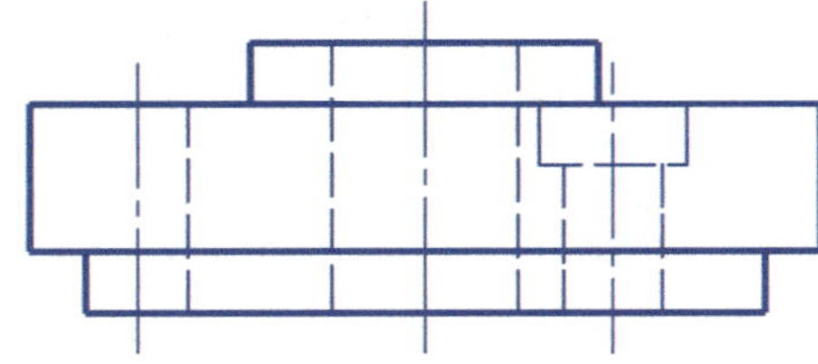

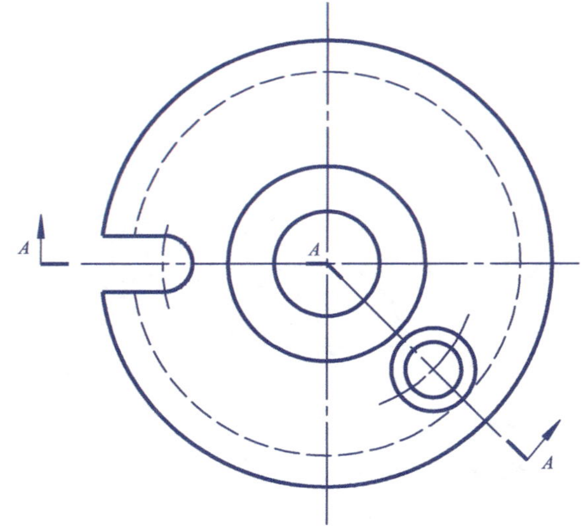

8-20 In the space given, draw the full sectional view according to the cutting position *A*-*A* as shown.
在中间空白处按给出的剖切位置，画出*A*-*A*全剖视图。

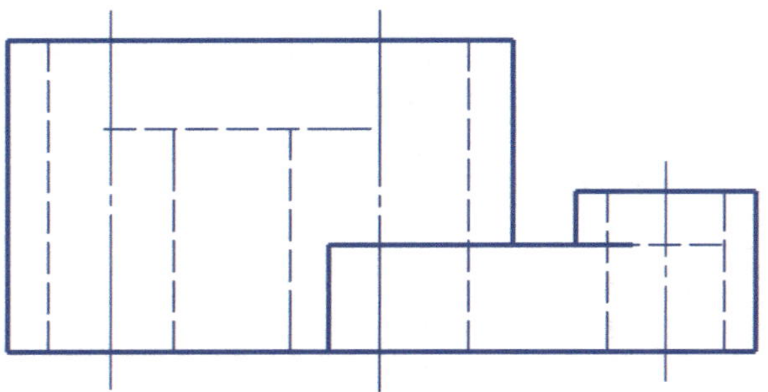

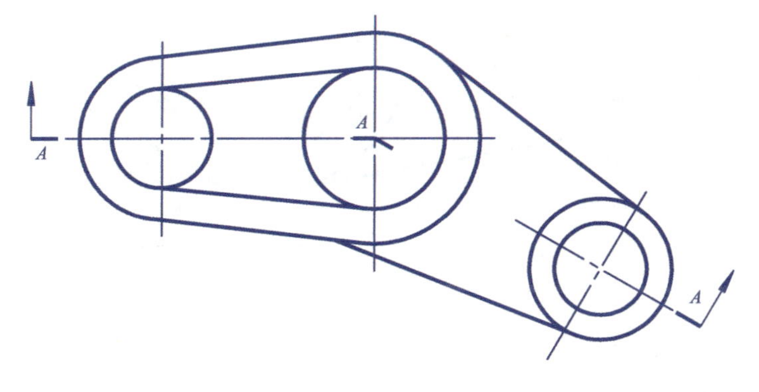

8-21 Complete the full sectional view of the front view by two intersecting cutting planes and add marks.
补全两相交剖切平面剖切的主视图，并加标注。

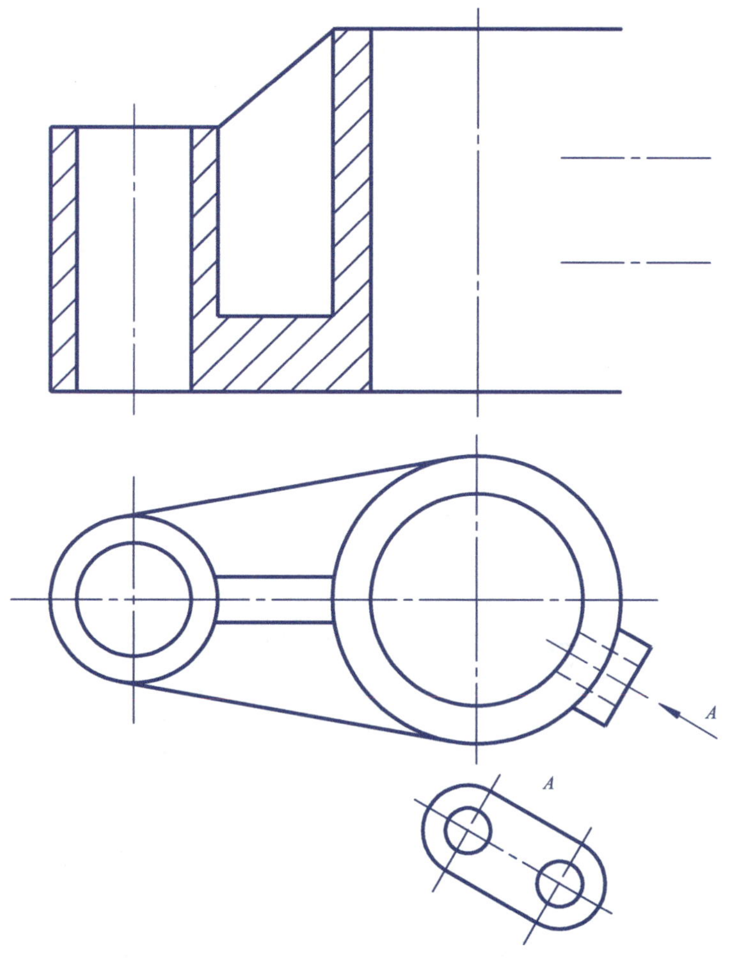

8-22 In the space given, change the front view into the compound full sectional view according to the cutting planes *A-A* as shown.
在中间空白处将主视图画成*A-A*复合全剖视图。

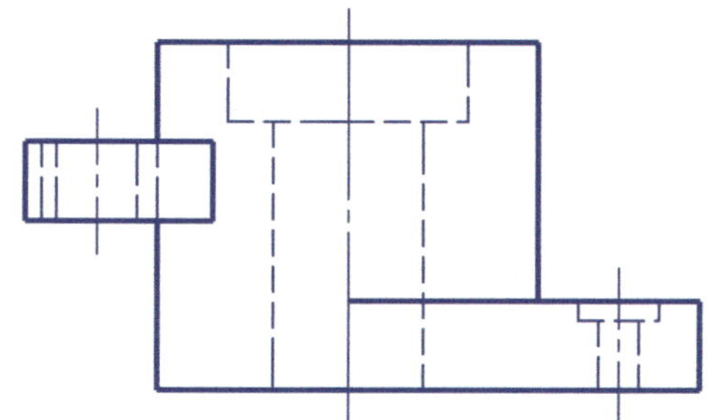

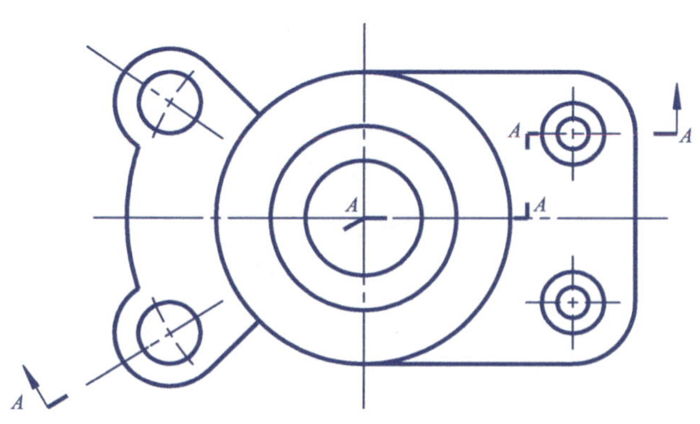

 Class 班级： Name 姓名： ID Number 学号：

8-23 Draw the removed cross-section view from the front view. The structure of left-side $\phi22$ axis is a symmetrical plane. The keyways of the middle $\phi30$ axis and the right-side $\phi22$ axis are both unilateral keyways, and the depth is 3.5 mm.

根据主视图，画出移出断面图。其中左端$\phi22$轴结构为前后对称平面，中间$\phi30$轴、右端$\phi22$轴上的键槽为单侧键槽，深3.5mm。

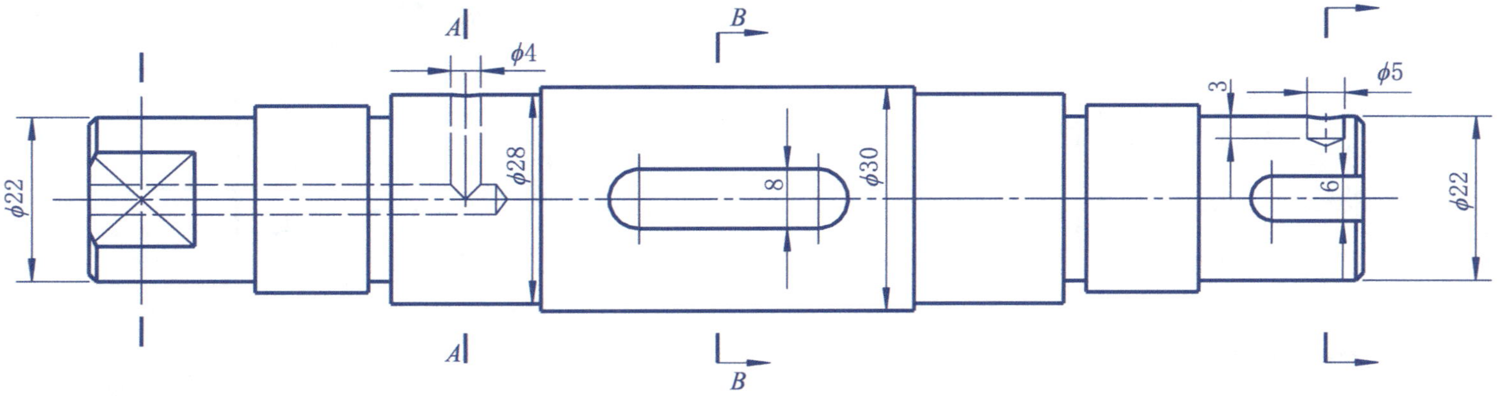

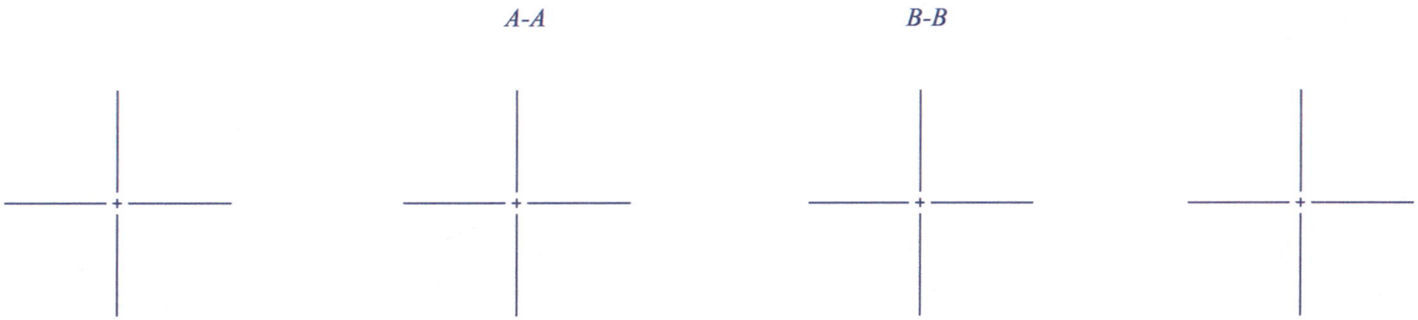

A
$\phi4$
B
3
$\phi5$
$\phi22$
$\phi28$
8
$\phi30$
6
$\phi22$
A
B

A-A

B-B

8-24 Choose the correct cross-section view and write the answer in the bracket.
选择正确的断面图，将答案写在括号内。

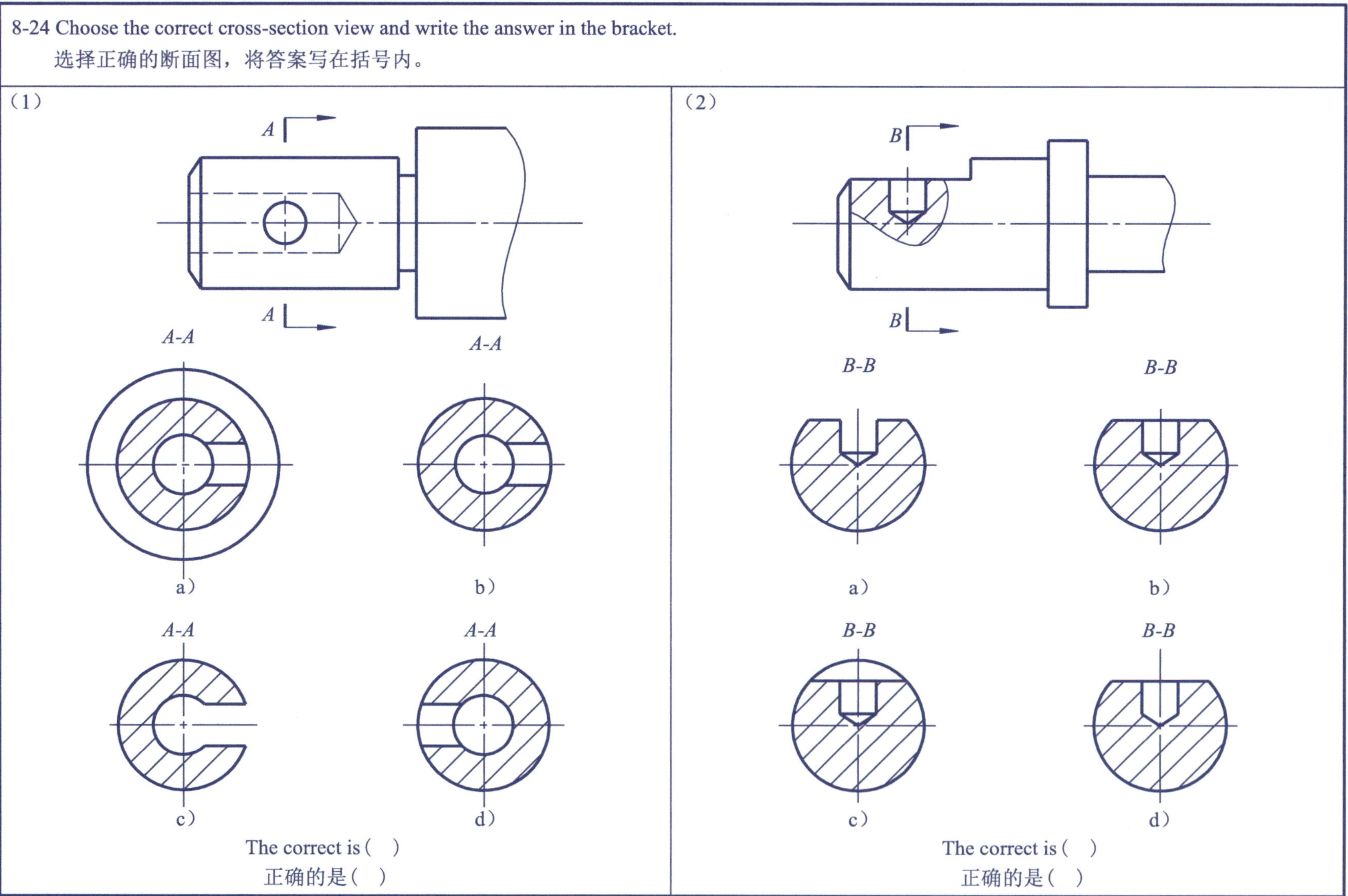

(1) The correct is ( )
正确的是( )

(2) The correct is ( )
正确的是( )

 Class 班级: Name 姓名: ID Number 学号:

8-24 Continued.（续）

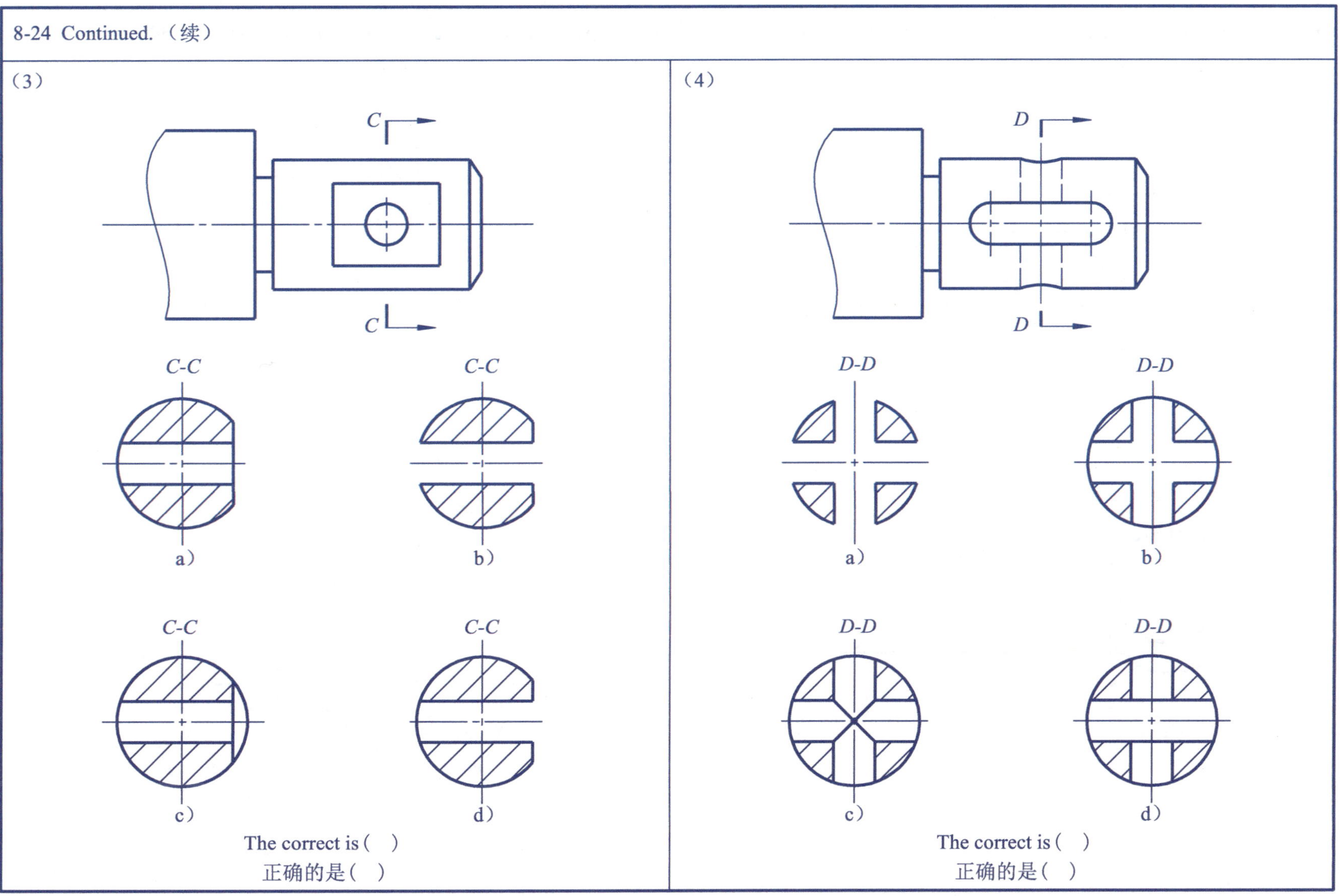

(3) The correct is ( )
正确的是（ ）

(4) The correct is ( )
正确的是（ ）

8-25 Draw the superposed cross-section view at the specified position.
画出指定位置的重合断面图。

8-26 Draw the front view as a full sectional view in the middle space.
在中间空白处将主视图画成全剖视图。

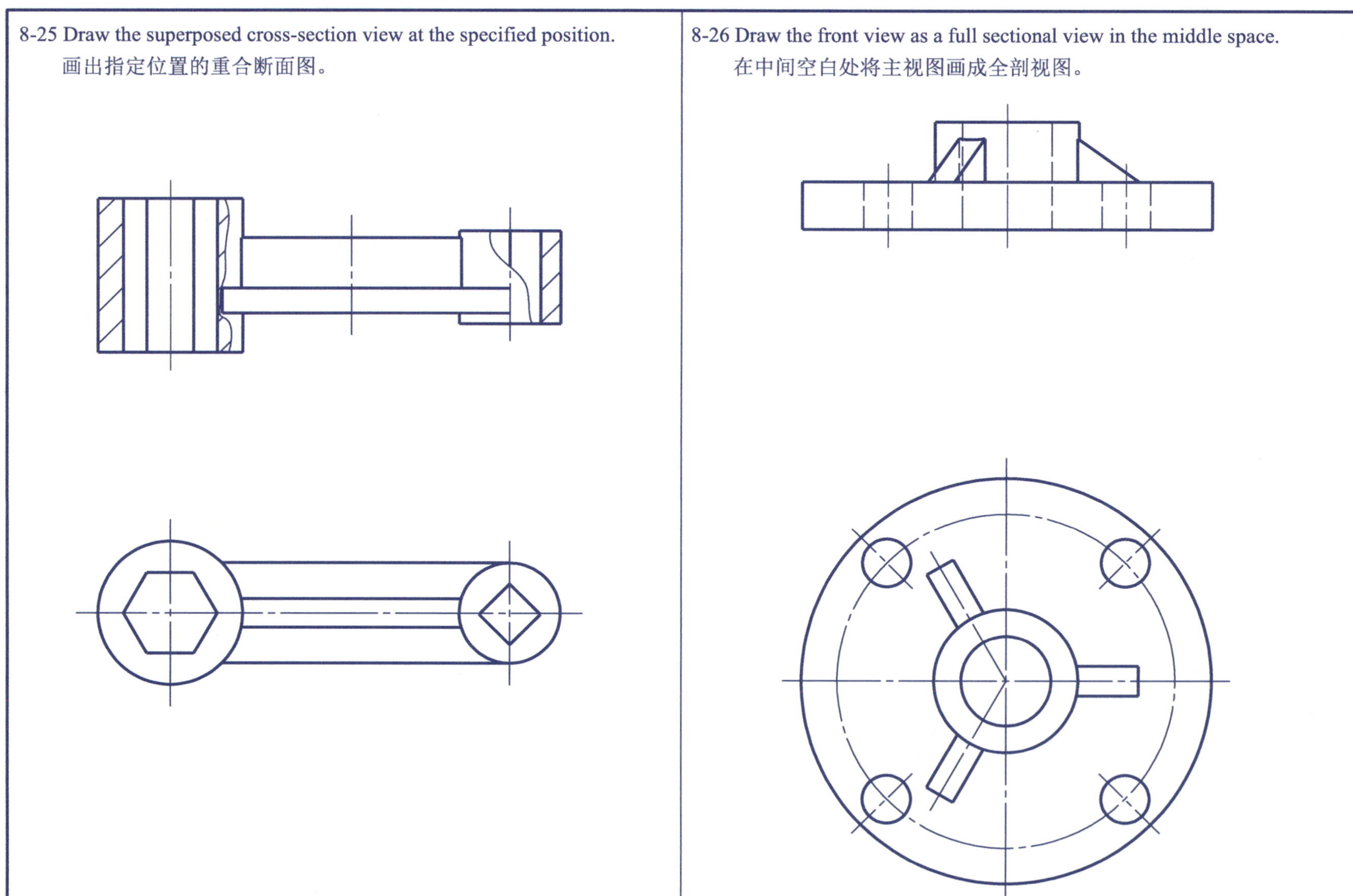

Class 班级：　　Name 姓名：　　ID Number 学号：

8-27 Draw the *C-C* sectional view in the blank.
在空白处画出*C-C*剖视图。

8-28 Given the three views of the part, choose the best scheme to express it.
已知机件三视图，选择最优的方案表达该机件。

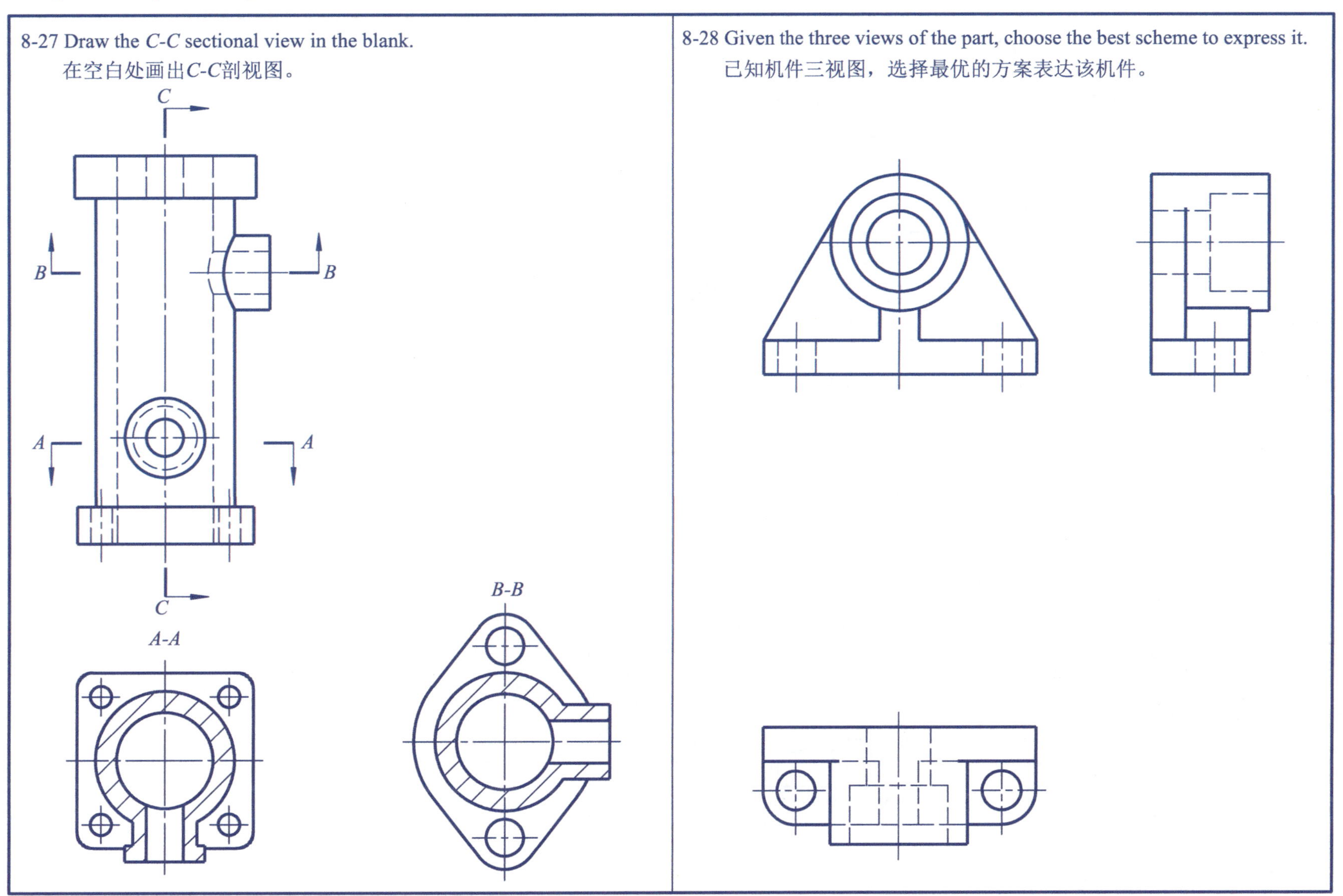

8-29 Based on the front view and top view given, select the best scheme to express the part. Draw it on the A3 drawing with a scale of 1∶1 and mark the dimensions.

已知主、俯两个视图，选择最优的方案表达该机件。在A3图纸上按1∶1比例绘图，并标注尺寸。

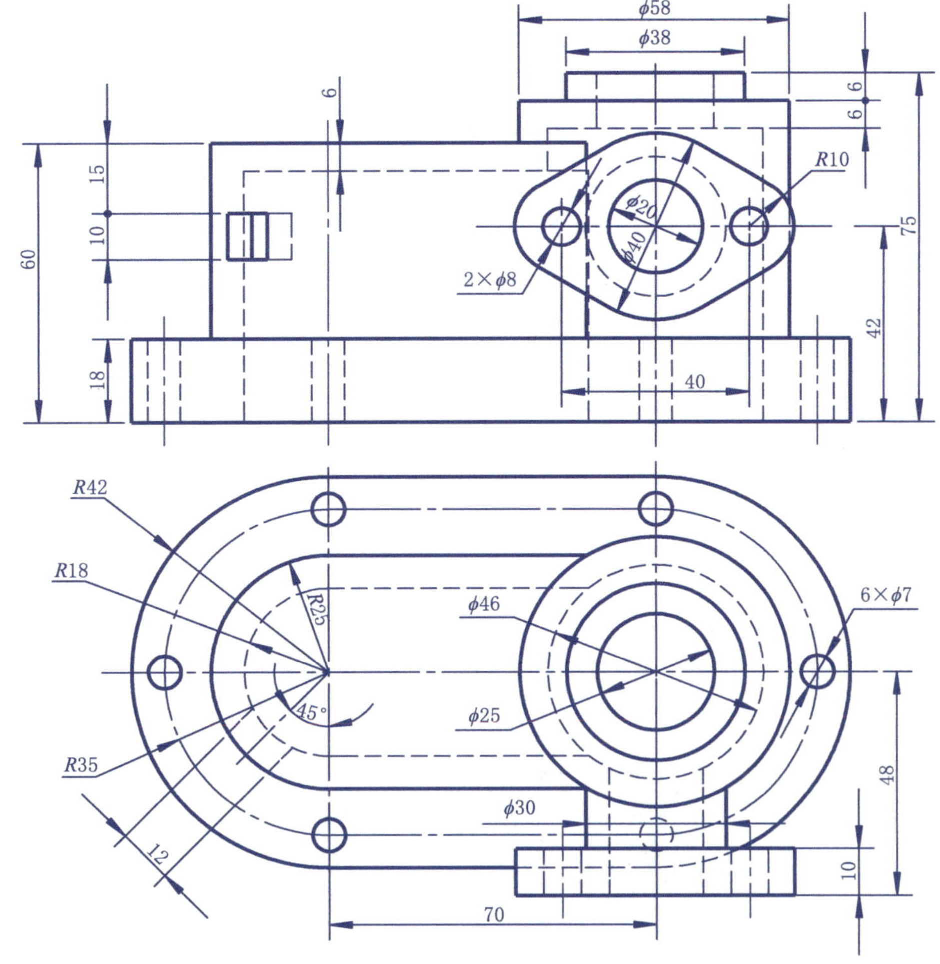

8-30 Based on the front view and top view given, select the best scheme to express the part. Draw it on the A3 drawing with a scale of 1∶1 and mark the dimensions.

已知主、俯两个视图，选择最优的方案表达该机件。在A3图纸上按1∶1比例绘图，并标注尺寸。

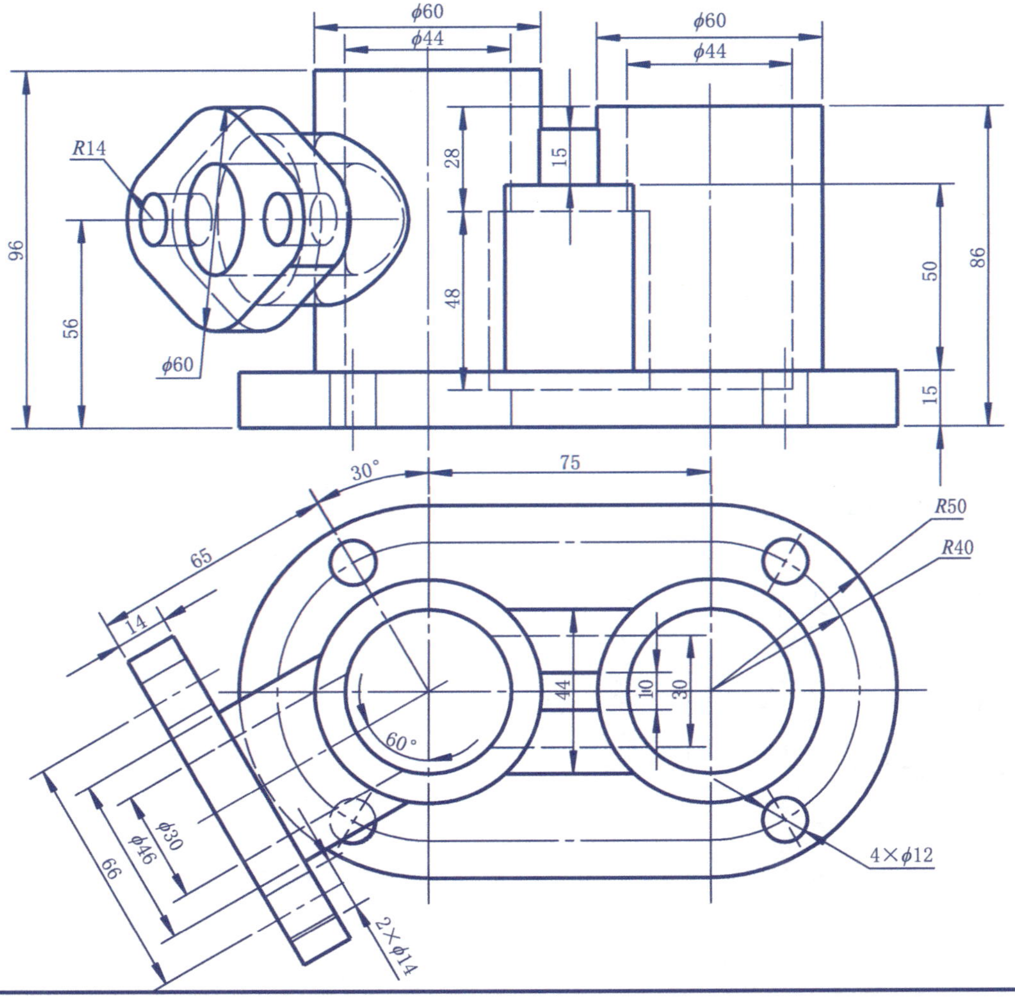

Class 班级：　　Name 姓名：　　ID Number 学号：

8-31 Based on the front view and top view given, select the best scheme to express the part. Draw it on the A3 drawing with a scale of 1∶1 and mark the dimensions.

已知主、俯两个视图，选择最优的方案表达该机件。在A3图纸上按1∶1比例绘图，并标注尺寸。

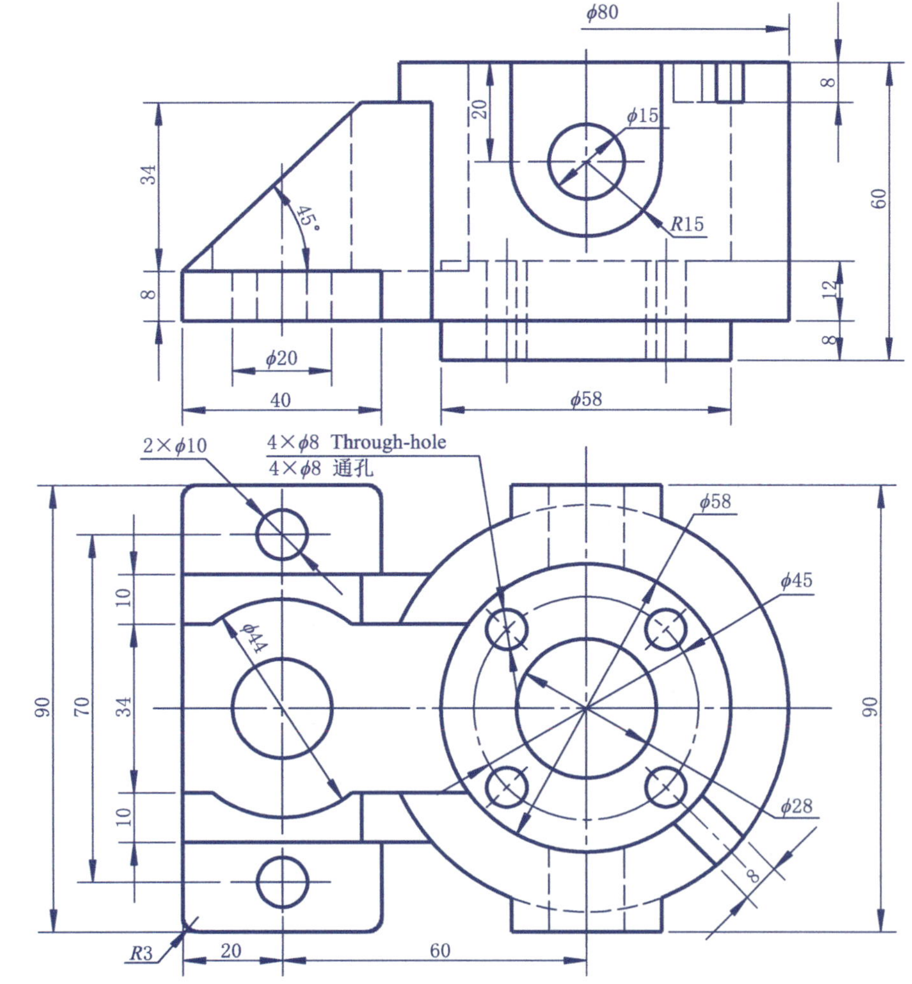

 Class 班级： Name 姓名： ID Number 学号：

8-32 Based on the axonometric projection given, select the optimal scheme to express the part.
根据给出的轴测图，选择最优的方案表达该机件。

（1）

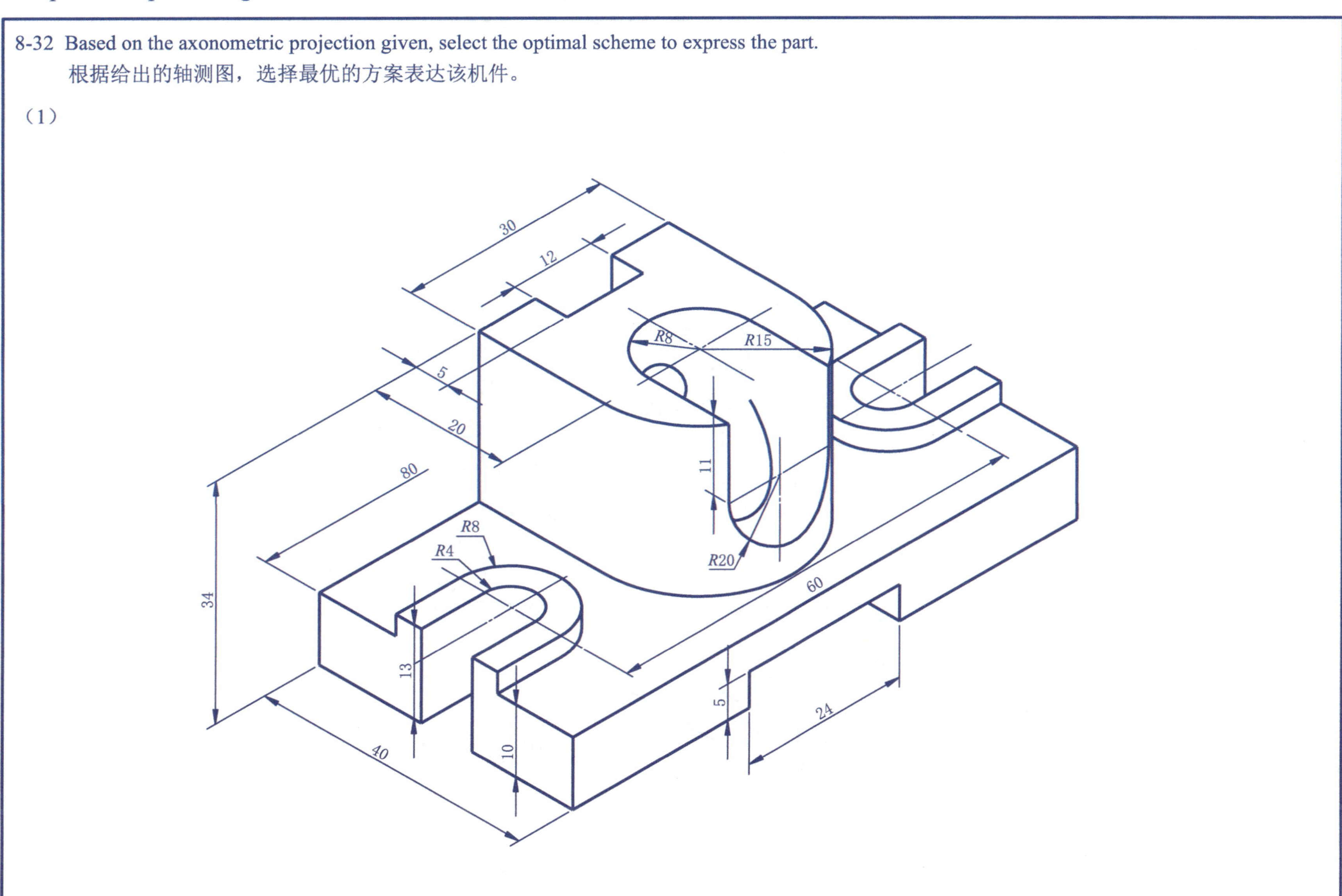

Class 班级： Name 姓名： ID Number 学号：

8-32 Continued.（续）

（2）

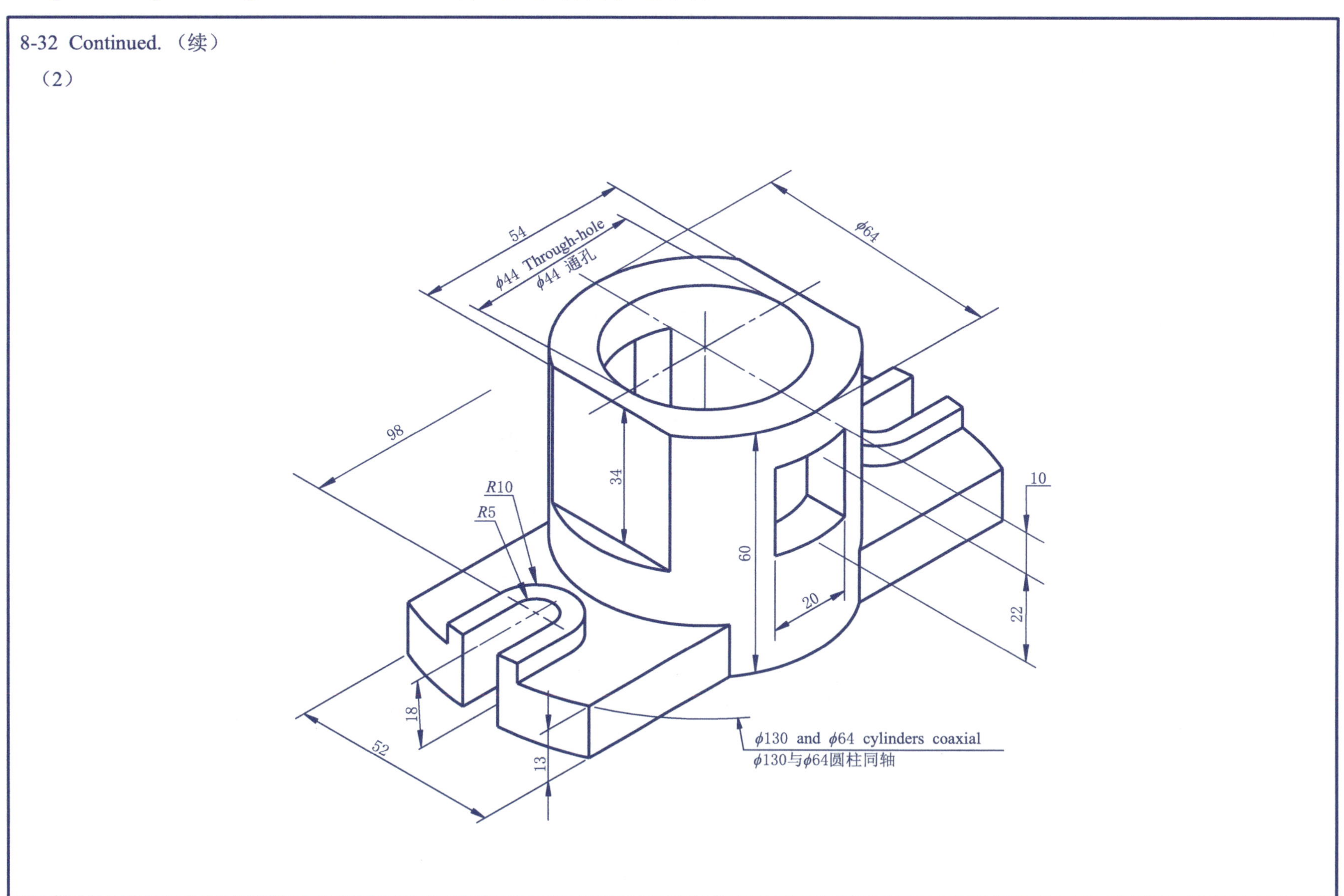

 Class 班级： Name 姓名： ID Number 学号：

8-32 Continued. （续）

（3）

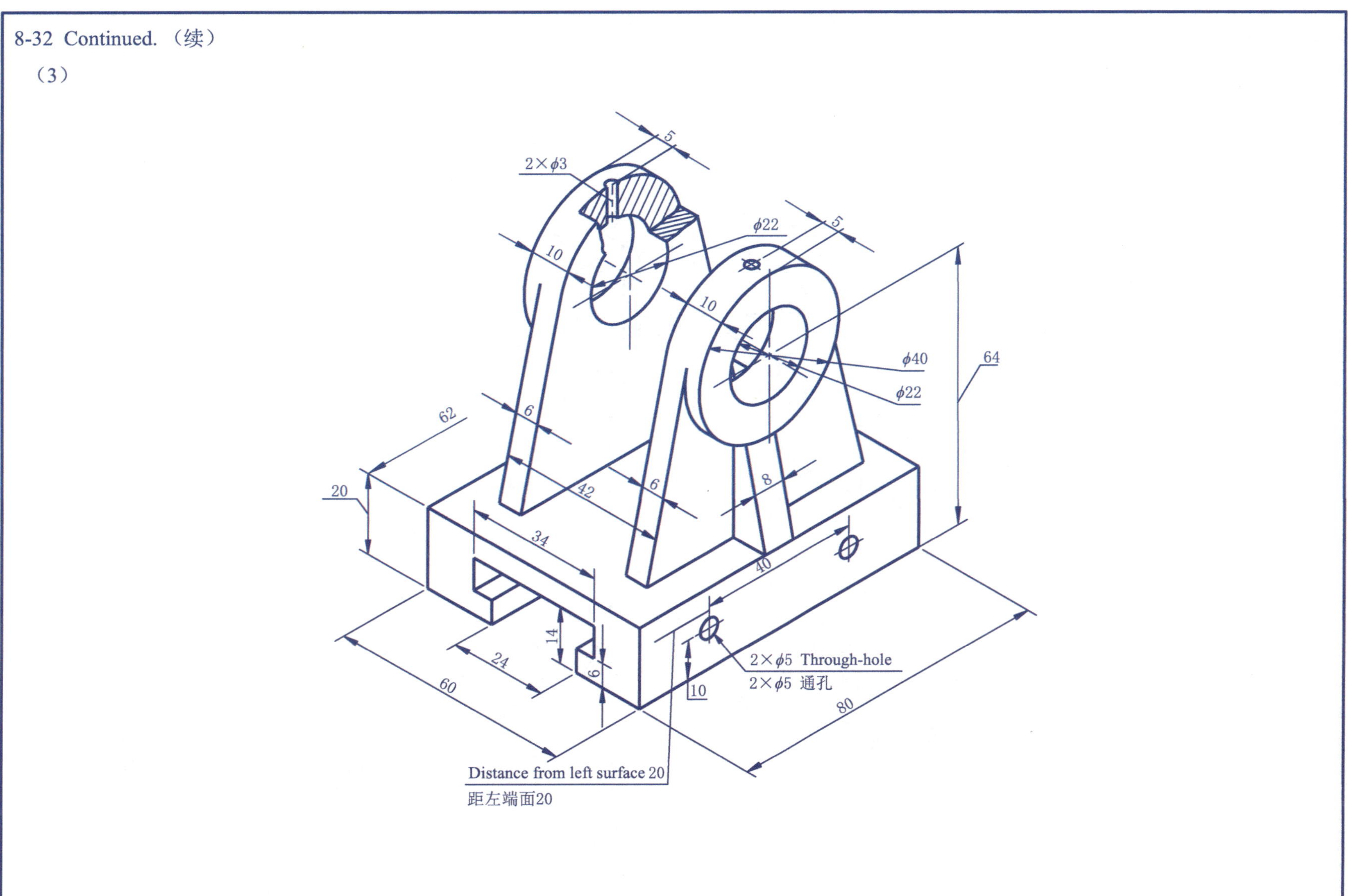

8-32 Continued. （续）

（4）

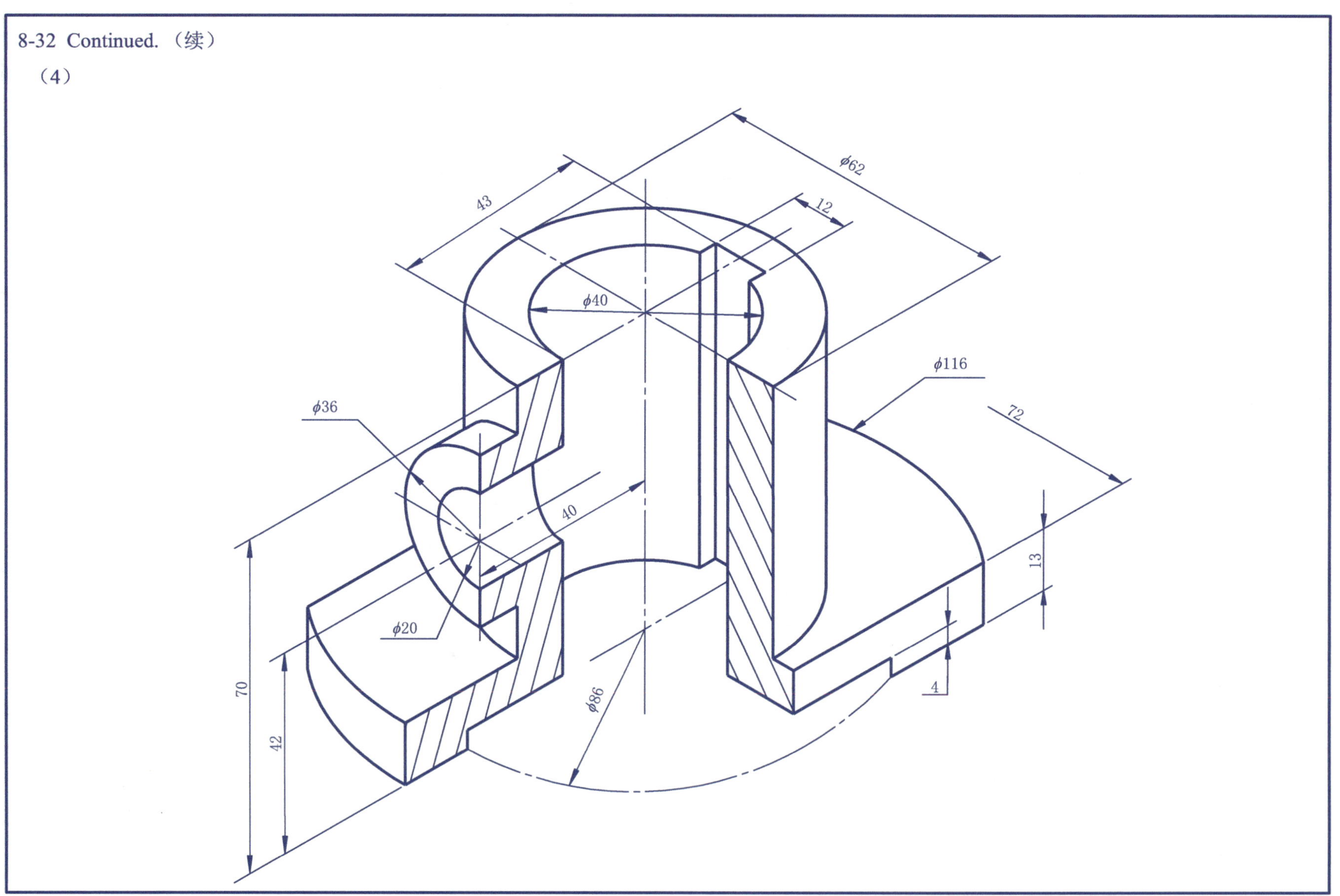

Class 班级： Name 姓名： ID Number 学号：

8-32 Continued.（续）

（5）

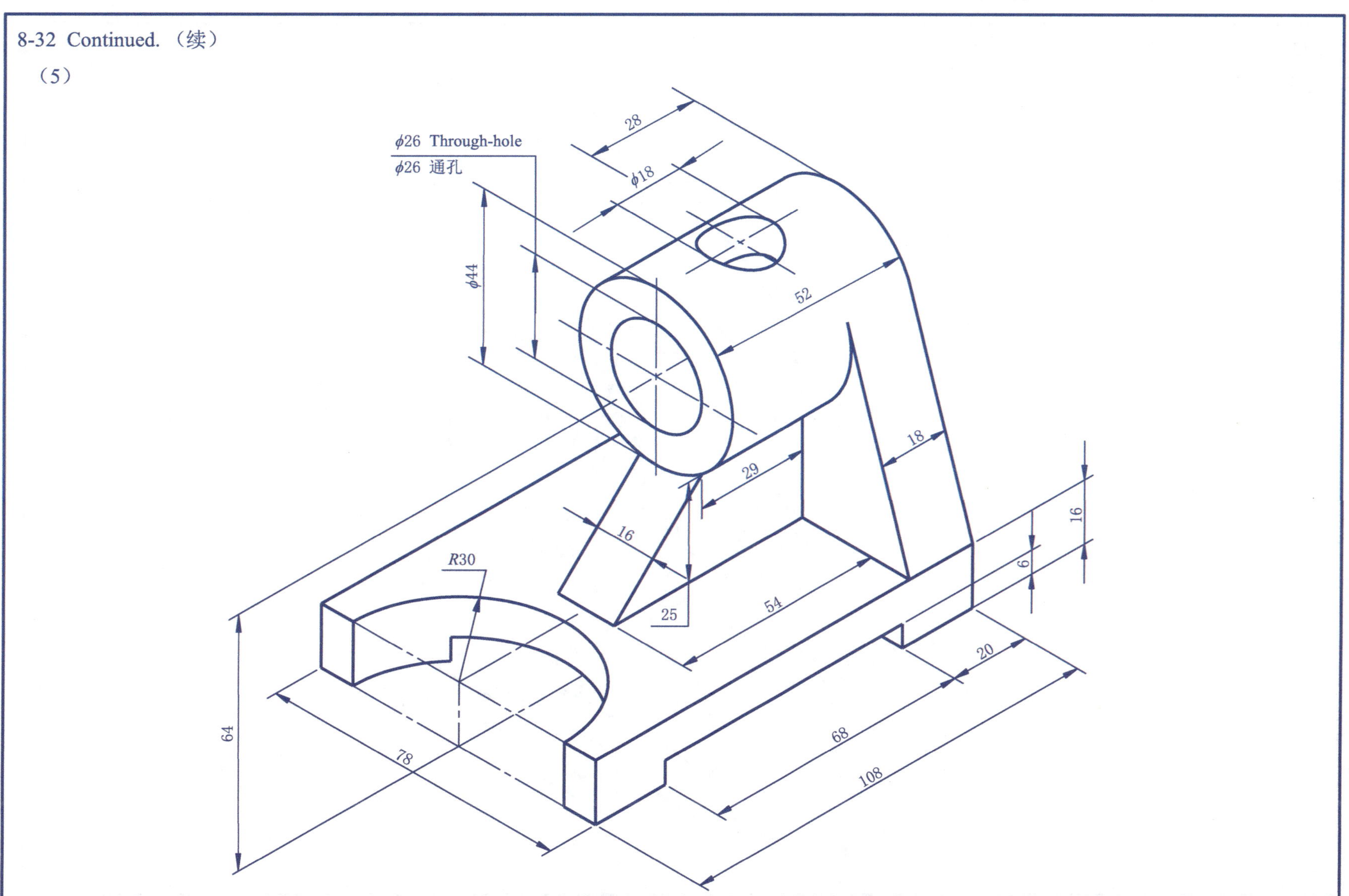

8-32 Continued. （续）

（6）

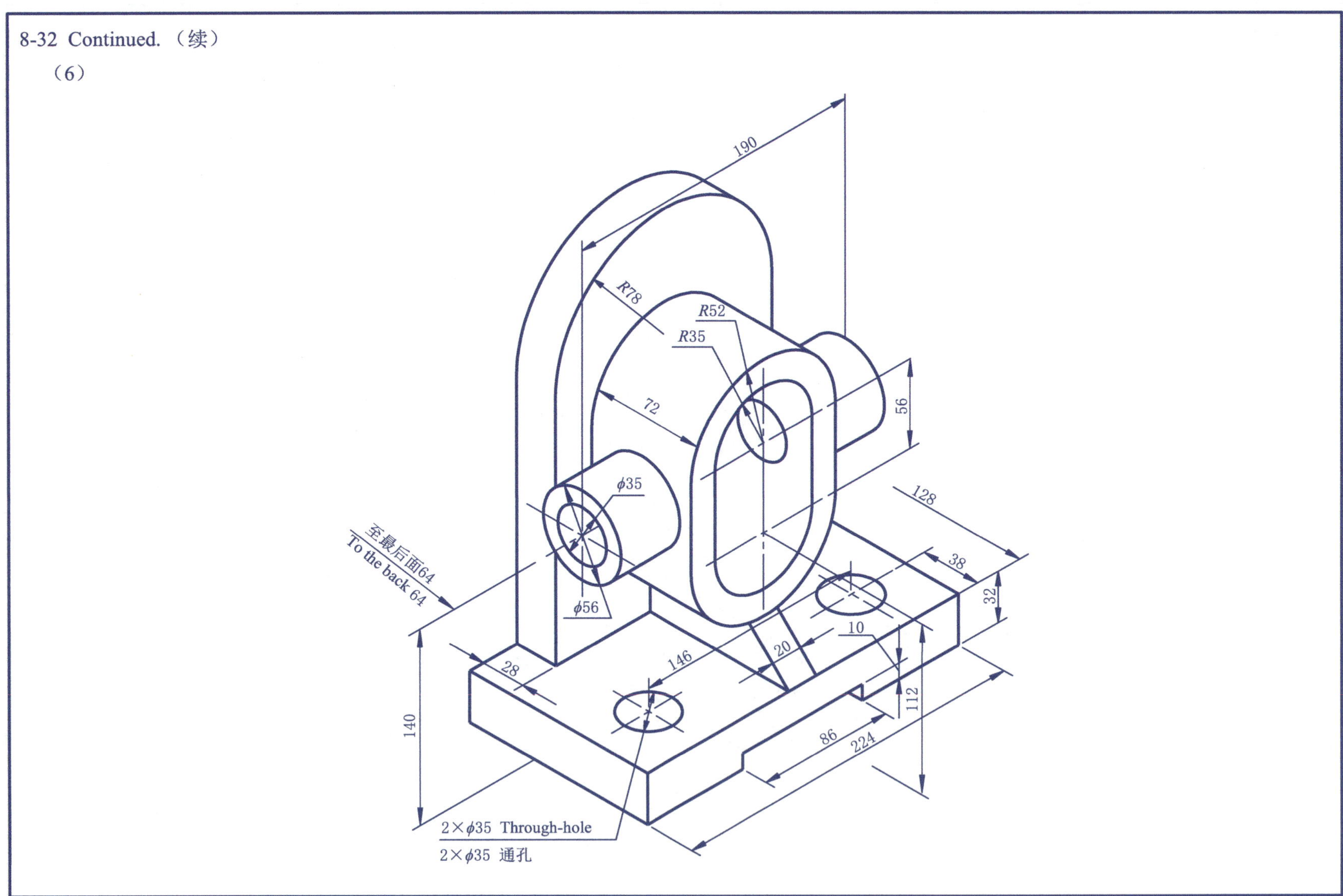

Class 班级： Name 姓名： ID Number 学号：

8-32 Continued.（续）

（7）

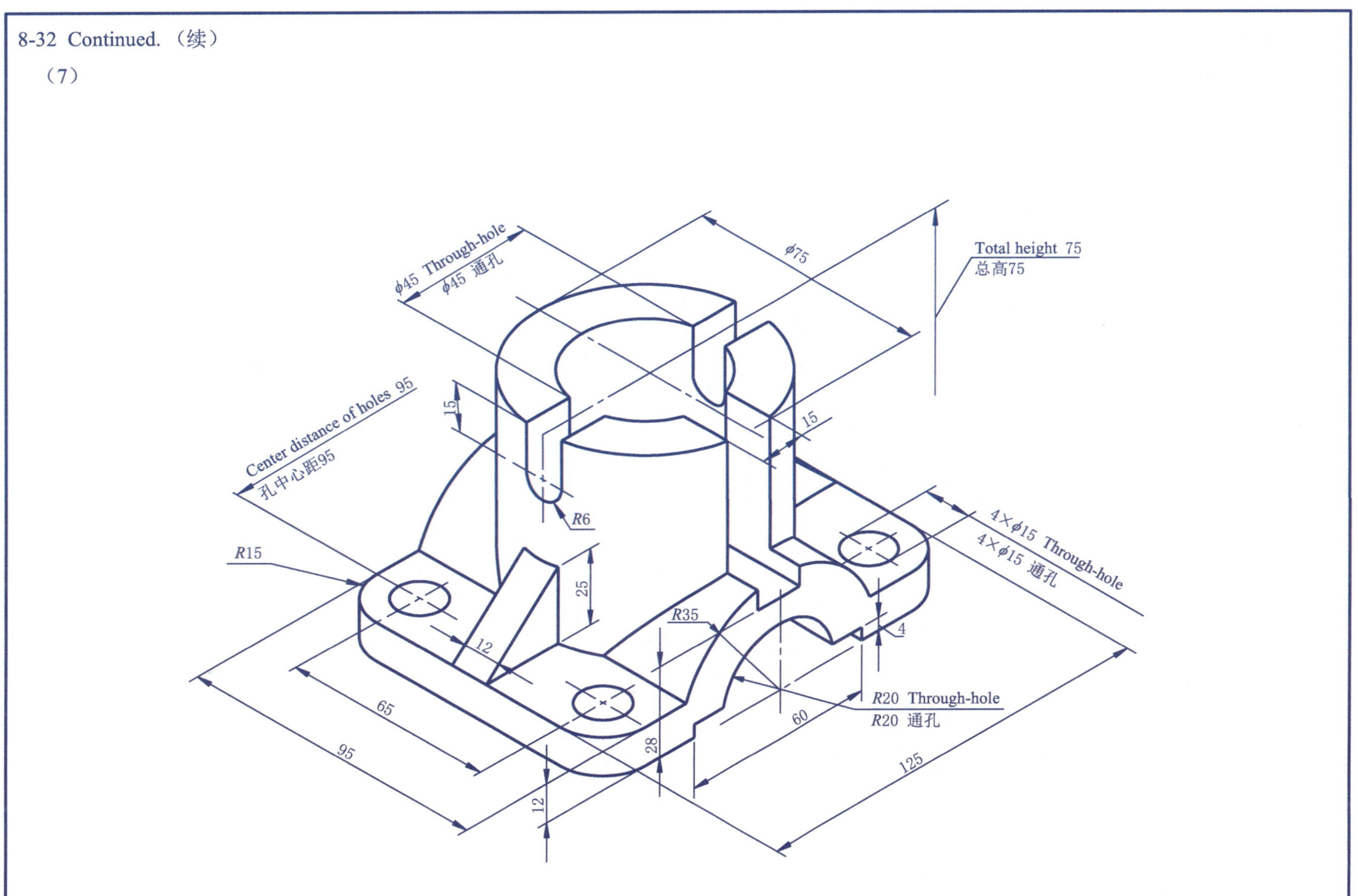

8-32 Continued.（续）

（8）

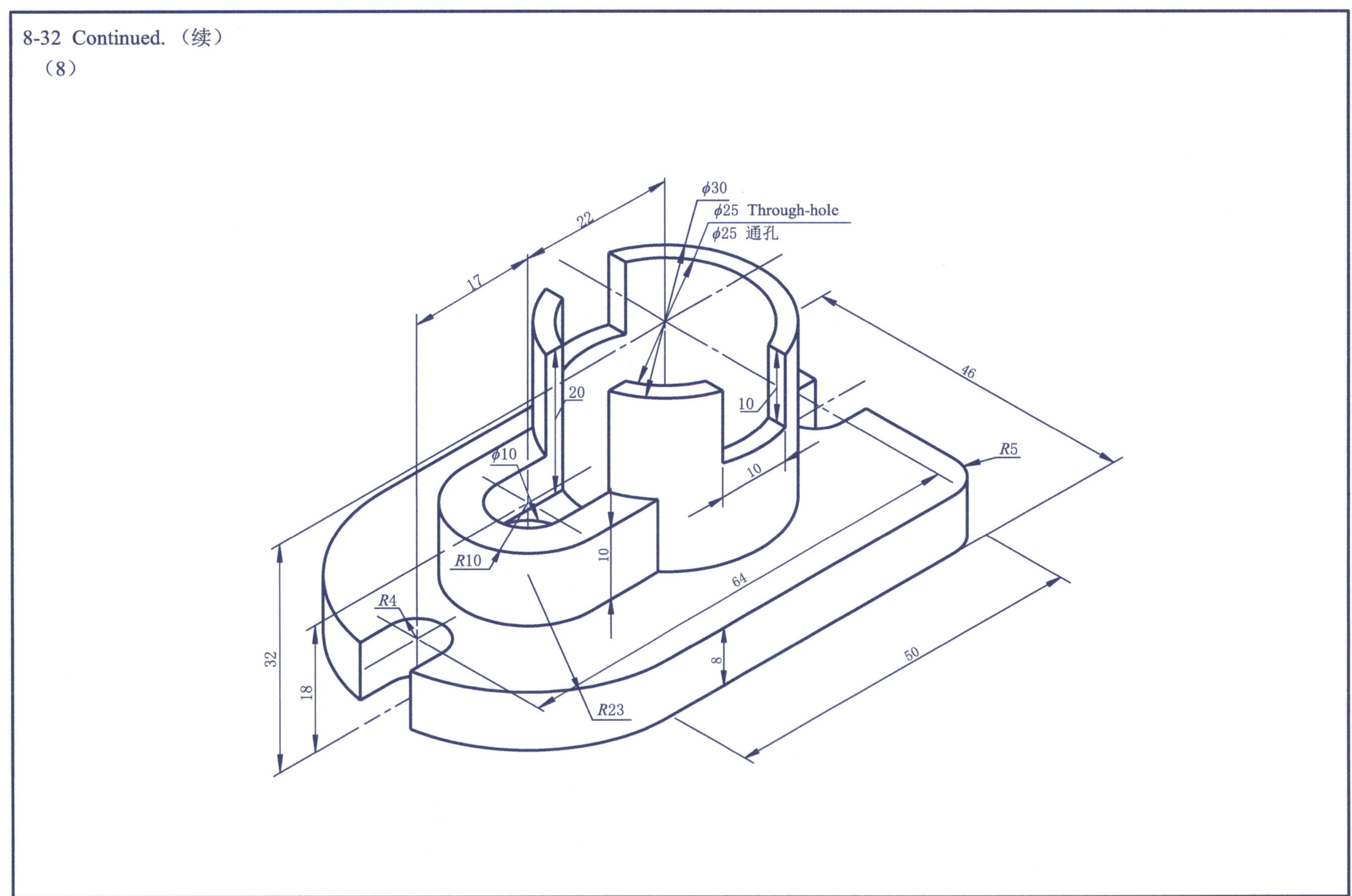

8-33 Based on the views given, select the optimal scheme to express the part.
根据给出的视图，选择最优的方案表达该机件。

（1）

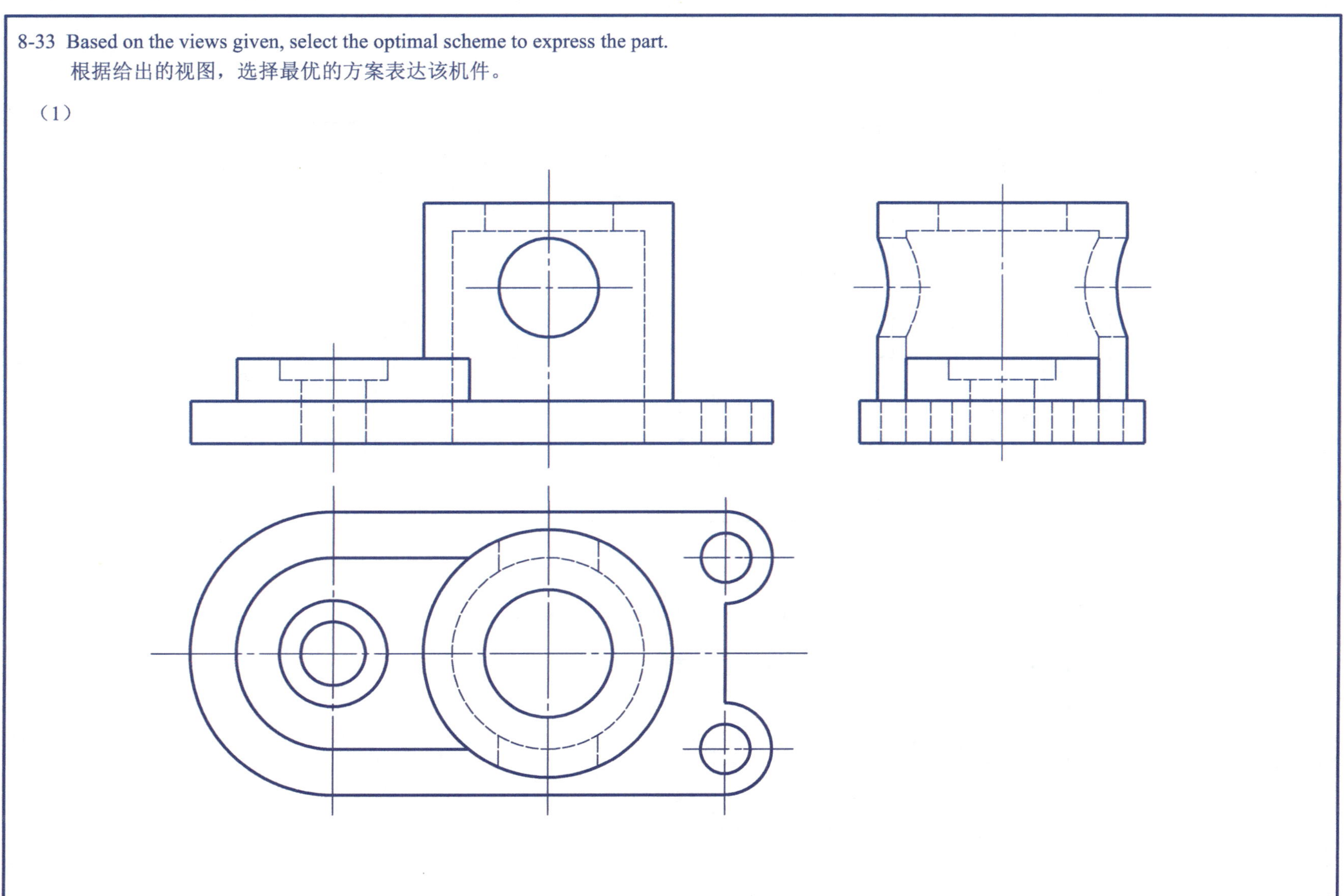

8-33 Continued.（续）

（2）

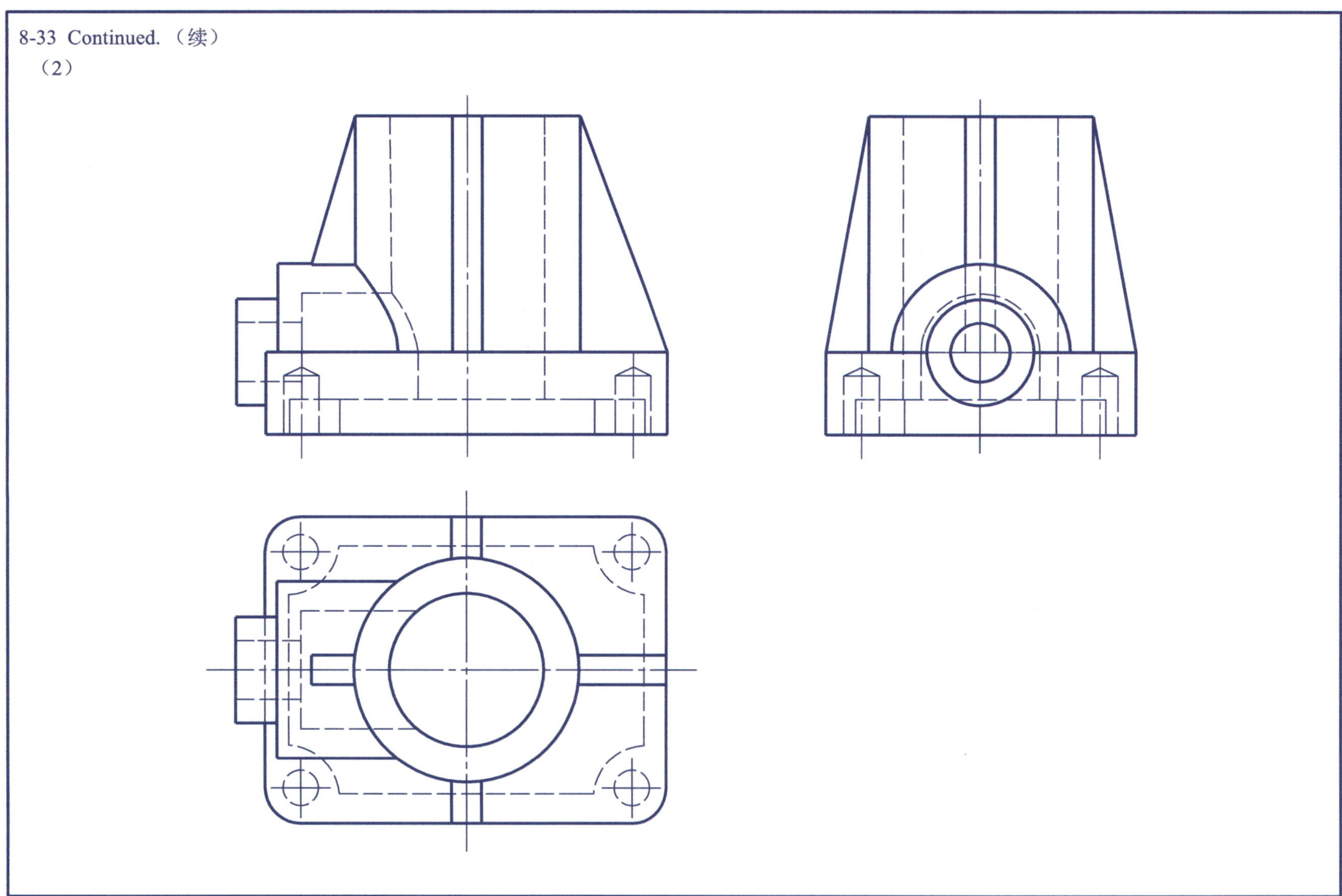

 Class 班级： Name 姓名： ID Number 学号：

8-33 Continued.（续）

（3）

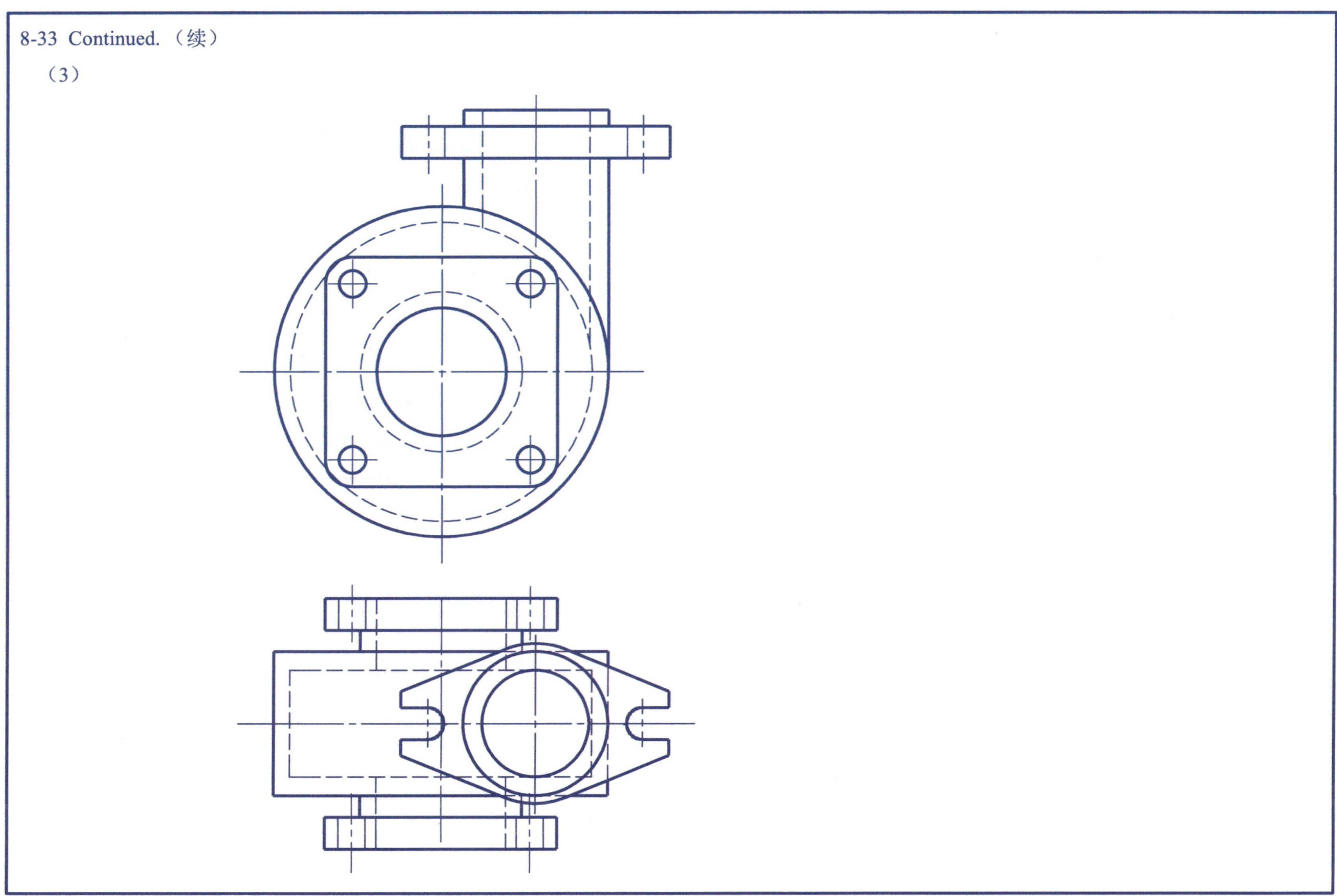

8-33 Continued. （续）

（4）

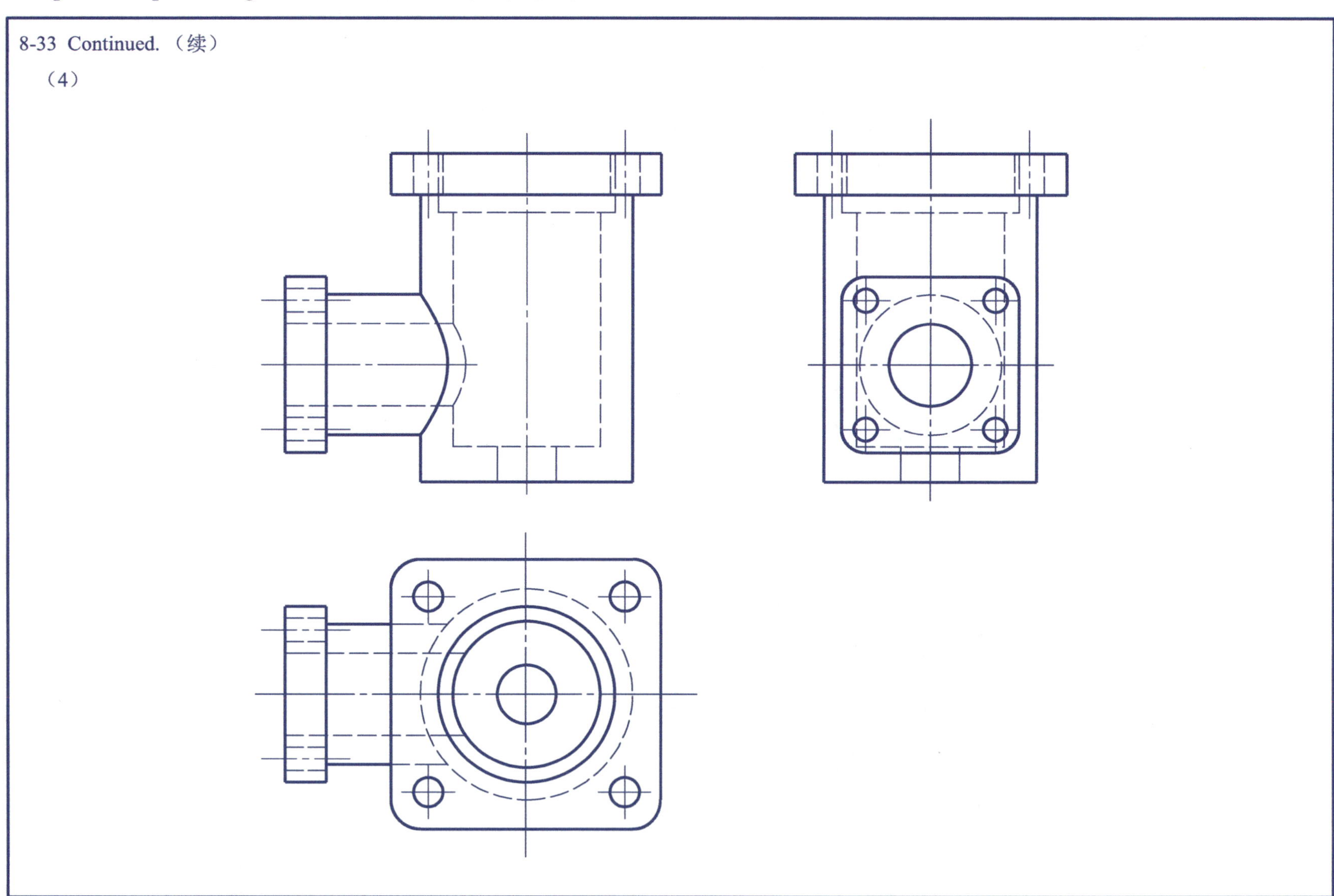

Class 班级： Name 姓名： ID Number 学号：

8-33 Continued. （续）

（5）

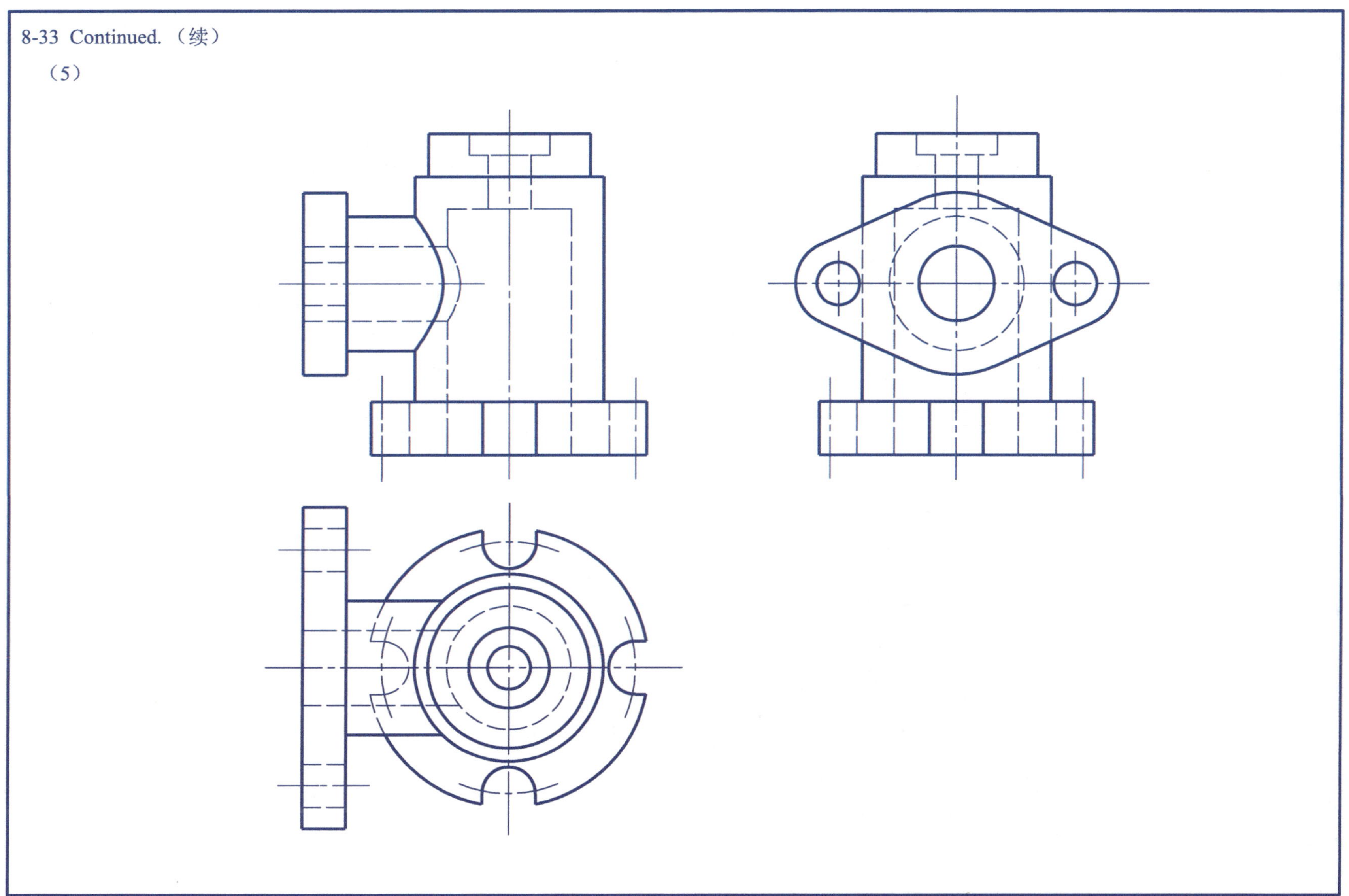

8-33 Continued. （续）

（6）

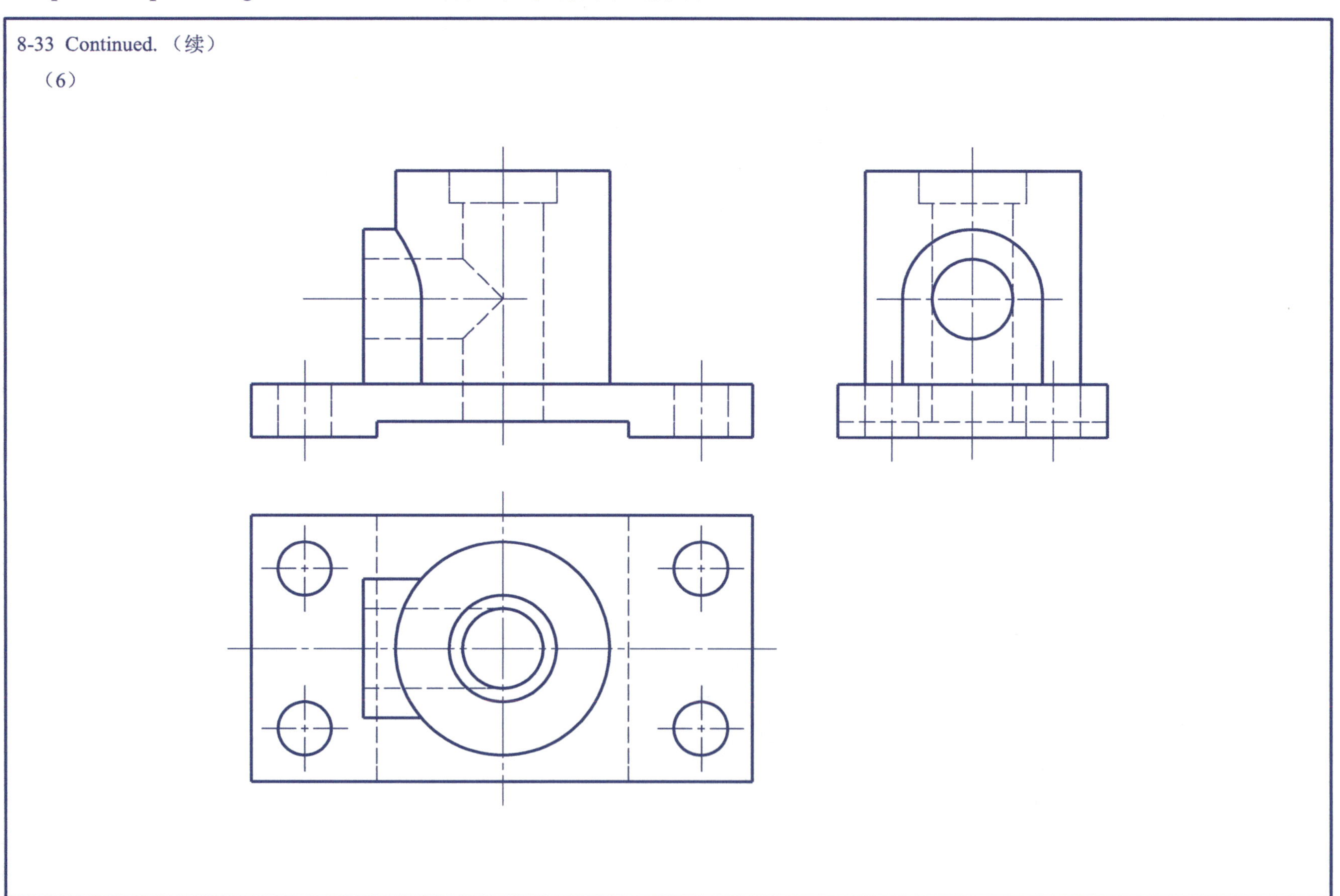

 Class 班级： Name 姓名： ID Number 学号：

8-33 Continued.（续）

（7）

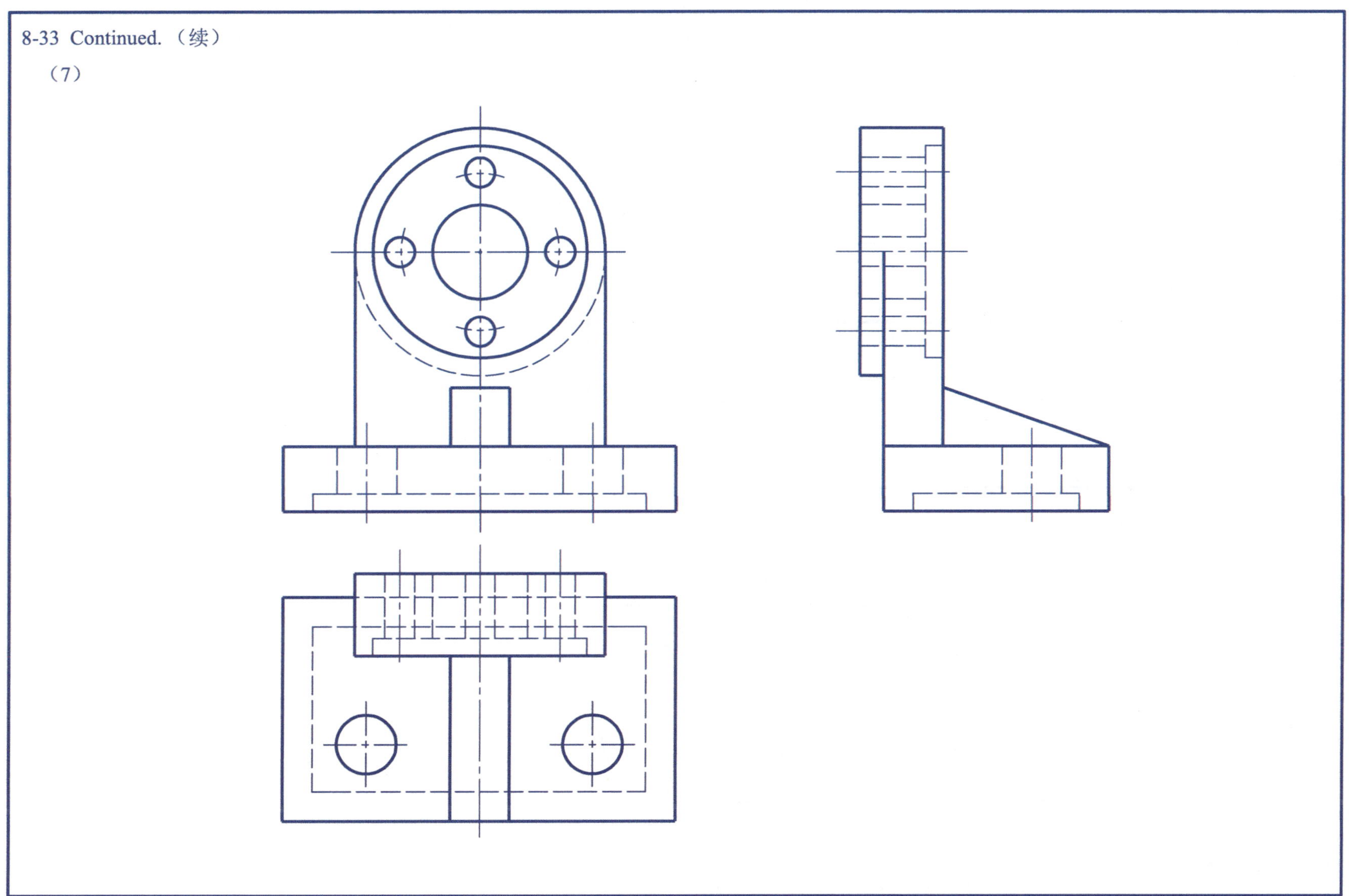

8-33 Continued.（续）

（8）

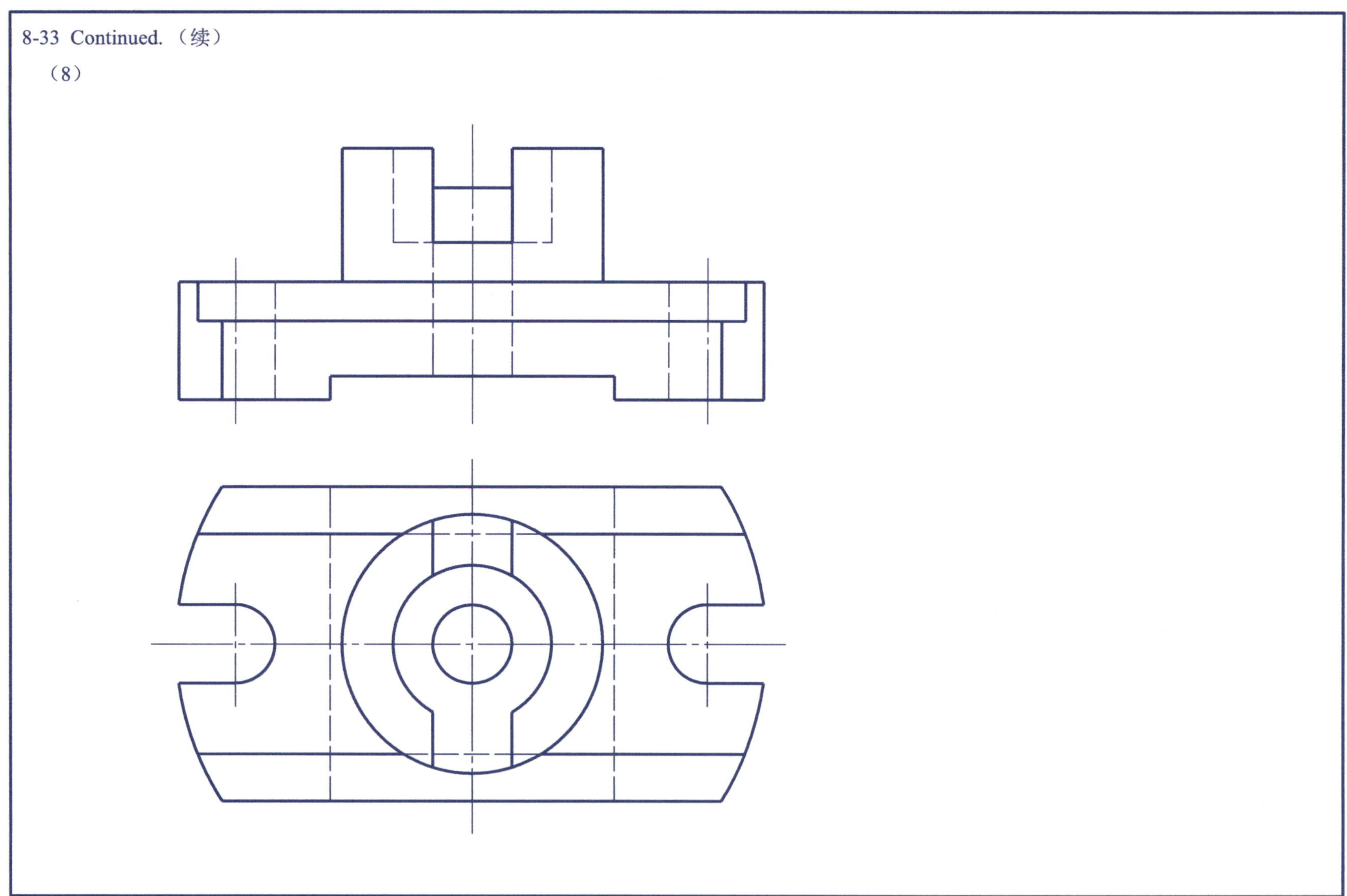

Class 班级：　　Name 姓名：　　ID Number 学号：

8-33 Continued.（续）

（9）

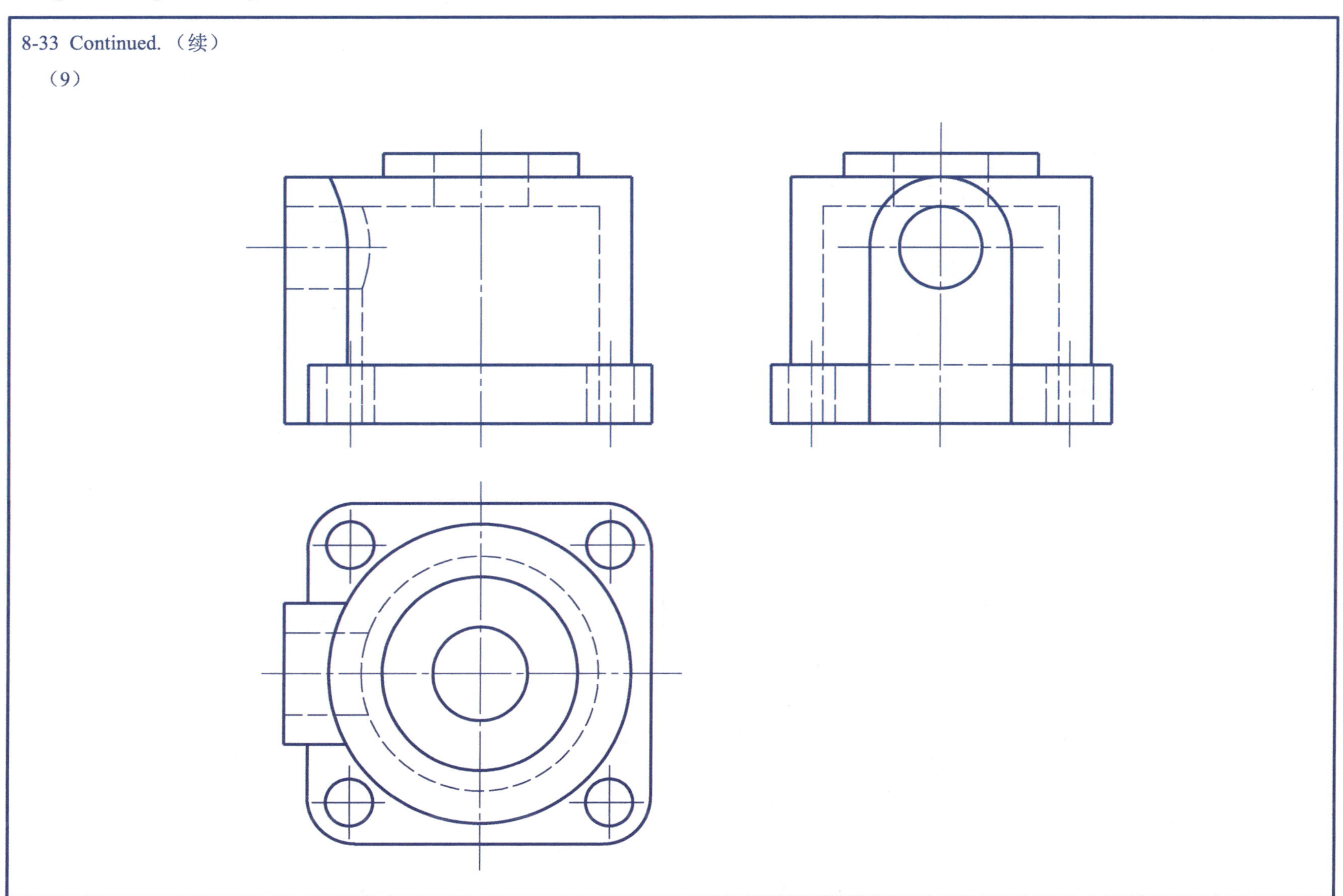

8-33 Continued.（续）

（10）

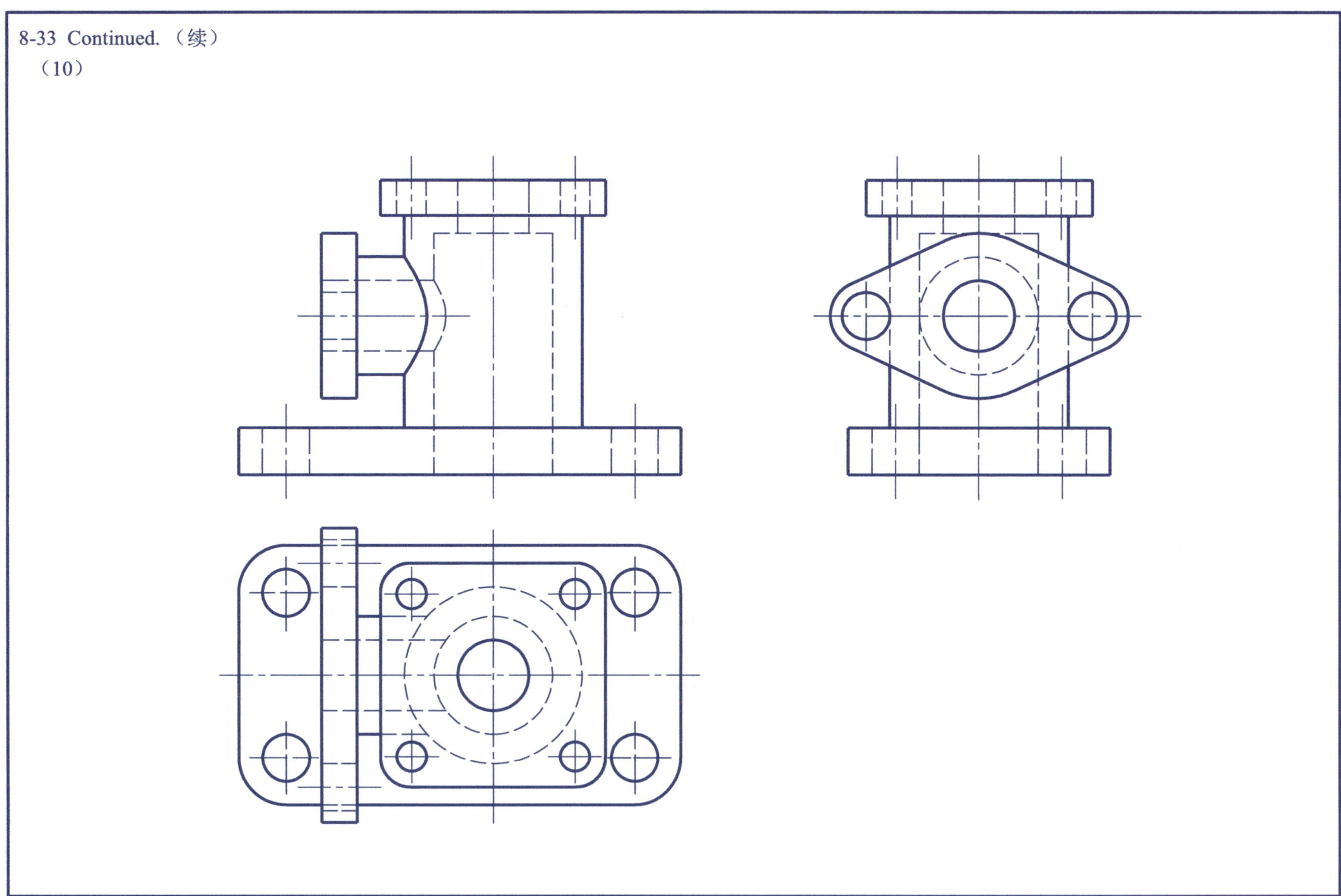

 Class 班级： Name 姓名： ID Number 学号：

8-33 Continued.（续）

（11）

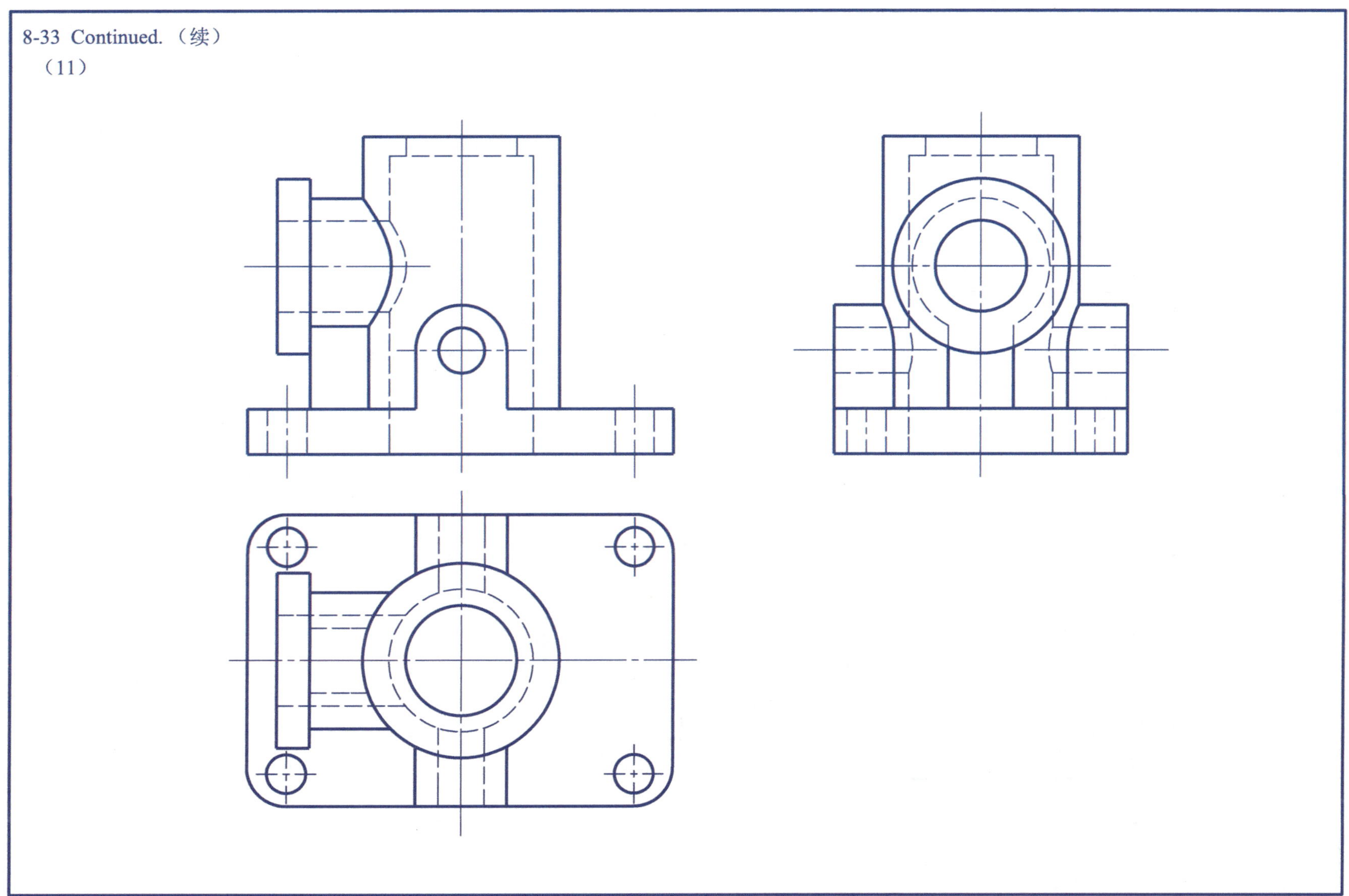

8-33 Continued.（续）

（12）

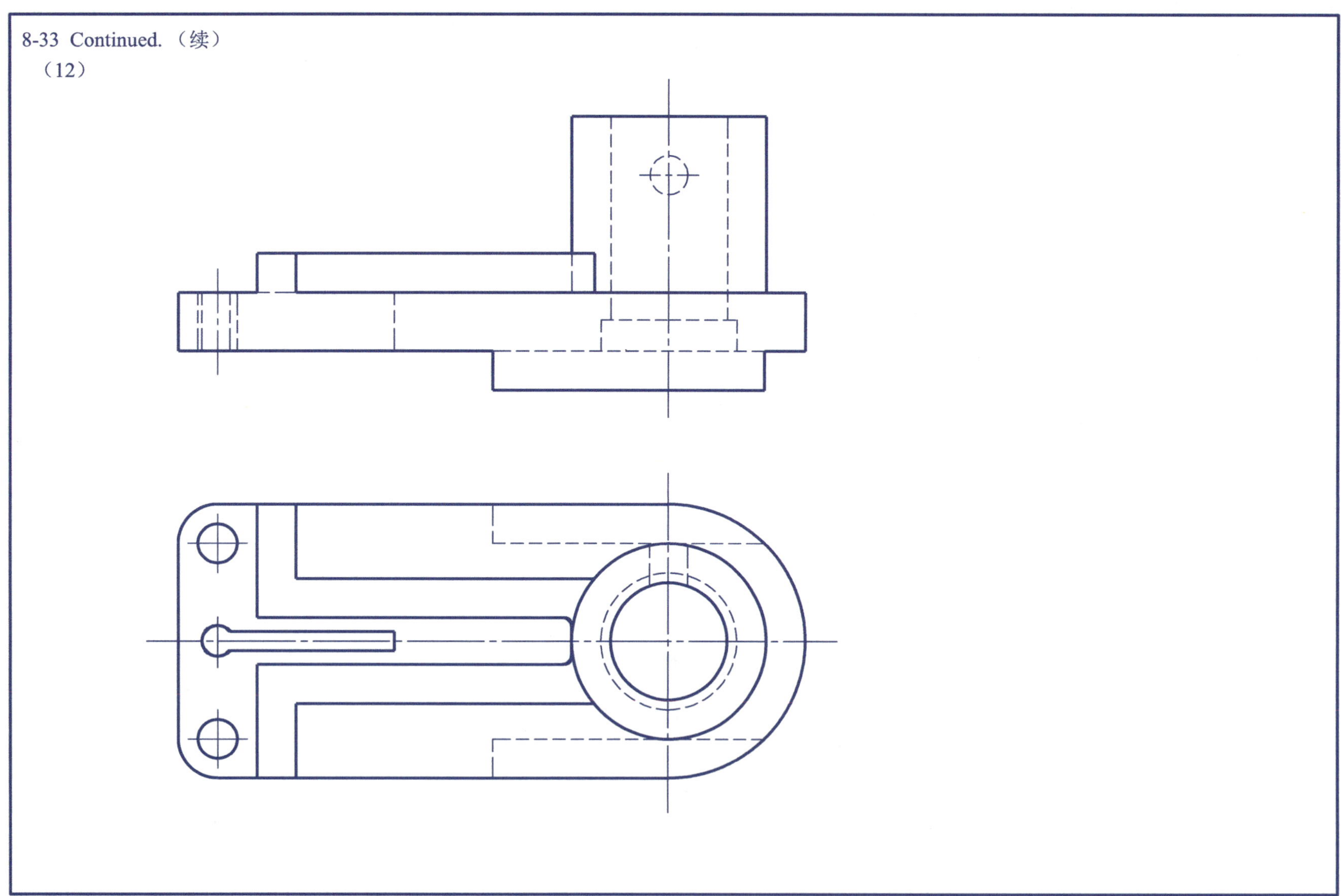

 Class 班级： Name 姓名： ID Number 学号：

8-34 Based on the axonometric projection and front view given, select the optimal scheme to express the part.
根据给出的轴测图和主视图，选择最优的方案表达该机件。

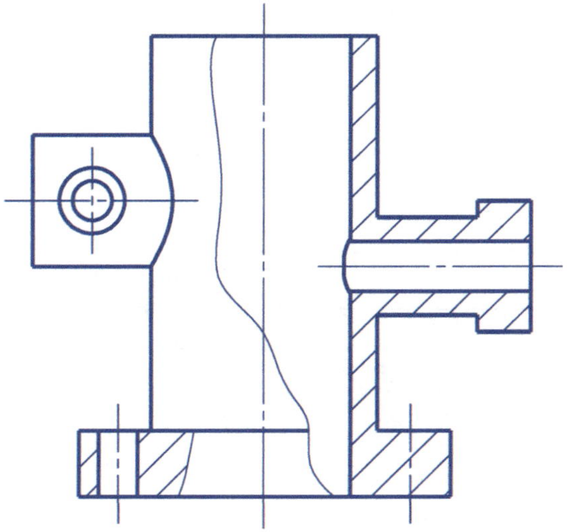

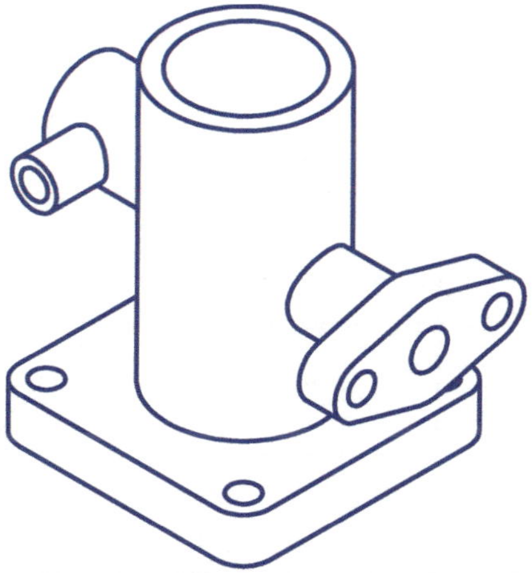

9-1 Draw the front view and left view of the thread as specified representation (Draw the chamfer of the internal and external threads) .
按规定画法画出螺纹的主、左视图（画出内外螺纹倒角）。

（1） The major diameter of external thread is M20, thread length is 30mm, and screw length is 40mm.
外螺纹大径M20，螺纹长度30mm，螺杆长度40mm。

（2） The major diameter of internal thread is M20, thread length is 25mm, and drilling depth is 35mm.
内螺纹大径M20，螺纹长度25mm，钻孔深度35mm。

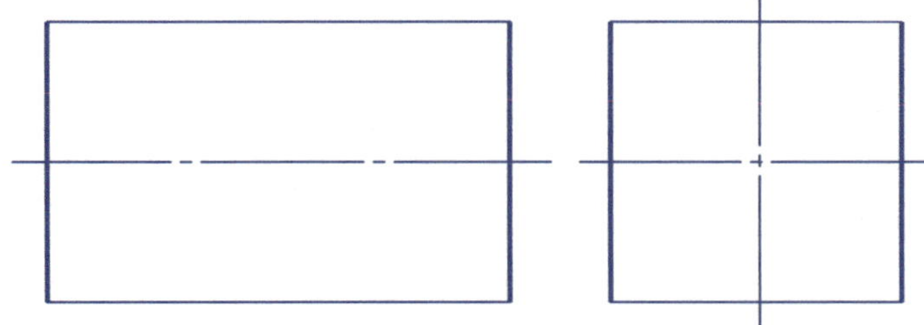

9-2 Find the drawing errors in the thread connection diagram and draw the correct thread connection diagram below.
找出螺纹连接图中的画法错误，并在下方画出正确的螺纹连接图。

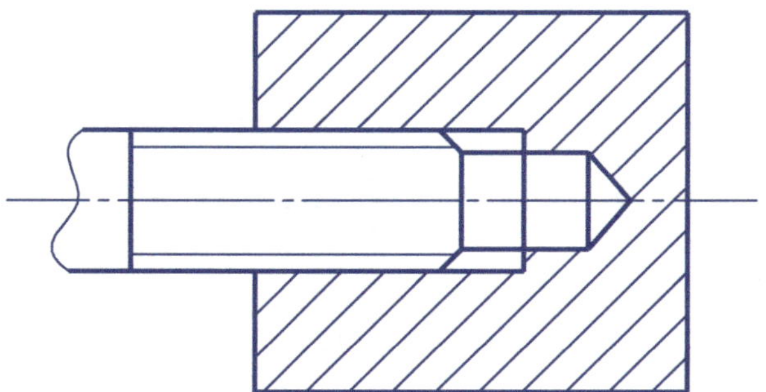

Class 班级： Name 姓名： ID Number 学号：

9-3 According to the thread elements, mark the thread code.
根据螺纹要素，标注螺纹代号。

（1） One coarse plain thread with a nominal diameter of 22mm, and a tolerance zone symbol of 6H.
粗牙普通螺纹，公称直径22mm，公差带代号为6H。

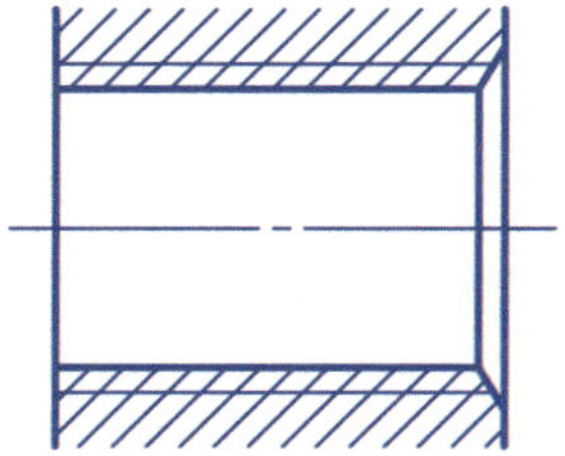

（2） One fine plain thread with a nominal diameter of 18mm, a thread pitch of 2mm, and with a central diameter tolerance zone symbol of 5g and a major diameter tolerance zone symbol of 6g.
细牙普通螺纹，公称直径18mm，螺距2mm，公差带代号为中径5g、顶径6g。

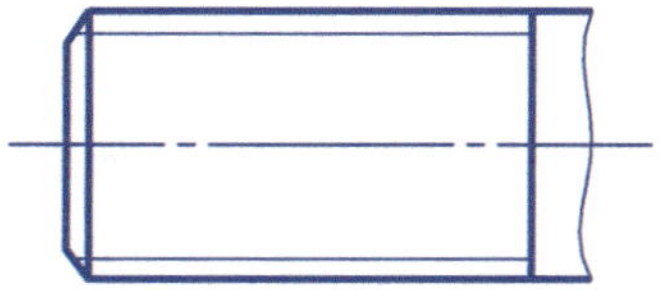

（3） One non-screw-sealed pipe thread, with a dimention symbol of 3/4 and a tolerance grade of A.
非螺纹密封管螺纹，尺寸代号3/4，公差等级A。

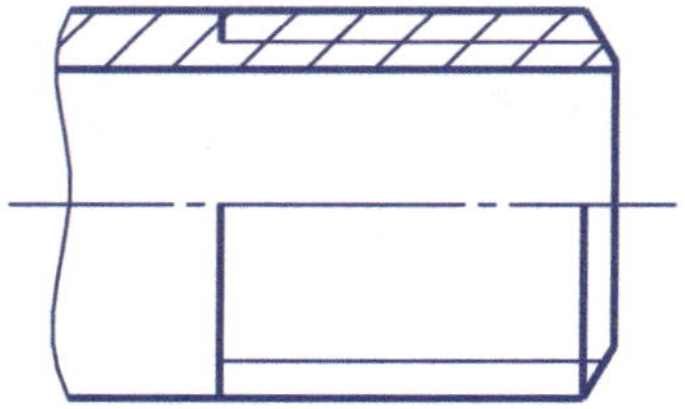

（4） One trapezoidal thread with a nominal diameter of 28mm, with a single-start, left-hand rotation and a thread pitch of 5mm.
梯形螺纹公称直径28mm，单线左旋，螺距5mm。

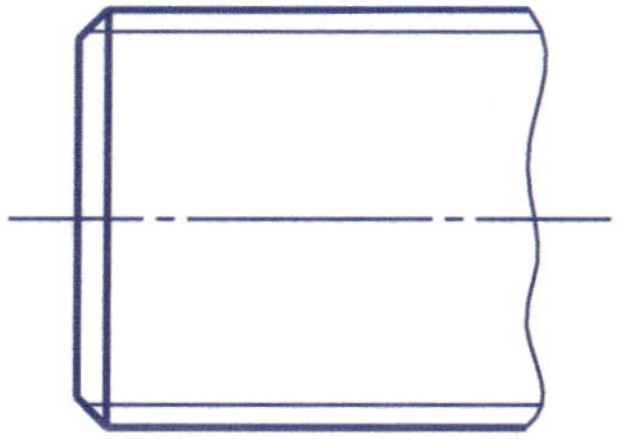

9-4 Write down the marking of threaded fasteners.
写出螺纹紧固件的标记。

(1) Hexagon head bolt 六角头螺栓

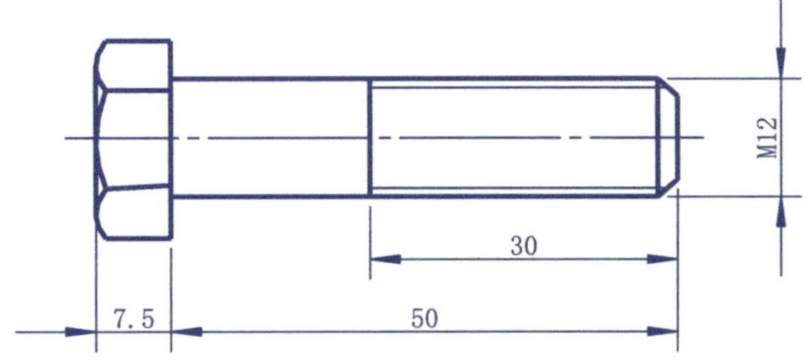

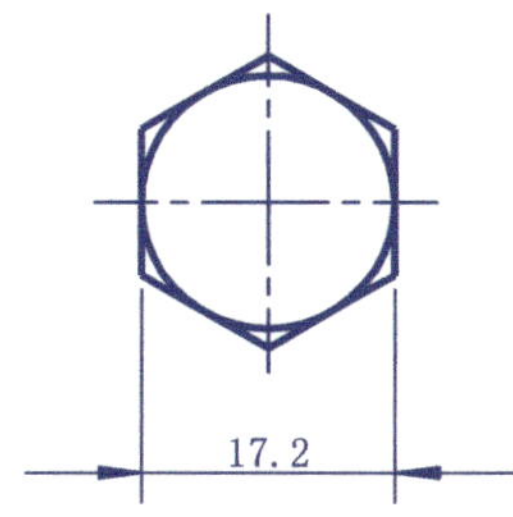

Specified mark 规定标记：

____

(2) Double-end stud 双头螺柱

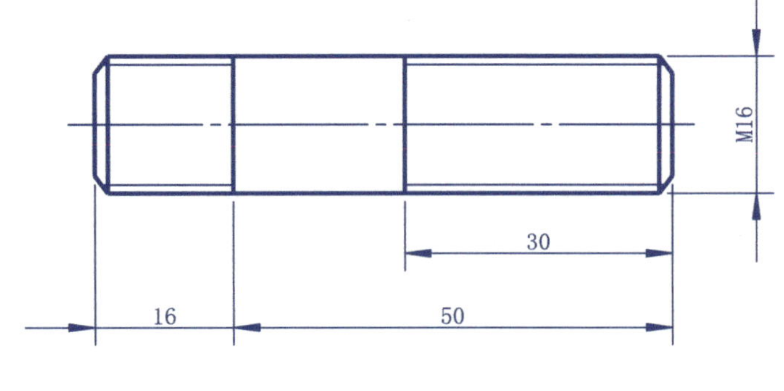

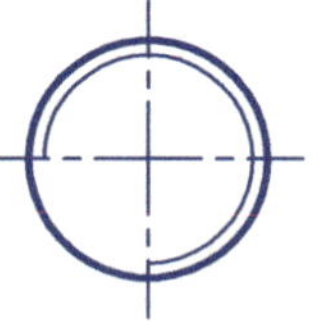

Specified mark 规定标记：

____

9-5 Draw the assembly drawing of the thread fastener connections by scale drawing (Draw it on the A3 drawing with a scale of 2∶1).

用比例画法画出螺纹紧固件连接装配图（用尺规在A3图纸上进行绘图, 比例2∶1）。

（1） Bolted connection 螺栓连接

Bolt GB/T 5782—2016 M12×*l*

Nut GB/T 6170—2015 M12

Washer GB/T 97.1—2002 12

Each plate is 23mm thick and 35mm wide.

每块板厚23mm，板宽35mm。

（2） Stud connection 螺柱连接

Stud GB/T 897—88 M12×*l*

Nut GB/T 6170—2015 M12

Washer GB/T 93—87 12

The upper plate is 15mm thick, the lower cast steel base is 35mm thick and the plates are both 35mm wide.

上面板厚15mm，下面铸钢底座厚35mm，板宽35mm。

（3） Screw connection 螺钉连接

Screw GB/T 65—2016 M10×*l*

The upper plate is 15mm thick, the lower cast iron base is 35mm thick and the plates are both 30mm wide.

上面板厚15mm，下面铸铁底座厚35mm，板宽30mm。

Class 班级： Name 姓名： ID Number 学号：

9-6 Common flat keys and connections 普通平键及联结

(1) Check the size of the keyway according to the diameter of the shaft and mark it in the drawing.

按轴的直径查表确定键槽的尺寸，并在图中标注出来。

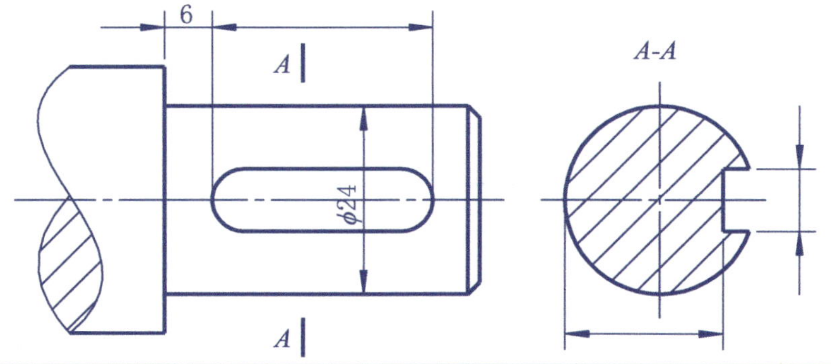

(2) Check the size of the keyway according to the diameter of the hole and mark it in the drawing.

按孔的直径查表确定键槽的尺寸，并在图中标注出来。

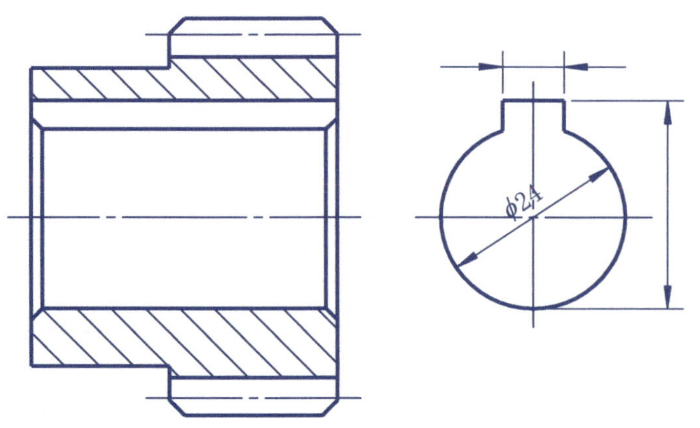

(3) Connect the shaft and gear in question (1) and (2) with a common flat key and draw the whole assembly drawing.

将(1)(2)两题中的轴和齿轮用普通平键联结起来，画全装配图。

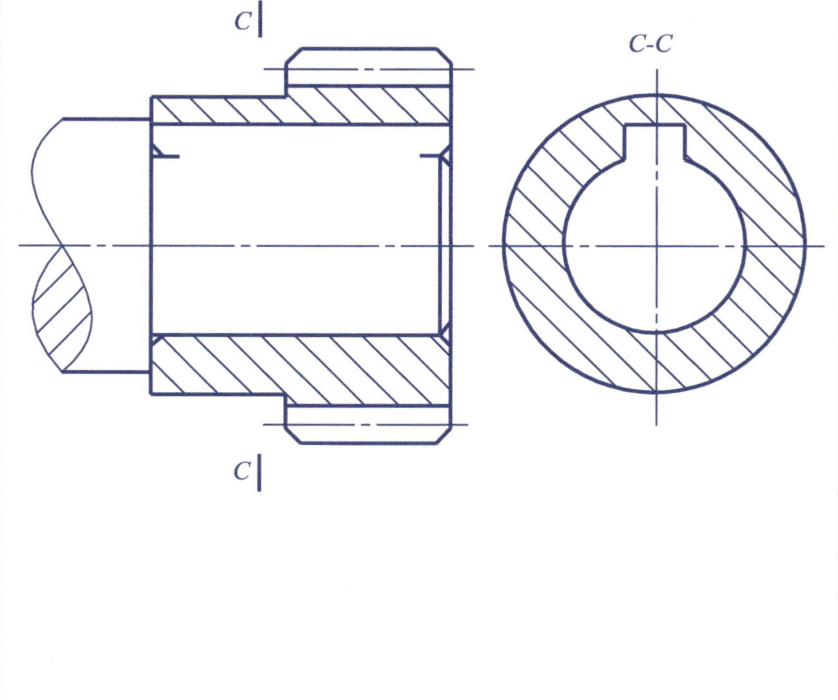

 Class 班级： Name 姓名： ID Number 学号：

9-7 Complete the assembly drawing of cylindrical pin connection.
完成圆柱销连接的装配图。

Pin GB/T 119.1—2000 10×40

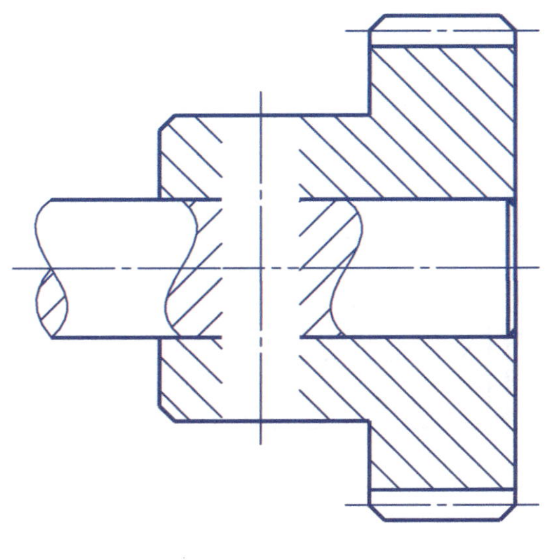

9-8 Known a cylindrical helical compression spring with wire diameter d =5 mm, pitch diameter D =36mm, pitch t =12mm, active coils number n= 8, supporting coils number n_z =2.5, and right hand, draw the sectional view of the spring and mark the dimensions of d, D, t and H_0.
已知圆柱螺旋压缩弹簧的钢丝直径d =5mm，中径D =36mm，节距t =12mm，有效圈数n =8，支撑圈数n_z =2.5，右旋，画出此弹簧的剖视图，并标注出d、D、t 及H_0的尺寸。

Class 班级： Name 姓名： ID Number 学号：

9-9 Draw the rolling bearings in the assembly drawing as specified representation (scale 1∶2). Bearing 1 is GB/T 276—2013 6409 (inner diameter of 45 mm, outer diameter of 120mm, width of 29mm) , bearing 2 is GB/T 276—2013 6413 (inner diameter of 65mm, outer diameter of 160mm, width of 37 mm) .

用规定画法（比例1∶2）画出装配图中的滚动轴承。其中轴承1为GB/T 276—2013 6409 （内径为45mm，外径为120mm，宽为29mm），轴承2为GB/T 276—2013 6413（内径为65mm，外径为160mm，宽为37mm）。

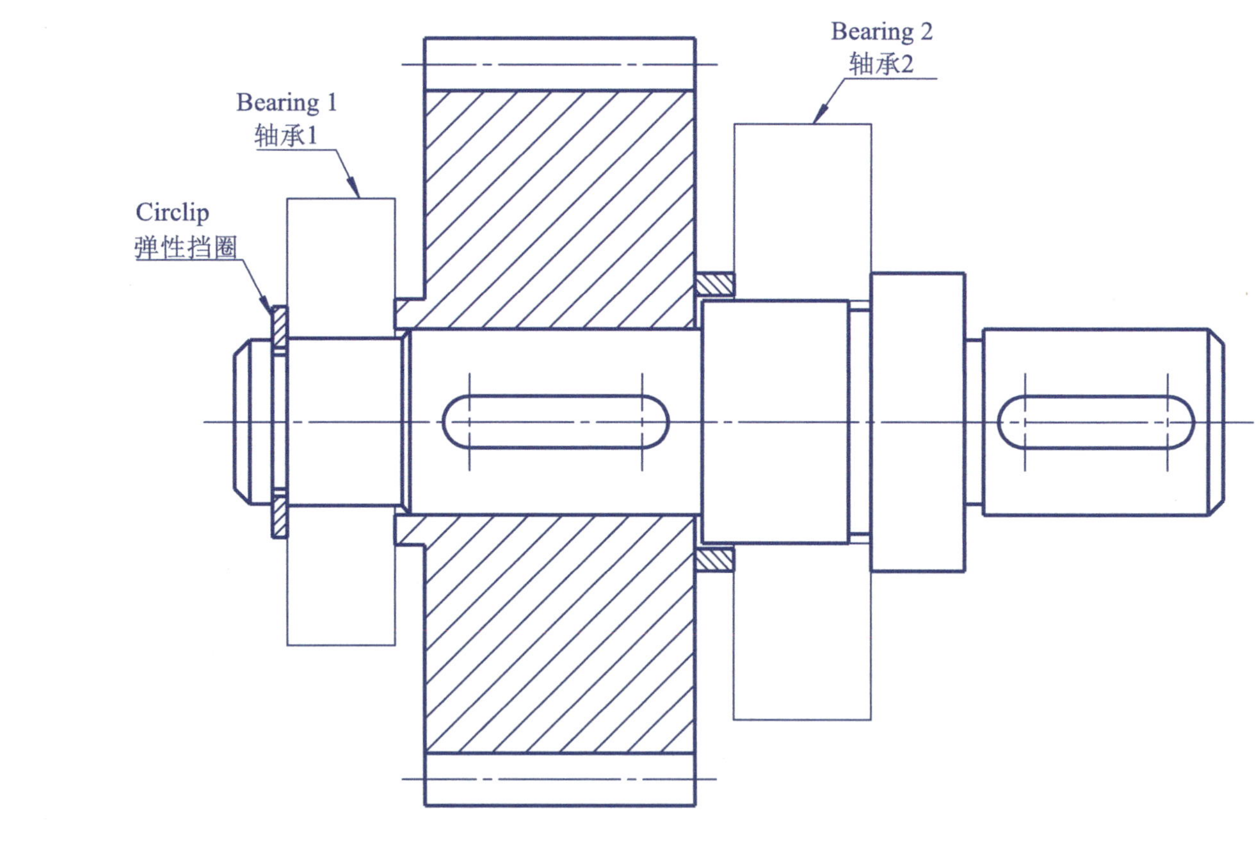

 Class 班级： Name 姓名： ID Number 学号：

9-10 The modulus m =3, the number of teeth z =25, the pressure angle α =20° of a cylindrical spur gear are given, complete two views of the gear and mark the dimensions of d, d_a and d_f.

已知一圆柱直齿轮的模数m =3, 齿数z =25，压力角α =20°，完成齿轮的两个视图，并标注出d、d_a、d_f的尺寸。

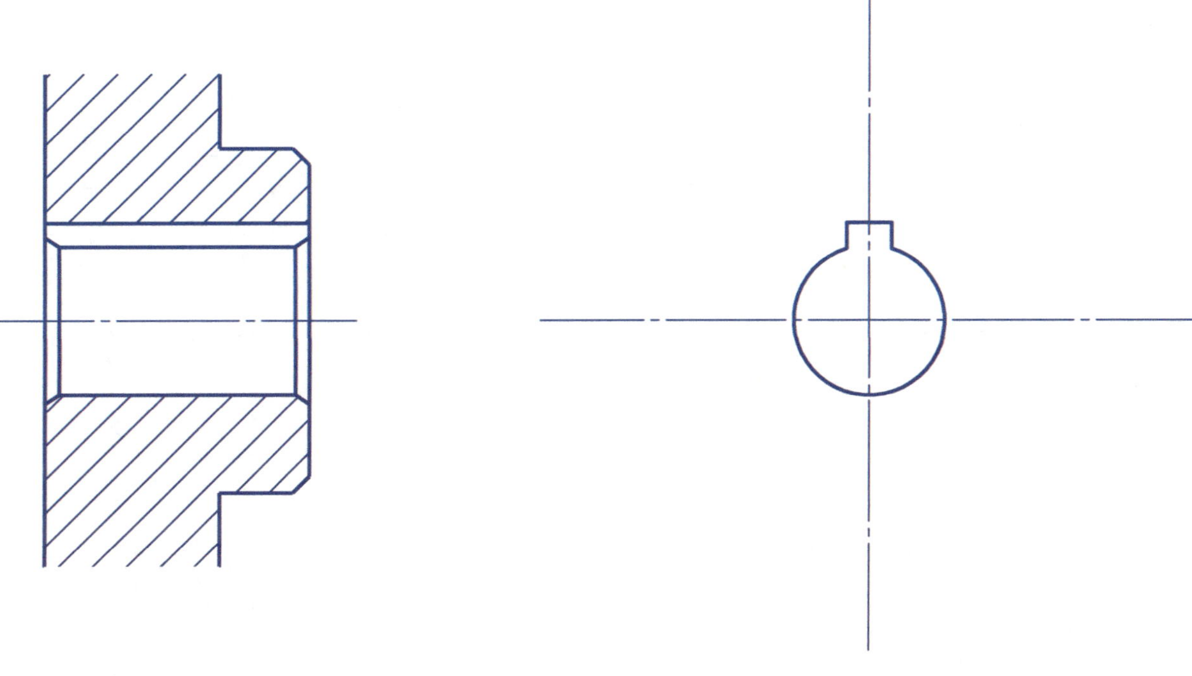

Class 班级： Name 姓名： ID Number 学号：

9-11 Known that the modulus of two gears is $m=2$, the number of teeth is $z_1=18$ for the small gear and $z_2=32$ for the large gear, draw the views of gear engagement according to the specified representation with a scale of 1 ：1.

已知两个齿轮的模数$m=2$，小齿轮的齿数$z_1=18$，大齿轮的齿数$z_2=32$，用1：1的比例按规定画法画出齿轮啮合图。

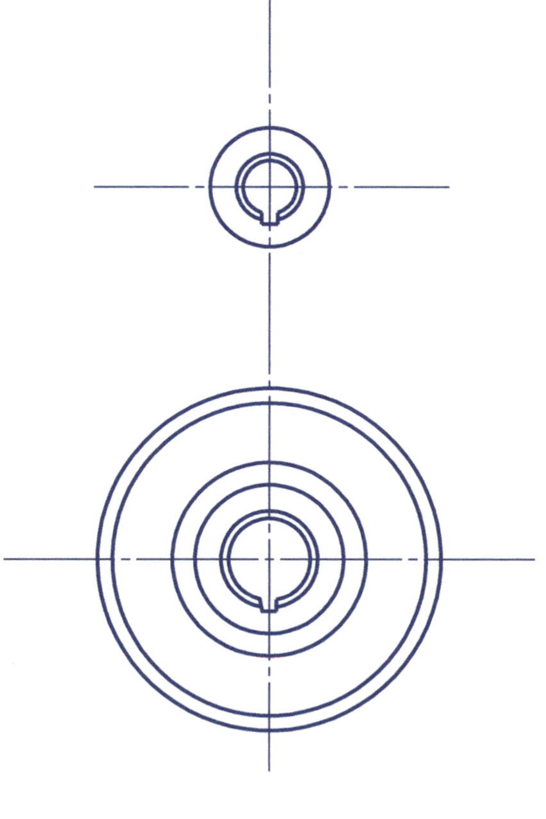

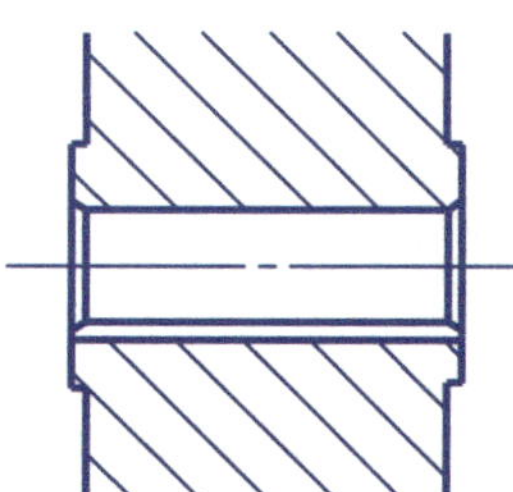

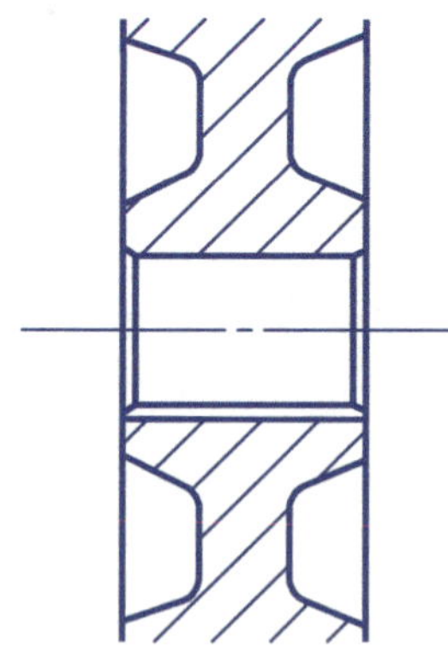

10-1 Draw a detail drawing according to the axonometric drawing. Title: shaft. Material: 45 steel (Check the sizes of the thread undercut, grinding undercut and keyway according to the national standard).

根据轴测图绘制零件图。零件名称：轴。材料：45 钢（要求查表确定退刀槽、越程槽、键槽的尺寸）。

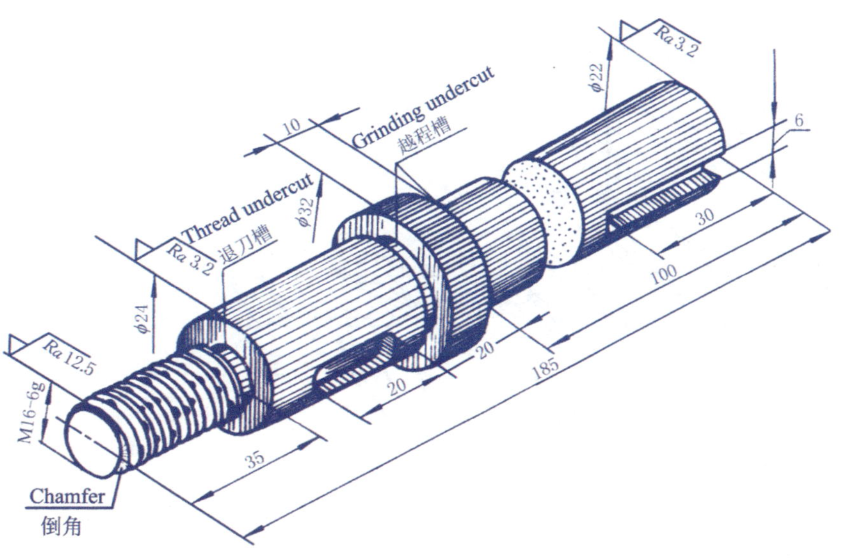

TECHNICAL REQT 技术要求

The shaft should be tempered. 轴须经调质处理。

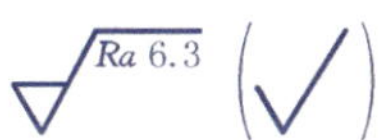

10-2 Draw a detail drawing according to the axonometric drawing. Title: Bearing cover. Material: HT200.
根据轴测图绘制零件图。零件名称：轴承盖。材料：HT200。

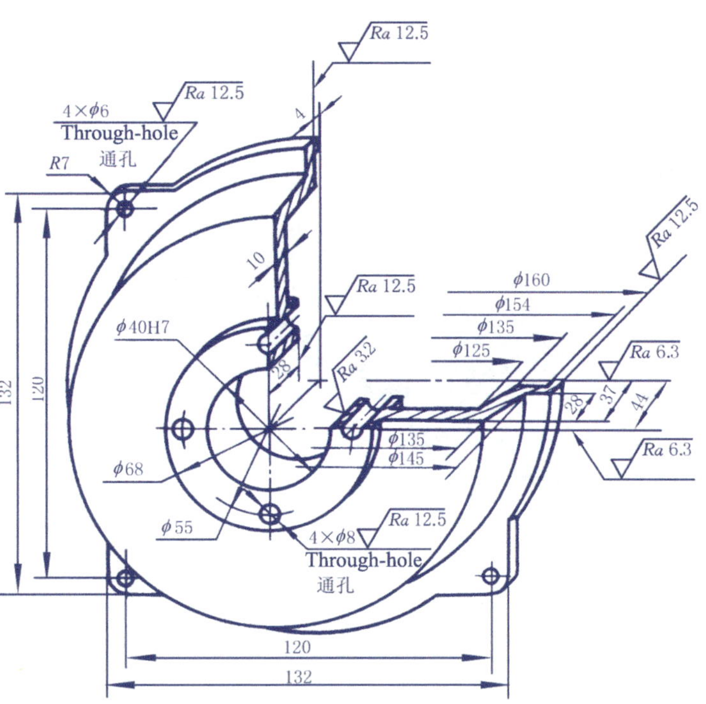

TECHNICAL REQT　　技术要求
Unspecified casting fillets are $R3$.　未注铸造圆角为 $R3$。

　Class 班级：　Name 姓名：　ID Number 学号：

10-3 Draw a detail drawing according to the axonometric drawing. Tile : Tube joint. Material: HT230.

根据轴测图绘制零件图。零件名称：管接头。材料：HT230。

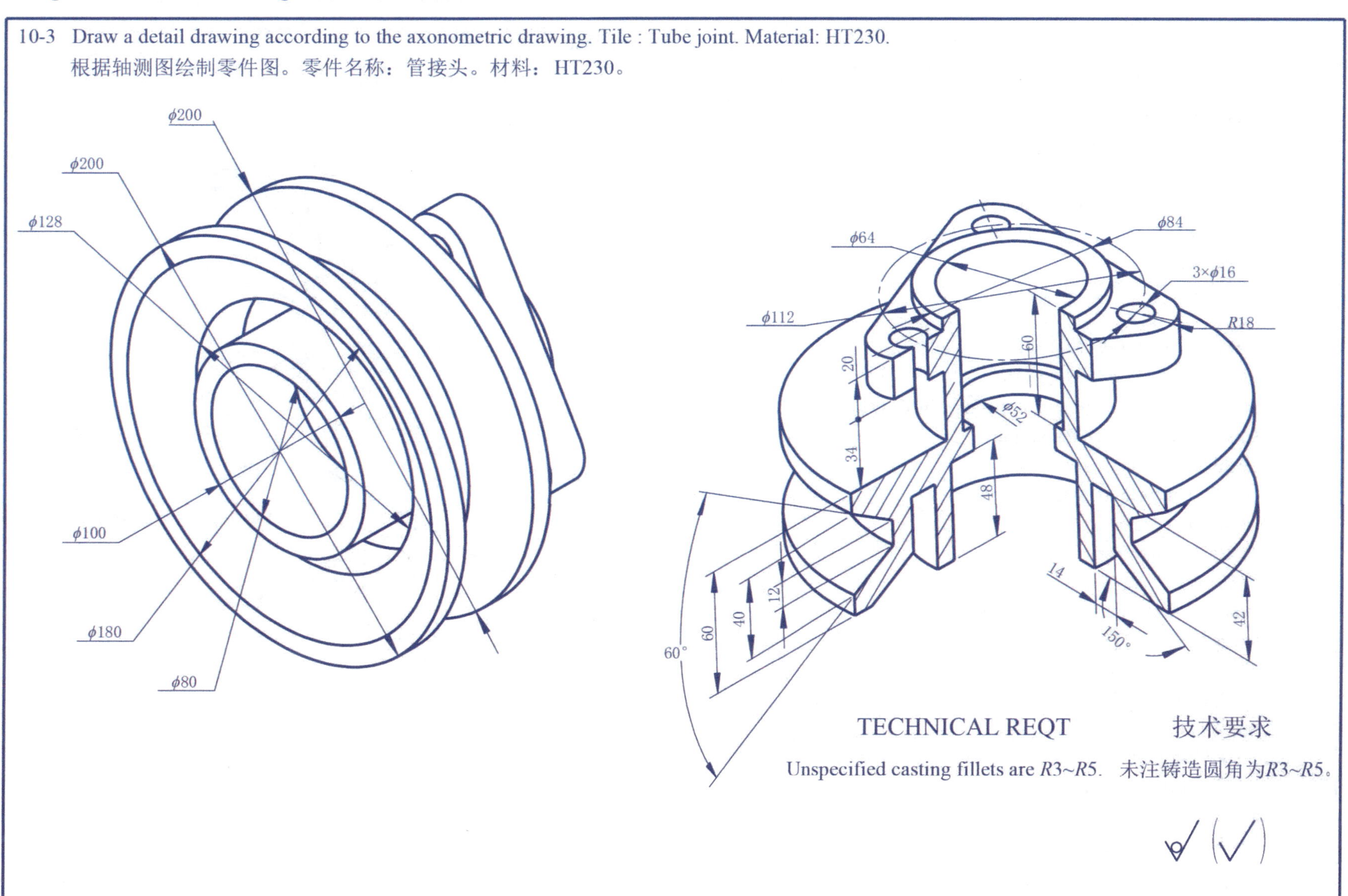

TECHNICAL REQT 技术要求

Unspecified casting fillets are *R*3~*R*5. 未注铸造圆角为*R*3~*R*5。

10-4 Draw a detail drawing according to the axonometric drawing. Title: Engine base. Material: HT200.
根据轴测图绘制零件图。零件名称：机座。材料：HT200。

Work requirements:

1. Mark the dimensions and surface roughness of the part.
2. Fill in the technical requirements:
   (1) Castings should undergo aging treatment.
   (2) Unspecified casting fillets are *R*3~*R*5.

作业要求：

1. 标注零件尺寸和表面粗糙度。
2. 填写技术要求：
   (1) 铸件应进行时效处理。
   (2) 未注铸造圆角为 *R*3~*R*5。

Mark the surface roughness of the machined surfaces:

*ϕ*30 inner hole surface *Ra* 6.3μm;
*ϕ*20 inner hole surface *Ra* 3.2μm;
140 lower surface *Ra* 6.3μm;
*ϕ*50 upper surface *Ra* 6.3μm;
*ϕ*40 upper and lower surface *Ra* 6.3μm;
65×30 surface *Ra* 6.3μm;
Other machined surfaces *Ra* 12.5μm.

标注加工表面的表面粗糙度：

*ϕ*30 内孔面 *Ra* 6.3μm;
*ϕ*20 内孔面 *Ra* 3.2μm;
140 下端面 *Ra* 6.3μm;
*ϕ*50 上端面 *Ra* 6.3μm;
*ϕ*40 上下端面 *Ra* 6.3μm;
65×30 平面 *Ra* 6.3μm;
未注加工面 *Ra* 12.5μm。

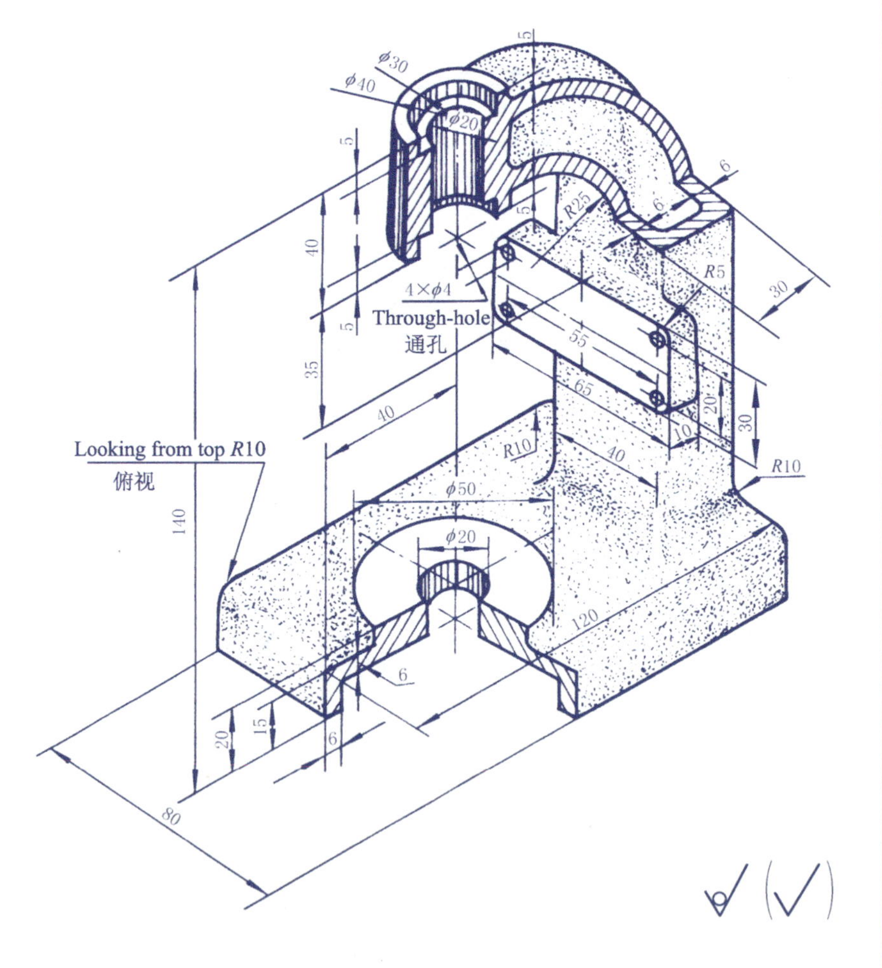

10-5 Draw a detail drawing according to the axonometric drawing. Title: Support. Material: HT200.
根据轴测图绘制零件图。零件名称：支座。材料：HT200。

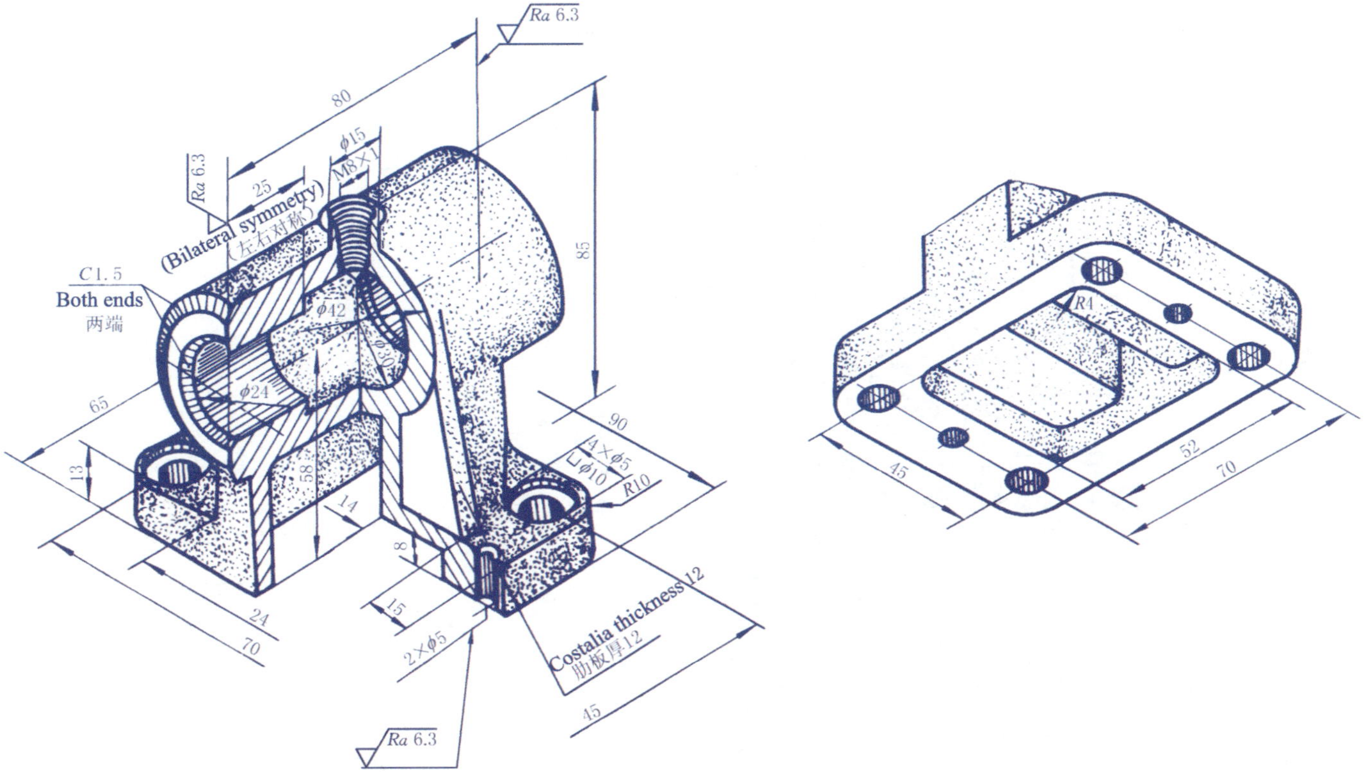

TECHNICAL REQT

1. Unspecified casting fillets are *R*2~*R*3.
2. The unidirectional upper limit for the surface roughness of the unmarked machined surfaces is *Ra*12.5.

技术要求

1. 未注铸造圆角为 *R*2~*R*3。
2. 未注加工面表面粗糙度的单向上限值 *Ra* 为 12.5。

10-6 Draw a detail drawing according to the axonometric drawing. Title: Valve body. Material: HT150.
根据轴测图绘制零件图。零件名称：阀体。材料：HT150。

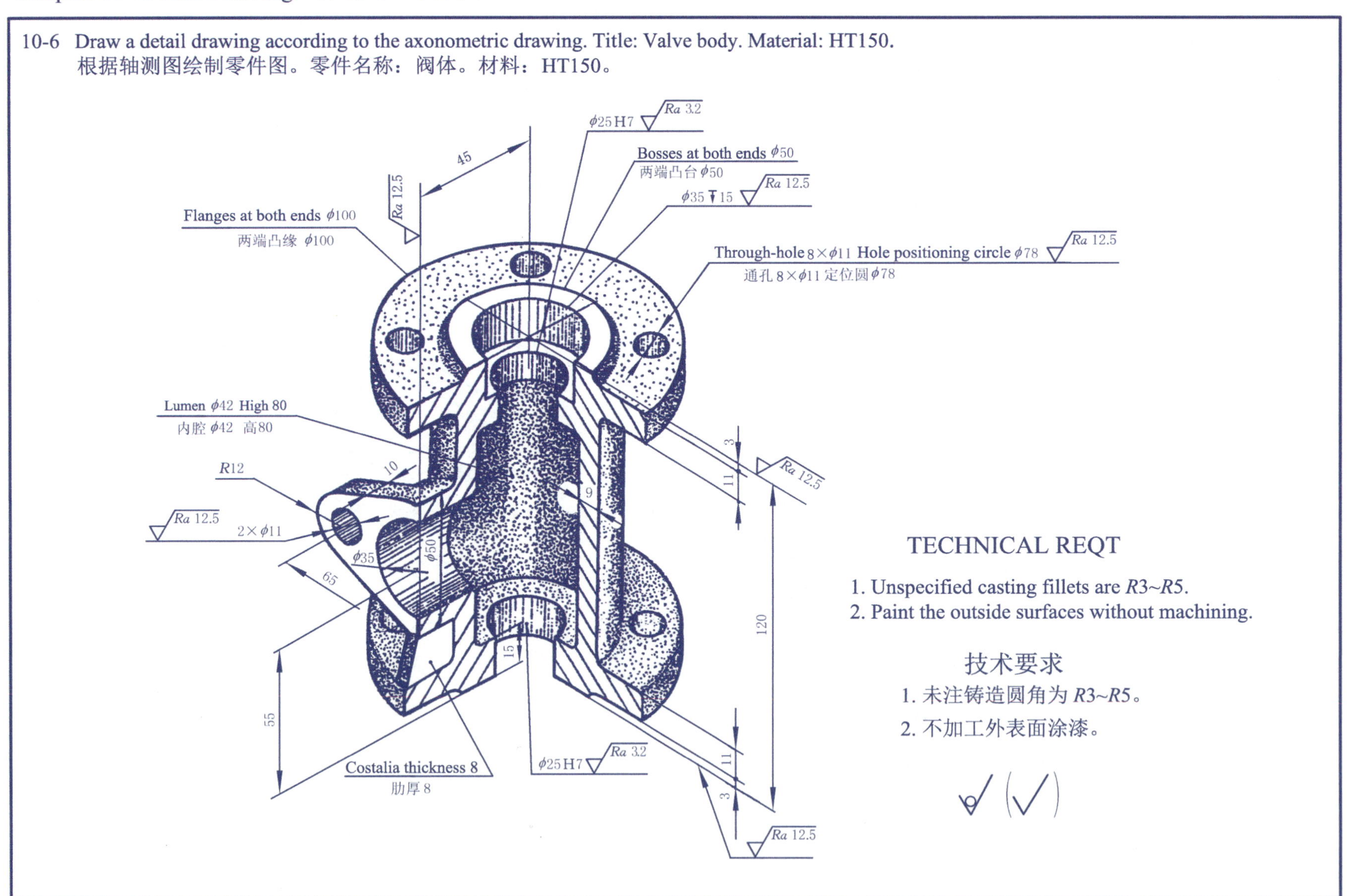

TECHNICAL REQT

1. Unspecified casting fillets are R3~R5.
2. Paint the outside surfaces without machining.

技术要求

1. 未注铸造圆角为R3~R5。
2. 不加工外表面涂漆。

$\sqrt{}$ ($\sqrt{}$)

Class 班级： Name 姓名： ID Number 学号：

10-7 Draw a detail drawing according to the axonometric drawing. Title: Pylon. Material: HT150.
根据轴测图绘制零件图。零件名称：挂架。材料：HT150。

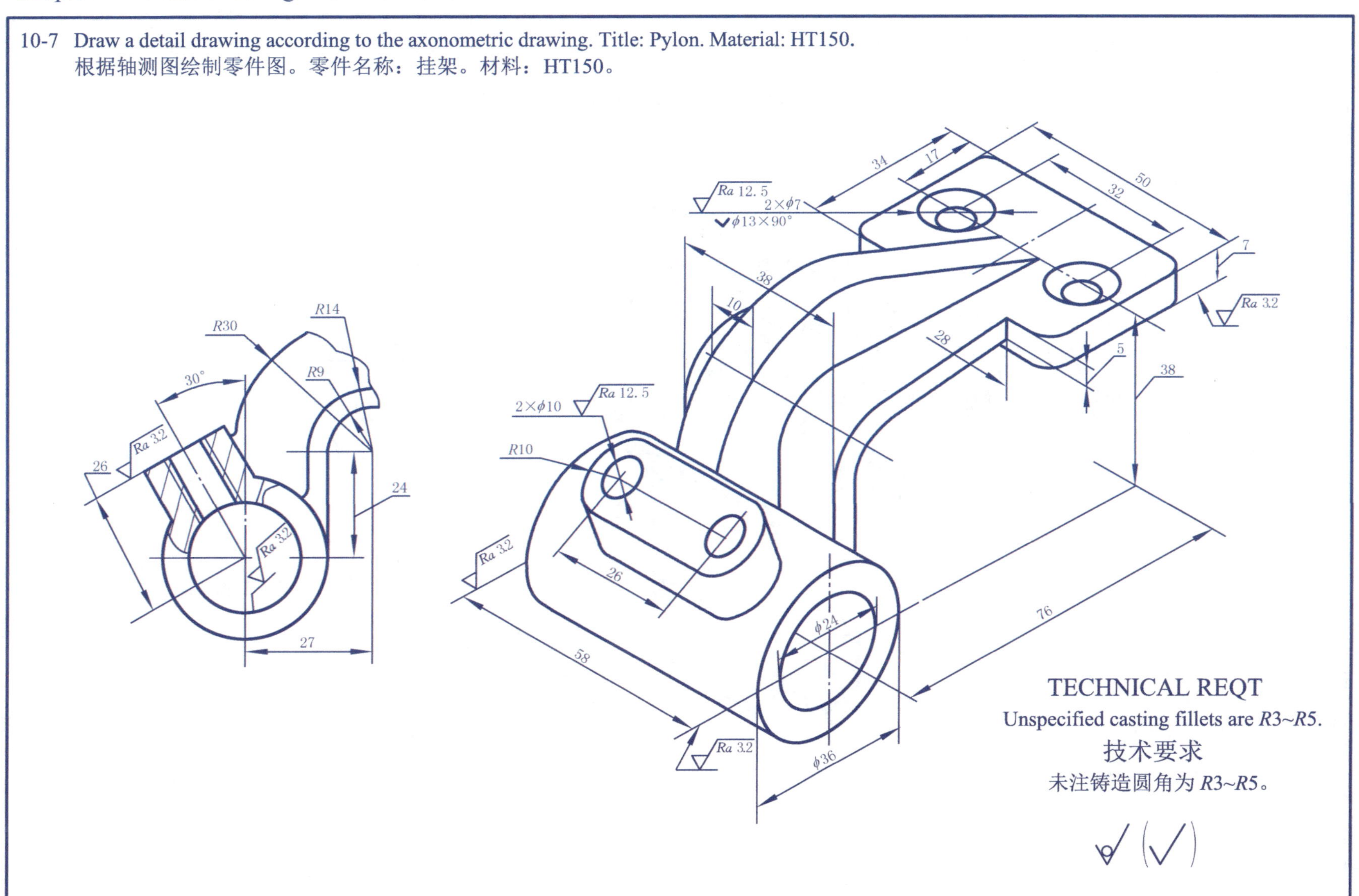

TECHNICAL REQT
Unspecified casting fillets are *R*3~*R*5.
技术要求
未注铸造圆角为*R*3~*R*5。

10-8 Draw a detail drawing according to the axonometric drawing. Title: Underframe. Material: HT250. ▽ is the symbol of surface roughness (The parameter value is self-determined).

根据轴测图绘制零件图。零件名称：底架。材料：HT250。▽ 为表面粗糙度符号（参数值自定）。

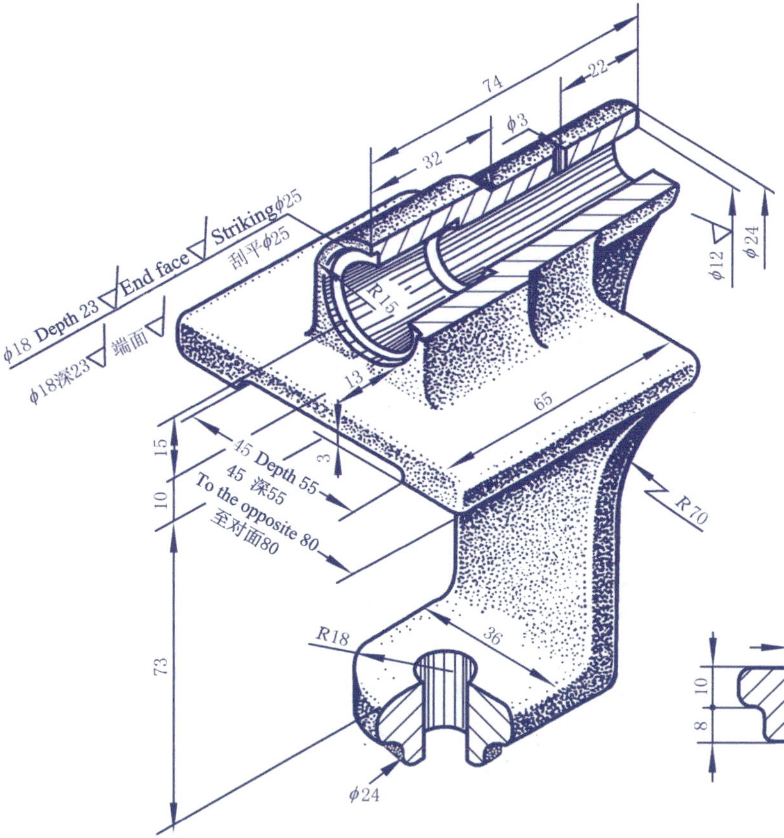

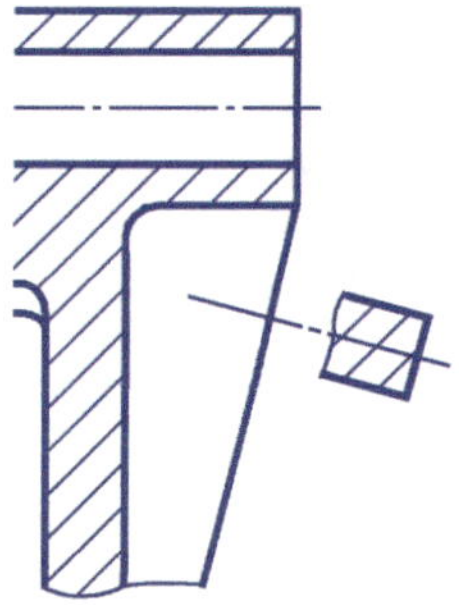

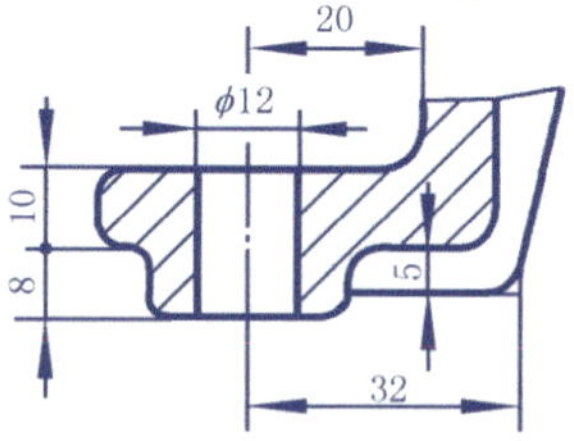

TECHNICAL REQT

Unspecified casting fillets are *R*3~*R*5.

技术要求

未注铸造圆角为 *R*3~*R*5。

Class 班级： Name 姓名： ID Number 学号：

10-9 Draw a detail drawing according to the axonometric drawing. Title: Base body. Material: ZL5. ▽ is the symbol of surface roughness (The parameter value is self-determined).

根据轴测图绘制零件图。零件名称：座体。材料：ZL5。▽为表面粗糙度符号（参数值自定）。

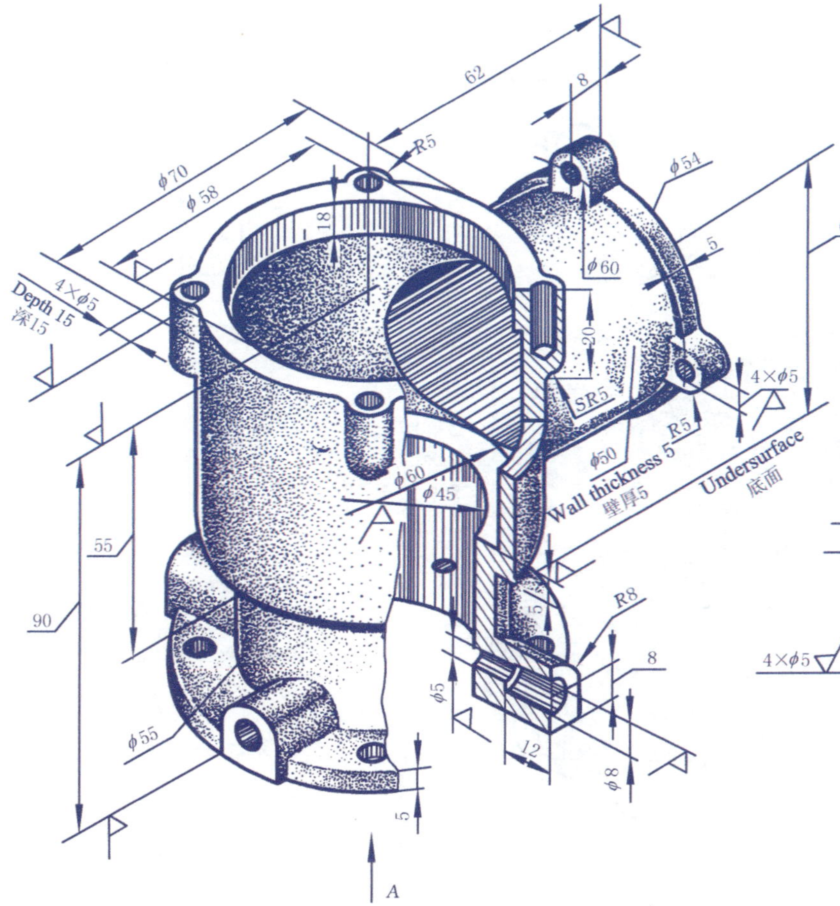

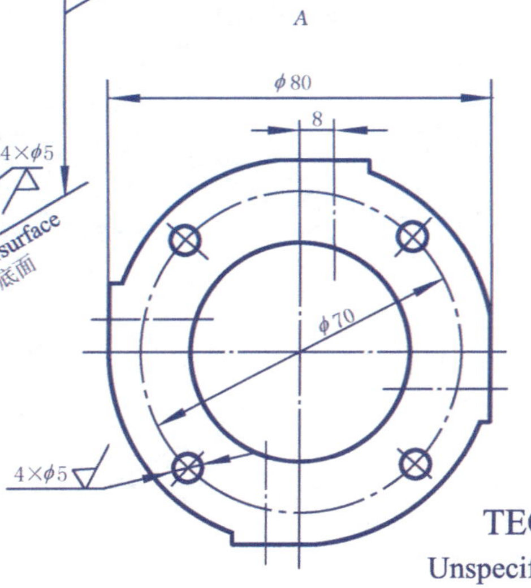

TECHNICAL REQT

Unspecified casting fillets are $R3$~$R5$.

技术要求

未注铸造圆角为$R3$~$R5$。

10-10 Draw a detail drawing according to the axonometric drawing. Title: Gallows. Material: HT150. ▽/ is the symbol of surface roughness (The parameter value is self-determined).

根据轴测图绘制零件图。零件名称：挂架。材料：HT150。▽/为表面粗糙度符号（参数值自定）。

TECHNICAL REQT

Unspecified casting fillets are *R*3~*R*5.

技术要求

未注铸造圆角为*R*3~*R*5。

Class 班级： Name 姓名： ID Number 学号：

10-11 Draw a detail drawing according to the axonometric drawing. Title: Nozzle. Material: ZL3. √ is the symbol of surface roughness (The parameter value is self-determined).

根据轴测图绘制零件图。零件名称：喷嘴。材料：ZL3。√为表面粗糙度符号（参数值自定）。

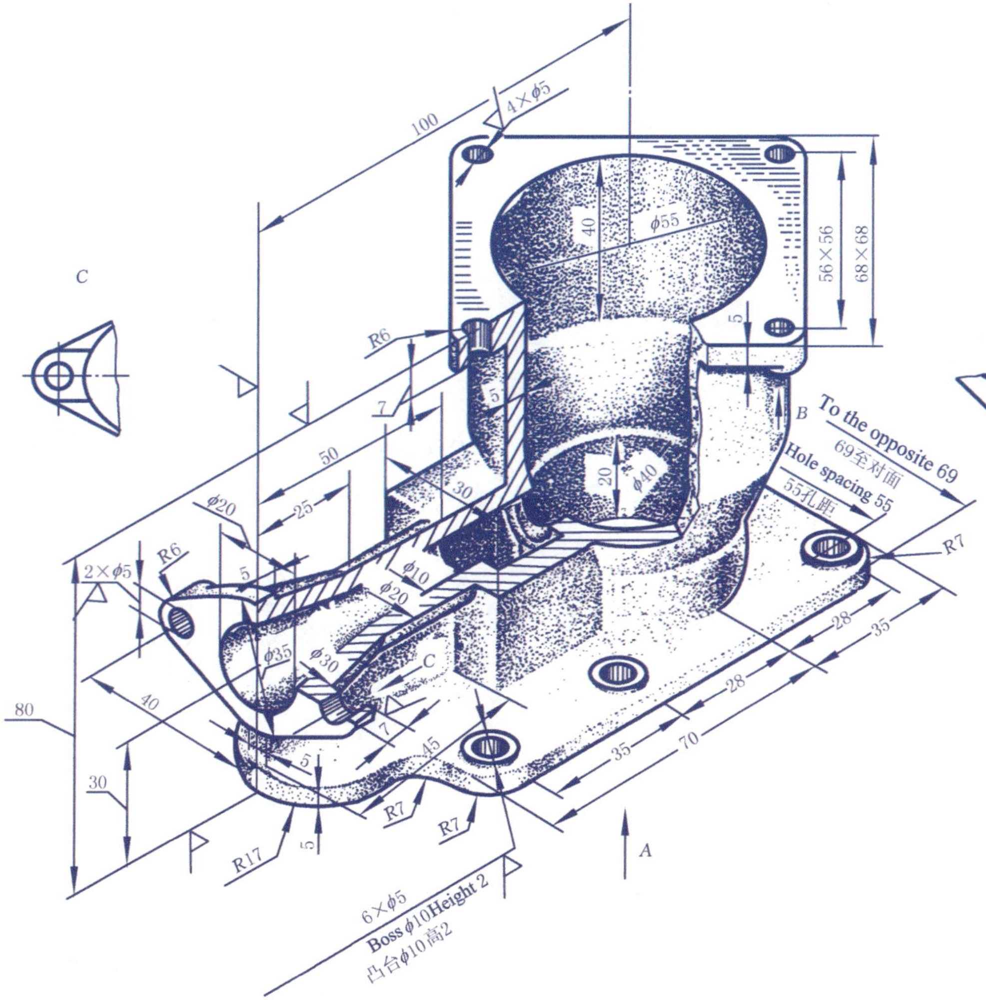

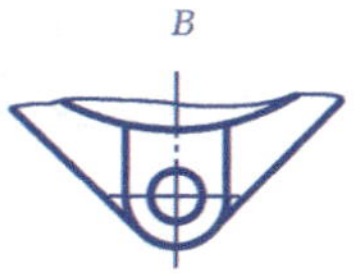

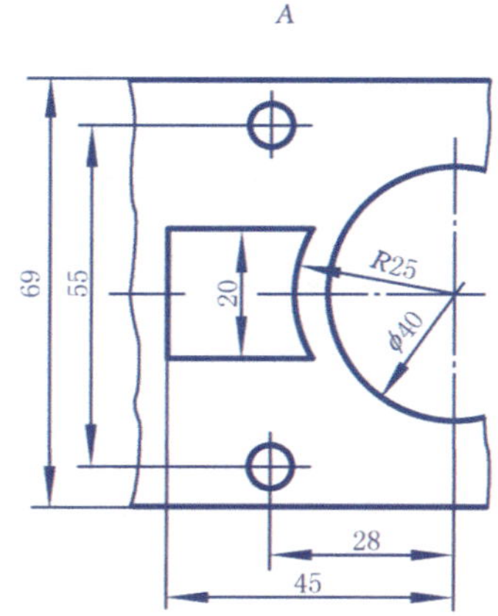

TECHNICAL REQT

Unspecified casting fillets are *R*3~*R*5.

技术要求

未注铸造圆角为 *R*3~*R*5。

√ (√)

10-12 Reasonably mark the dimensions on the detail drawing.
合理标注零件图上的尺寸。

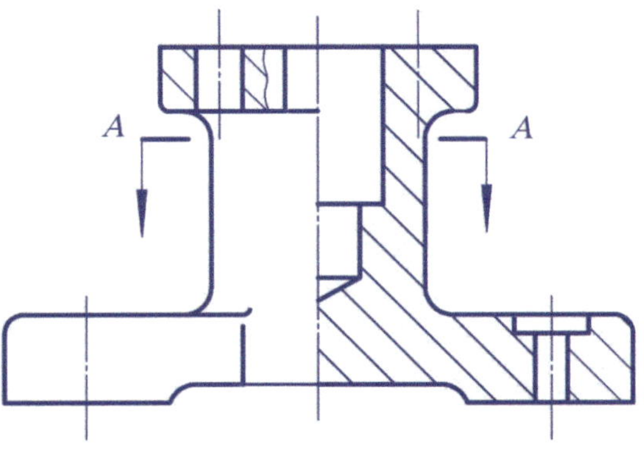

A-A

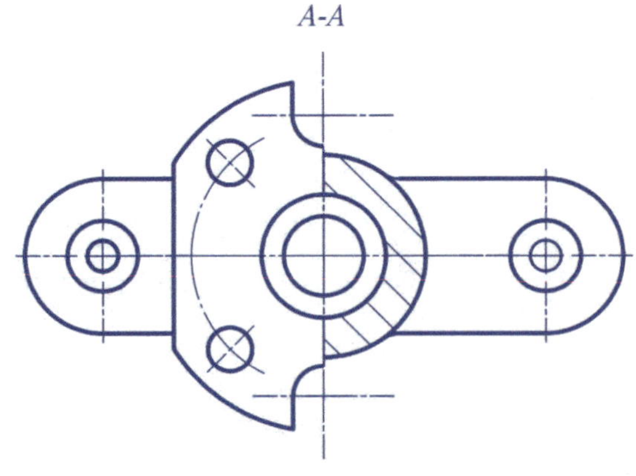

TECHNICAL REQT

1. Castings should undergo aging treatment.
2. Unspecified casting fillets are *R*3~*R*5.

技术要求

1. 铸件要经时效处理。
2. 未注铸造圆角为*R*3~*R*5。

Designed 设计			HT250				
Collated 校核							Abutment 支座
			Scale 比例				
Checked 审核			Class 班级		STU No. 学号		DRG No. 图号

10-13 Analyze the errors in the surface roughness in Figure a) and mark the correct ones in Figure b).

分析图a)中表面粗糙度标注中的错误，在图b)中作出正确的标注。

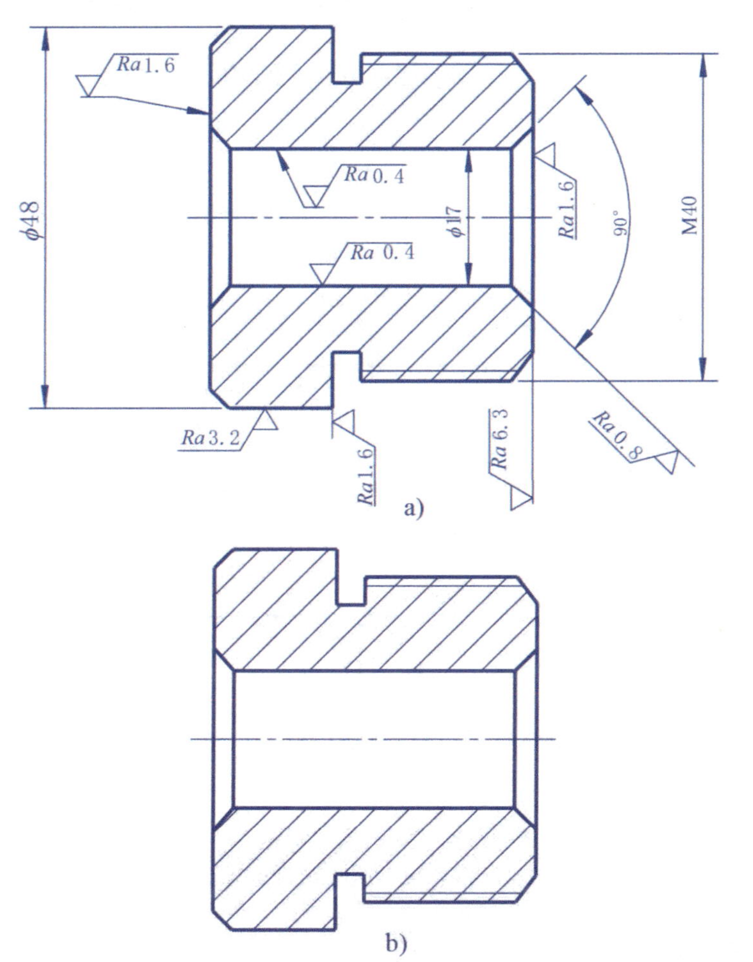

a)

b)

10-14 According to the surface roughness requirements given, mark the surface roughness on the corresponding surfaces below.

根据题中给出的表面粗糙度要求，把符号标注在下面相应表面上。

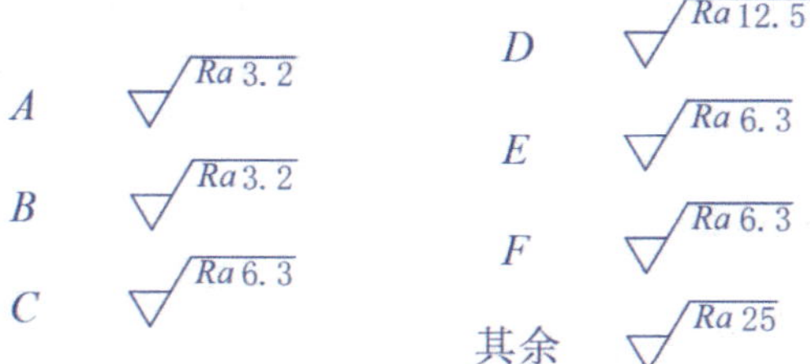

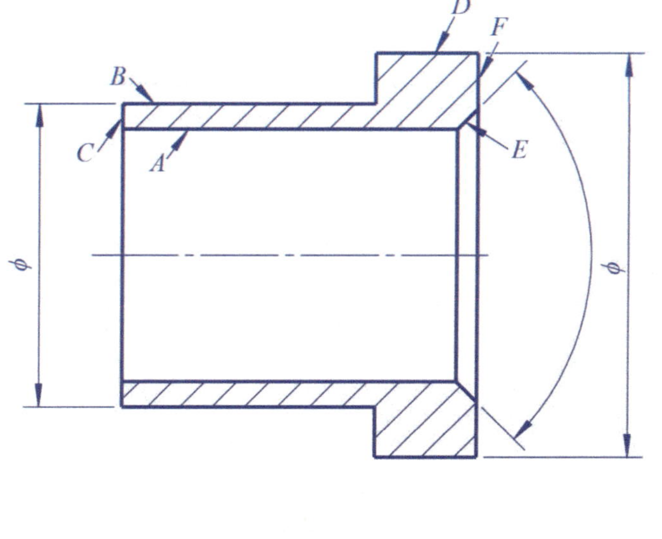

10-15 It is known that the nominal size of the shaft and hole is $\phi 32$, the base hole system is used, the standard tolerance grade of the hole is IT8, the basic deviation code of the shaft is g, and the standard tolerance grade is IT7.

已知轴和孔的公称尺寸为$\phi 32$，采用基孔制配合，孔的标准公差等级为IT8，轴的基本偏差代号为g，标准公差等级为IT7。

Requirements: (1) Note the nominal size, the tolerance zone symbol and the upper and lower limit deviation values on the corresponding detail drawing. (2) Note the nominal size and fit in the assembly drawing.

要求：（1）在相应的零件图上注出公称尺寸、公差带代号和上下极限偏差数值。（2）在装配图中注出公称尺寸和配合。

Shaft:
轴：

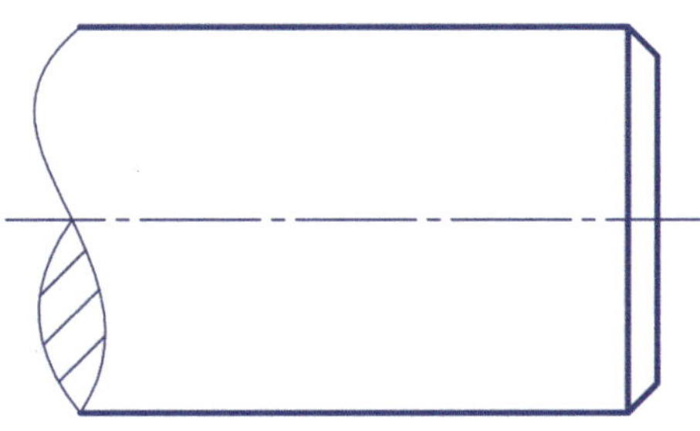

Hole:
孔：

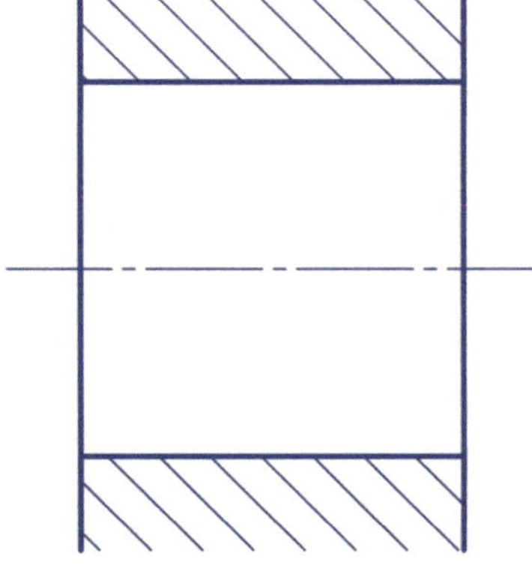

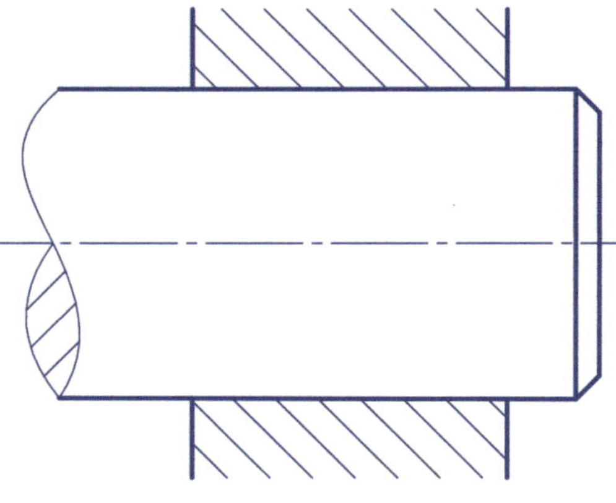

The fit of the hole and shaft belongs to the _______ fit.
孔和轴的配合属于 ________ 配合。

10-16 According to the matching code on the assembly drawing, explain the matching datum system and the matching type, and mark the nominal size and limit deviation values on the detail drawing respectively.

根据装配图上的配合代号，说明配合基准制和配合种类，并分别在零件图上标注出公称尺寸和极限偏差数值。

$\phi 20\frac{H8}{g7}$ is base ___ system ___ fit; $\phi 5\frac{R7}{h6}$ is base ___ system ___ fit.

$\phi 20\frac{H8}{g7}$是基 ____ 制 _____ 配合；$\phi 5\frac{R7}{h6}$是基 ____ 制 _____ 配合。

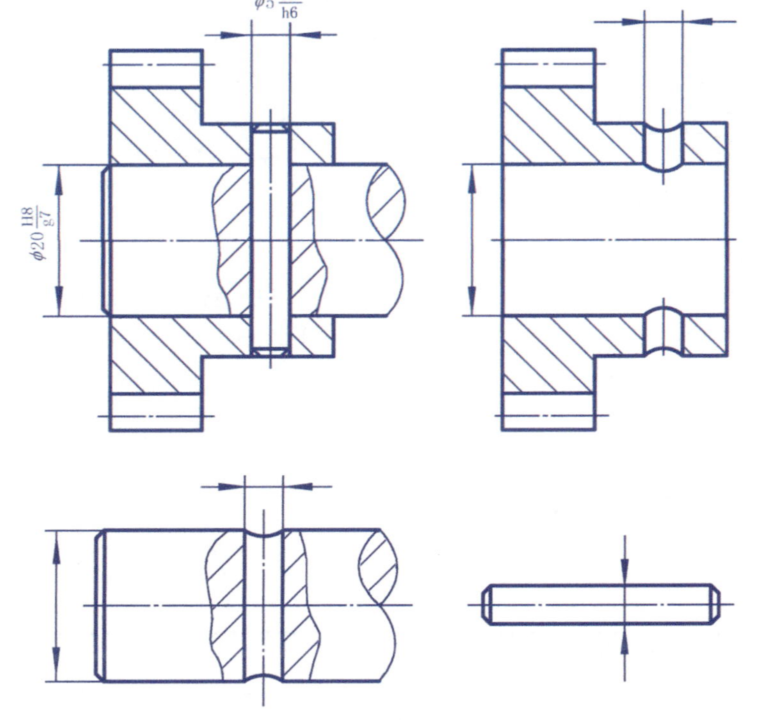

10-17 According to the matching code in the drawing, mark the diameter, tolerance zone code and limit deviation value of the shaft and hole on the detail drawing.

根据图中的配合代号，在零件图上分别标出轴和孔的直径、公差带代号和极限偏差值。

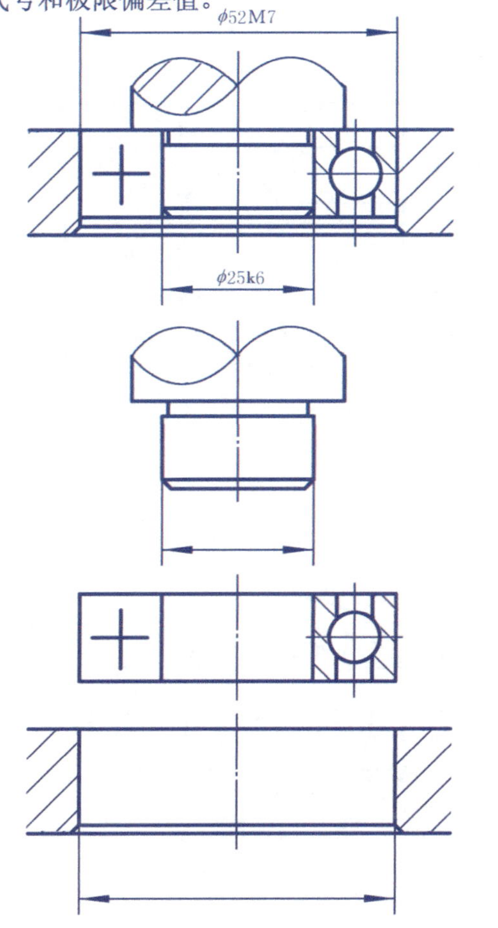

10-18 Read the detail drawing and answer the following questions.
读零件图，回答下列问题。

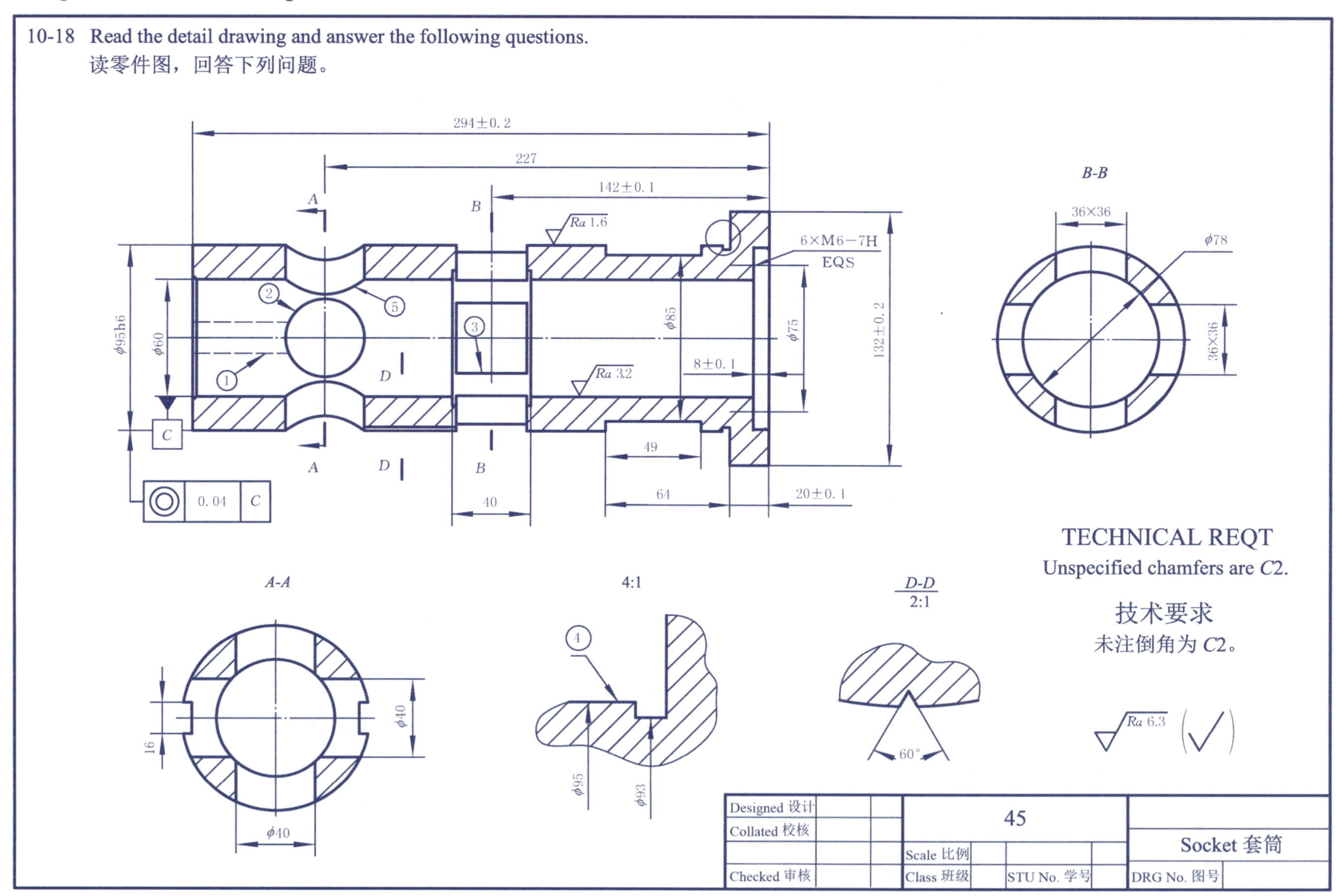

Class 班级: Name 姓名: ID Number 学号:

10-18 Continued. （续）

(1) The main axial dimension reference is ____________, and the main radial dimension reference is____________.
轴向主要尺寸基准是____________，径向主要尺寸基准是____________。

(2) The distance between the two dashed lines marked with ① in the drawing is ____________.
图中标有①的部位所指两条虚线间的距离为____________。

(3) The diameter dimension marked with ② in the drawing is ____________.
图中标有②所指的直径尺寸为____________。

(4) The shaping dimension of the wire frame marked with ③ in the drawing is ____________. It's positioning dimension is ____________.
图中标有③所指线框的定形尺寸为____________。定位尺寸为____________。

(5) The surface roughness of the leftmost end is ____________, and the surface roughness of the rightmost end is ____________.
最左端面的表面粗糙度为____________，最右端面的表面粗糙度为____________。

(6) The surface roughness of the ④ positions in the partial enlarged drawing is ____________.
局部放大图中④所指位置的表面粗糙度为____________。

(7) The curve marked with ⑤ in the drawing is ____________ formed by the intersection of ____________ and ____________.
图中标有⑤所指的曲线是由____________与____________相交而形成的________________________。

(8) The outer surface $\phi 132 \pm 0.2$ can be machined to the maximum ____________, the minimum ____________, and the tolerance is ____________.
外圆面$\phi 132 \pm 0.2$最大可加工成____________，最小可为____________，公差为____________。

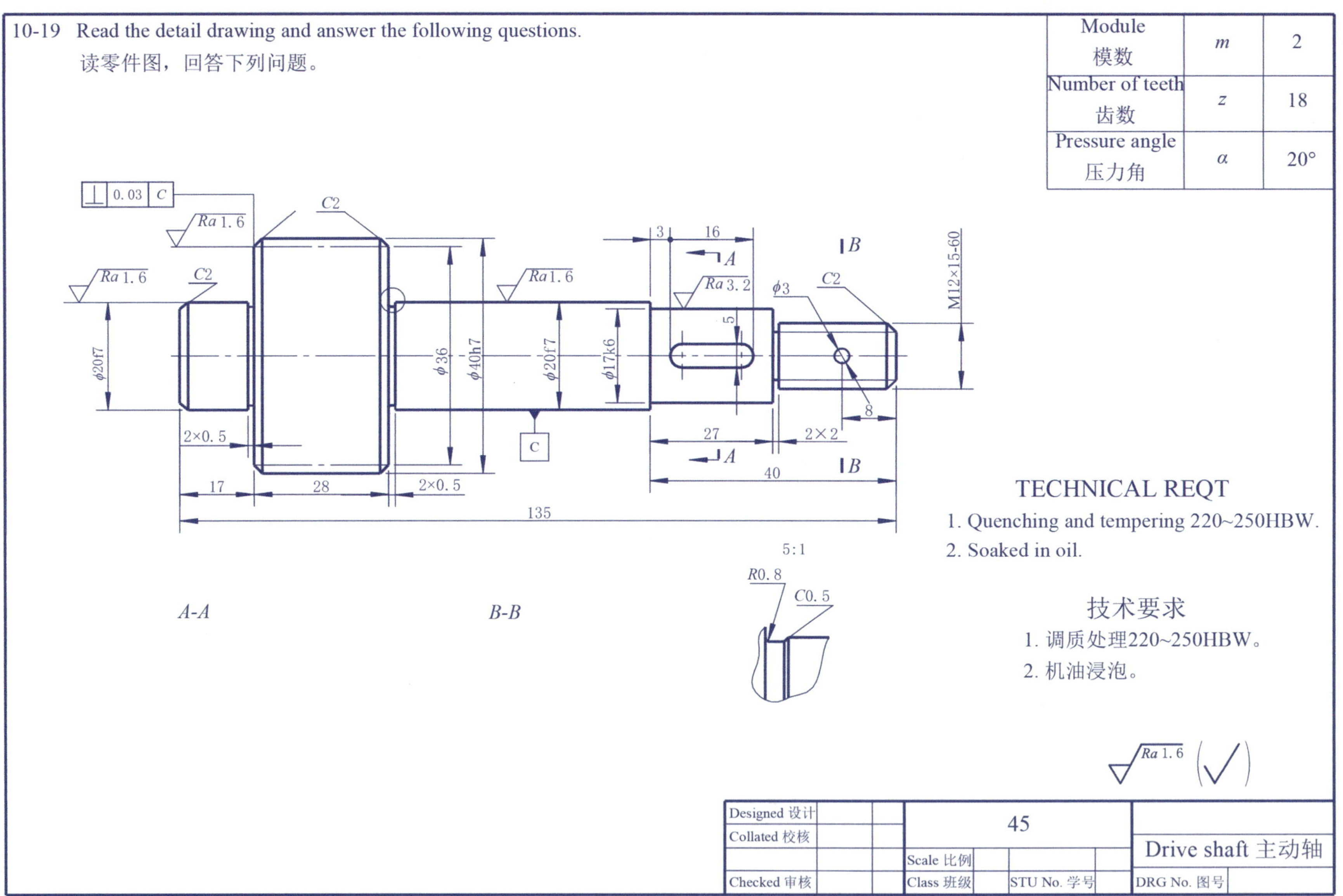

Class 班级：　　Name 姓名：　　ID Number 学号：

10-19 Continued. （续）

1. Read the detail drawing of drive shaft, and draw the removed cross-section views of *A-A* (Keyway depth is 3mm, and keyway surface roughness is *Ra* 3.2), and *B-B*.

   读主动轴零件图，画出*A-A*(键槽深3mm、键槽表面粗糙度为*Ra* 3.2)、*B-B*移出断面图。

2. Answer the following questions:

   回答下列问题：

(1) The structure of the drive shaft contains: ________shaft segments, ________undercuts, ________chamfers.

组成主动轴的结构有：________个轴段、________处退刀槽、________处倒角。

(2) The length of the keyway is________mm, and the width is________mm. The positioning dimension of the keyway is________.

键槽长为________mm，宽为________mm。键槽的定位尺寸是________。

(3) The surface roughness of the left end of ϕ20f7 is__________, and f7 is_________. Size 8 near the right end is the________.

左端ϕ20f7的表面粗糙度为__________，其中f7是_________。右端附近的尺寸8是________尺寸。

(4) The drawing at the mark 5：1 is the________.

标注5：1处的图为________图。

10-20 Read the detail drawing and answer the following questions.
读零件图，回答下列问题。

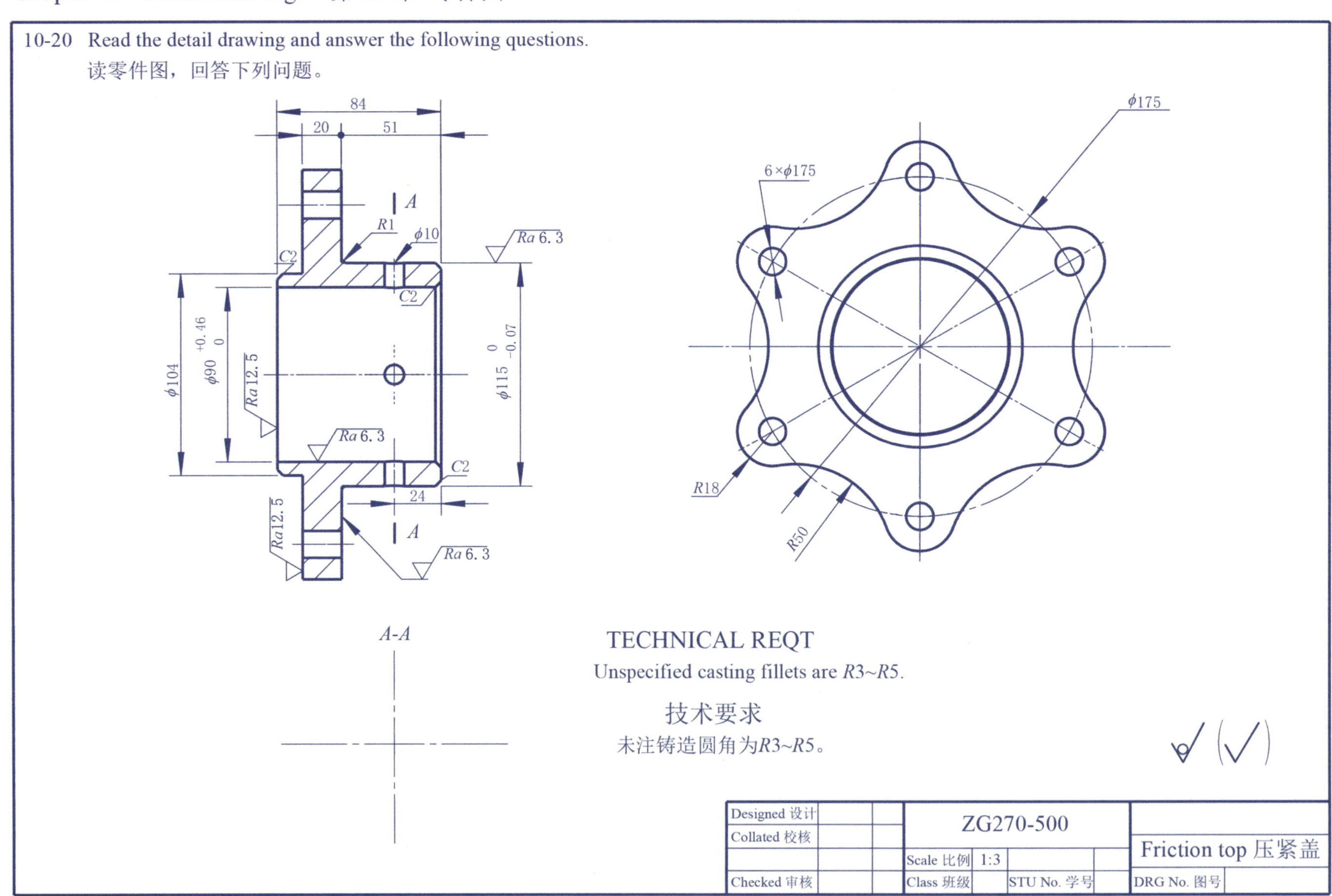

10-20 Continue.（续）

(1) The name of this part is _______ , the material is _______, and the scale is _______ , belonging to the _______scale

该零件的名称是_______， 材料是_______, 比例是_______, 属于_______比例。

(2) The contour line of the part is formed by _______ connecting arcs, the radius of the known arc is _______ , the location dimension for the six equal division of the circle is _______ , and the radius of the connecting arc is _______.

该零件的外形轮廓线由______段圆弧连接而成，其已知圆弧的半径是______, 圆周六等分的定位尺寸是______， 其连接圆弧的半径是______。

(3) Size $6\times\phi14$ indicates there are _______ through holes with a diameter of _______.

尺寸$6\times\phi14$表示有_______个直径是_______的通孔。

(4) The nominal size of dimension $\phi115^{\ 0}_{-0.07}$ is _______, the upper limit size is _______, the lower limit size is _______, and the tolerance value is _______.

尺寸$\phi115^{\ 0}_{-0.07}$的公称尺寸是_______， 上极限尺寸是_______， 下极限尺寸是_______， 公差值是_______。

(5) The meaning of chamfer dimension *C*2 is _______.

该零件上倒角尺寸*C*2的含义是_______。

(6) The minimum value of *Ra* required for the surface is _______, the maximum is _______, and the surface roughness code of the contour surface is _______ .

该零件表面*Ra*值要求最小的是_______， 最大的是_______， 外形轮廓表面的表面粗糙度代号是_______。

(7) The overall dimensions of the part: the length is _______, the width is _______.

该零件总体尺寸是：长 _______， 高 _______。

(8) The reference for the length direction of the part is _______ , the reference for the height direction is _______ .

该零件长度方向尺寸基准是_______， 高度方向尺寸基准是_______。

(9) Draw the cross-section view of *A*-*A*.

补画*A*-*A*断面图。

10-21 Read the detail drawing and answer the following questions.
读零件图，回答下列问题。

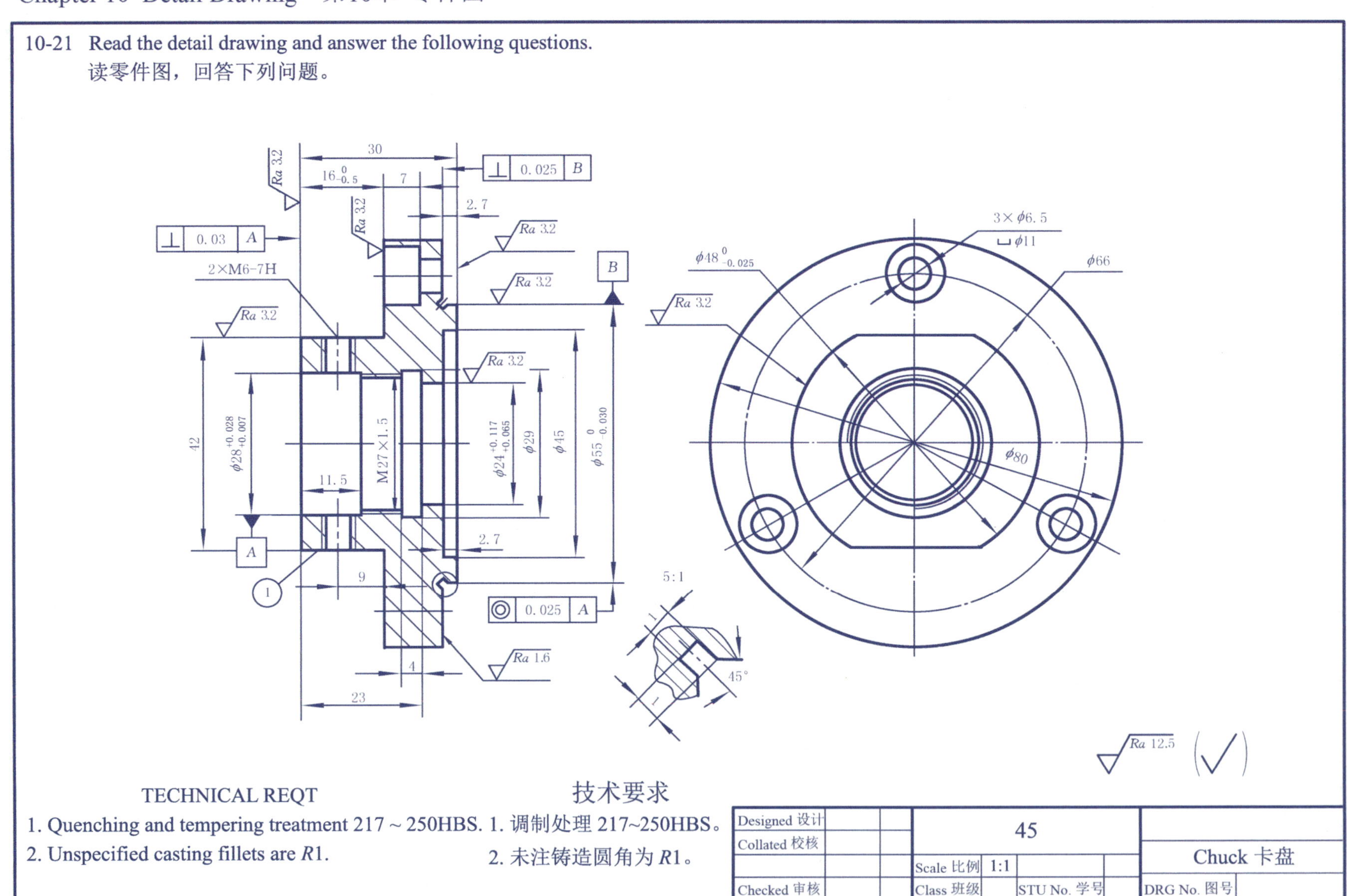

TECHNICAL REQT
1. Quenching and tempering treatment 217 ~ 250HBS.
2. Unspecified casting fillets are $R1$.

技术要求
1. 调制处理 217~250HBS。
2. 未注铸造圆角为 $R1$。

Designed 设计			45			
Collated 校核						Chuck 卡盘
			Scale 比例	1:1		
Checked 审核			Class 班级		STU No. 学号	DRG No. 图号

10-21 Continued. （续）

(1) The name of the part is ____________, the material is ____________ , and the scale is ____________.
该零件的名称是____________，材料是____________，比例是____________。

(2) The part shared ____________ basic views, and the front view used ____________.
该零件共用了____________个基本视图，其中主视图采用了____________。

(3) There is a grinding undercut in the part, whose shaping dimension is ____________ and positioning dimension is ____________.
该零件中有一个越程槽，其定形尺寸是____________，定位尺寸是____________。

(4) The surface referred to by ① in the front view is a ____________ (plane or cylindrical surface).
主视图中①所指的表面是____________（平面还是圆柱面）。

(5) The nominal size of dimension $\phi28^{+0.028}_{+0.007}$ is ____________, the upper limit size is ____________, the lower limit size is ____________, and the tolerance is ____________.
尺寸$\phi28^{+0.028}_{+0.007}$的公称尺寸是____________，上极限尺寸是____________，下极限尺寸是____________，公差是____________。

(6) Size 2×M6-7H, 2 represents ____________, M represents _________, 6 is _________, 7H is ____________, the rotation direction is ____________ , and the positioning dimension is ____________.
尺寸2×M6-7H中，2表示_________，M表示_________，6是_________，7H是 _________，旋向是__________，其定位尺寸是___________。

(7) The dimension reference of the length direction of the part is ___________, and the dimension reference of the width direction and the height direction is ____________.
该零件长度方向的尺寸基准是____________，宽度方向和高度方向的尺寸基准是____________。

10-22 Read the detail drawing and answer the following questions.

读零件图，回答下列问题。

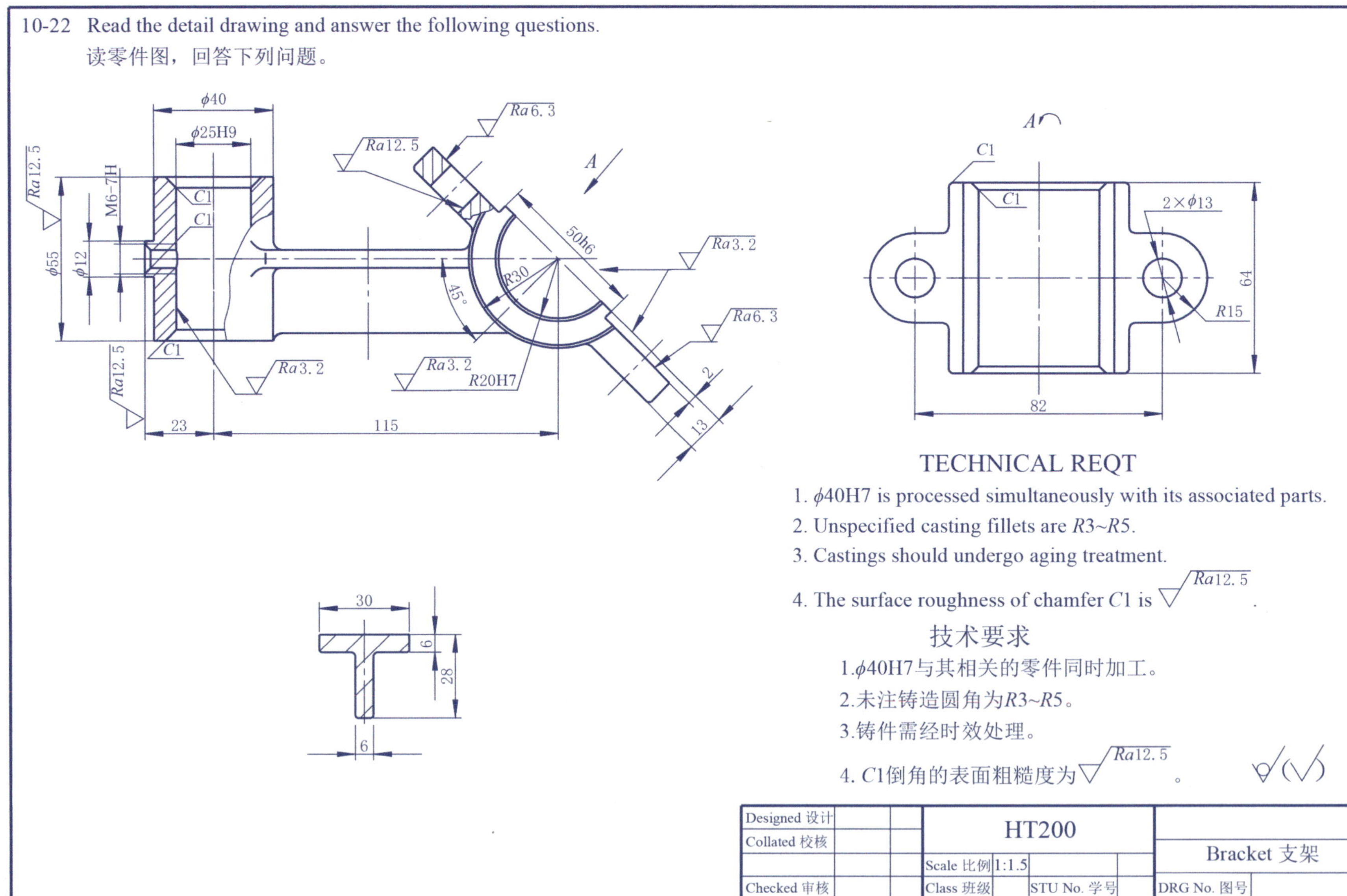

TECHNICAL REQT

1. ϕ40H7 is processed simultaneously with its associated parts.
2. Unspecified casting fillets are R3~R5.
3. Castings should undergo aging treatment.
4. The surface roughness of chamfer C1 is Ra12.5.

技术要求

1.ϕ40H7与其相关的零件同时加工。
2.未注铸造圆角为R3~R5。
3.铸件需经时效处理。
4. C1倒角的表面粗糙度为Ra12.5。

Designed 设计			HT200			
Collated 校核			Scale 比例	1:1.5		Bracket 支架
Checked 审核			Class 班级	STU No. 学号		DRG No. 图号

Class 班级： Name 姓名： ID Number 学号：

10-22 Continue.（续）

(1) The name of this part is _______ , the scale is _______ , and the material is _______.

该零件的名称是_______，比例是_______，材料是_______。

(2) This part is represented by _______ graphics, there are _______ sections in _______ sectional view in the front view, the *A*-direction rotation is a _______ drawing, and the other one is a _______ .

该零件共用了_______个图形来表达，主视图中有_______处作了_______剖视，*A*向旋转是_______图，还有一个是_______图。

(3) In dimension ϕ25H9, ϕ25 is _______ size, H represents _______, 9 represents _______, and H9 represents _______.

尺寸ϕ25H9中，ϕ25是_______尺寸，H表示_______，9表示_______，H9表示_______。

(4) In dimension M6-7H, M represents _______, 6 represents _______ size, and 7H represents _______.

尺寸M6-7H中，M表示_______，6表示_______尺寸，7H表示_______。

(5) There is/are _______ chamfere(s) of *C*1 on the bracket, there is/are _______ surface(s) with a surface roughness requirement of $\sqrt{Ra\ 12.5}$, and the symbol $\sqrt{\circ}$ represents _______。

支架上*C*1的倒角有_______处，表面粗糙度要求为$\sqrt{Ra\ 12.5}$的表面有_______个，符号$\sqrt{\circ}$表示_______。

(6) In the removed cross-section view, dimension 30 is the dimension in the _______ direction, and dimension 28 is the dimension in the _______ direction.

在移出断面图上，尺寸30是_______方向上的尺寸，28是_______方向上的尺寸。

(7) The shaping dimension of the cylinder on the left side of the bracket is _______, and the positioning dimension of the inclined plate on the right side of bracket is _______.

支架左侧圆筒的定形尺寸为_______，支架右侧斜板的定位尺寸是_______。

10-23 Read the detail drawing and answer the following questions.
读零件图，回答下列问题。

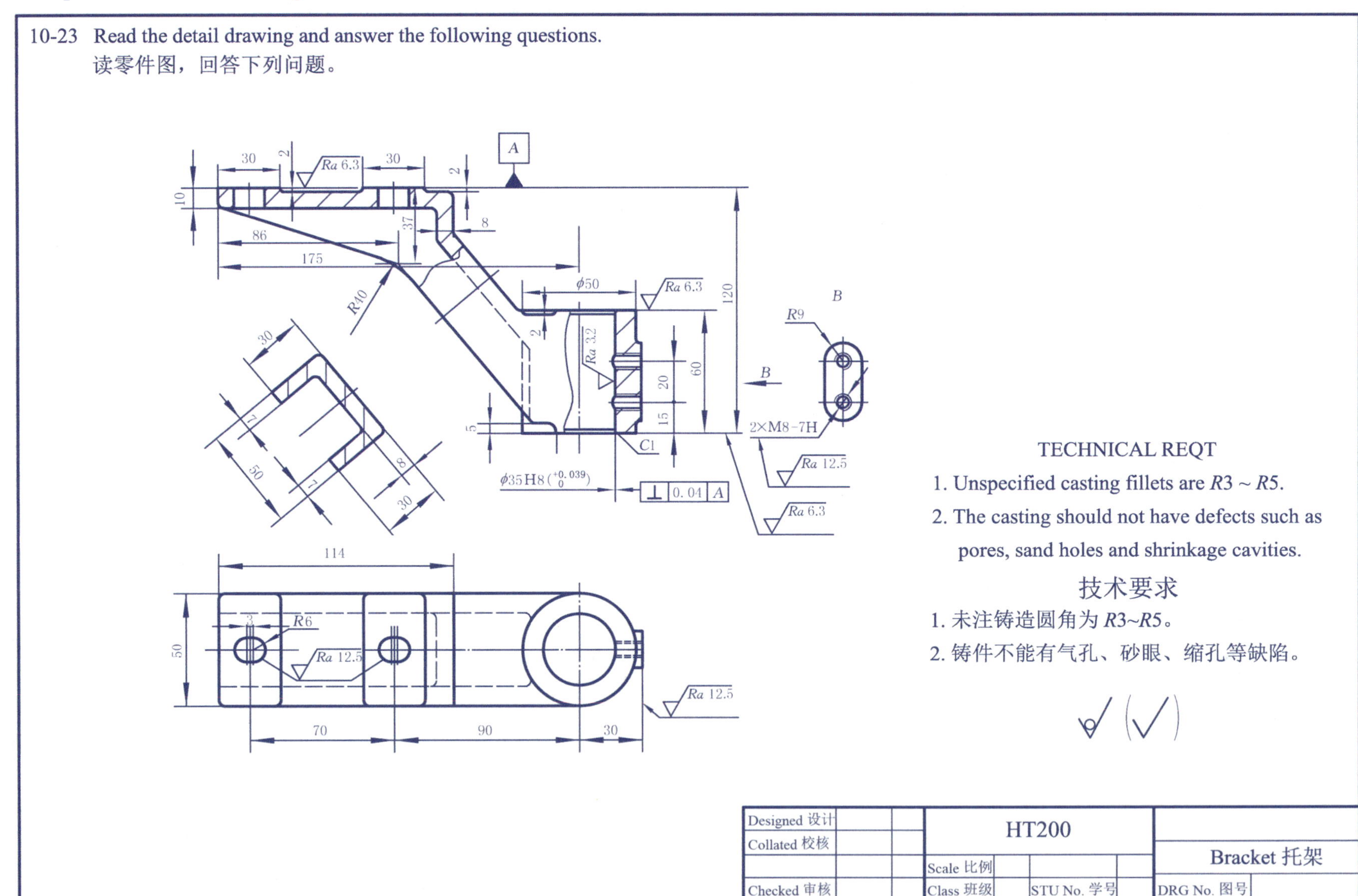

TECHNICAL REQT

1. Unspecified casting fillets are $R3 \sim R5$.
2. The casting should not have defects such as pores, sand holes and shrinkage cavities.

技术要求

1. 未注铸造圆角为 $R3 \sim R5$。
2. 铸件不能有气孔、砂眼、缩孔等缺陷。

Designed 设计			HT200			
Collated 校核						Bracket 托架
			Scale 比例			
Checked 审核			Class 班级		STU No. 学号	DRG No. 图号

Class 班级： Name 姓名： ID Number 学号：

10-23 Continued.（续）

(1) The part shared _____ graphic(s) to express, in the main view there is/are _____ made _____ section, the expression of the section shape of the ┌┐ shaped plate adopts ________ , *B*-direction view expresses the shape of the ________ .
该零件共用了_____个图形来表达，主视图中有_____处作了_____剖视，表达 ┌┐形板的截面形状采用了_____，*B*向表达了_____的形状。

(2) The shaping size of the ┌┐ shaped plate is _____.
┌┐ 形板截面的定形尺寸是_____。

(3) Two long circular holes in the top view, the shaping dimension is _, and the location dimension is _____.
俯视图中两个长圆孔，其定形尺寸是_____，定位尺寸是_____。

(4) In the bracket parts, the size reference in the length direction is _____, the size reference in the width direction is _____, and the size reference in the height direction is _____.
托架零件中，长度方向的尺寸基准是_____，宽度方向的尺寸基准是_____，高度方向的尺寸基准是_____。

(5) In the size 2×M8-7H: 2 denotes _____, M denotes _____, 8 denotes _____, 7H denotes _____, the positioning size is _____.
尺寸2×M8-7H中，2表示________，M表示________，8是________，7H是________，其定位尺寸是________。

(6) *R*40 is the size of _____ in the main view, 175 is the size of _____ Size.
主视图中尺寸*R*40是_____尺寸，175是_____尺寸。

(7) Dimension $\phi 35H8(^{+0.039}_{0})$, $\phi 35$is _____, H is _____, 8 is _____, + 0.039 is _____, 0 is _____, and the tolerance is _____.
尺寸$\phi 35H8(^{+0.039}_{0})$中，$\phi 35$是_____，H是_____，8是_____，+0.039是_____，0是_____，公差是_____。

10-24 Read the detail drawing and answer the following questions.

读零件图，回答下列问题。

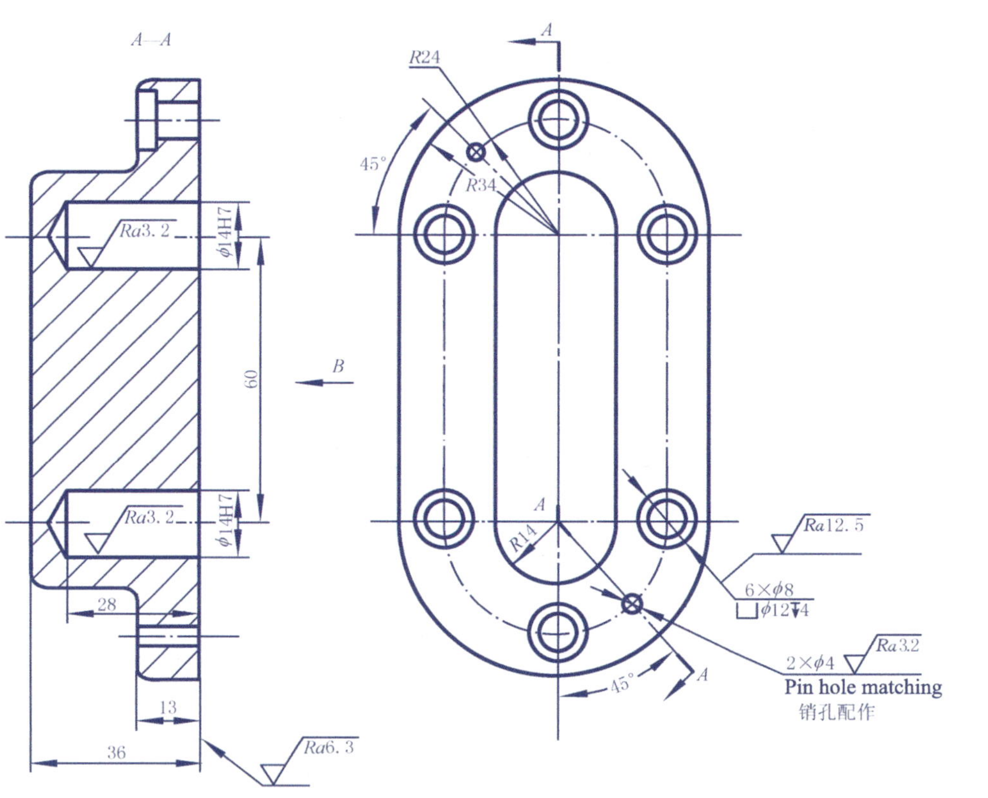

TECHNICAL REQT

1. Castings should not have defects such as pores and cracks.
2. Paint non-machined surfaces.
3. Unspecified casting fillets are *R*3.

技术要求

1. 铸件不能有气孔、裂纹等缺陷。
2. 非加工表面涂漆。
3. 未注铸造圆角为*R*3。

Designed 设计			HT200				
Collated 校核						Pump cover 泵盖	
			Scale 比例				
Checked 审核			Class 班级		STU No. 学号	DRG No. 图号	

Class 班级： Name 姓名： ID Number 学号：

10-24 Continued.（续）

(1) This part is expressed by ________ views, in which the front view is ________ sectional view.
本零件用了________个视图表达，其中主视图是________剖视图。

(2) The shaping dimension of the pin hole on the detail drawing is ________.
零件图上销孔的定形尺寸是________。

(3) Among all the machined surfaces of the parts, the surface roughness codes of the smoothest and roughest surfaces are ________________.
零件所有加工表面中，最光滑和最粗糙的表面粗糙度代号分别是________________。

(4) The roughness code of the non-machined surface of the part is ________.
零件非加工表面的粗糙度代号是________。

(5) The size ϕ14H7 in the figure, ϕ14 is called __________ size, H7 is __________.
图中尺寸ϕ14H7，ϕ14称为__________尺寸，H7是__________。

(6) The name of the part is __________, and the material used is __________.
零件的名称是__________，所用的材料是__________。

(7) Draw the *B*-direction view in the blank.
在空白处补画*B*向视图。

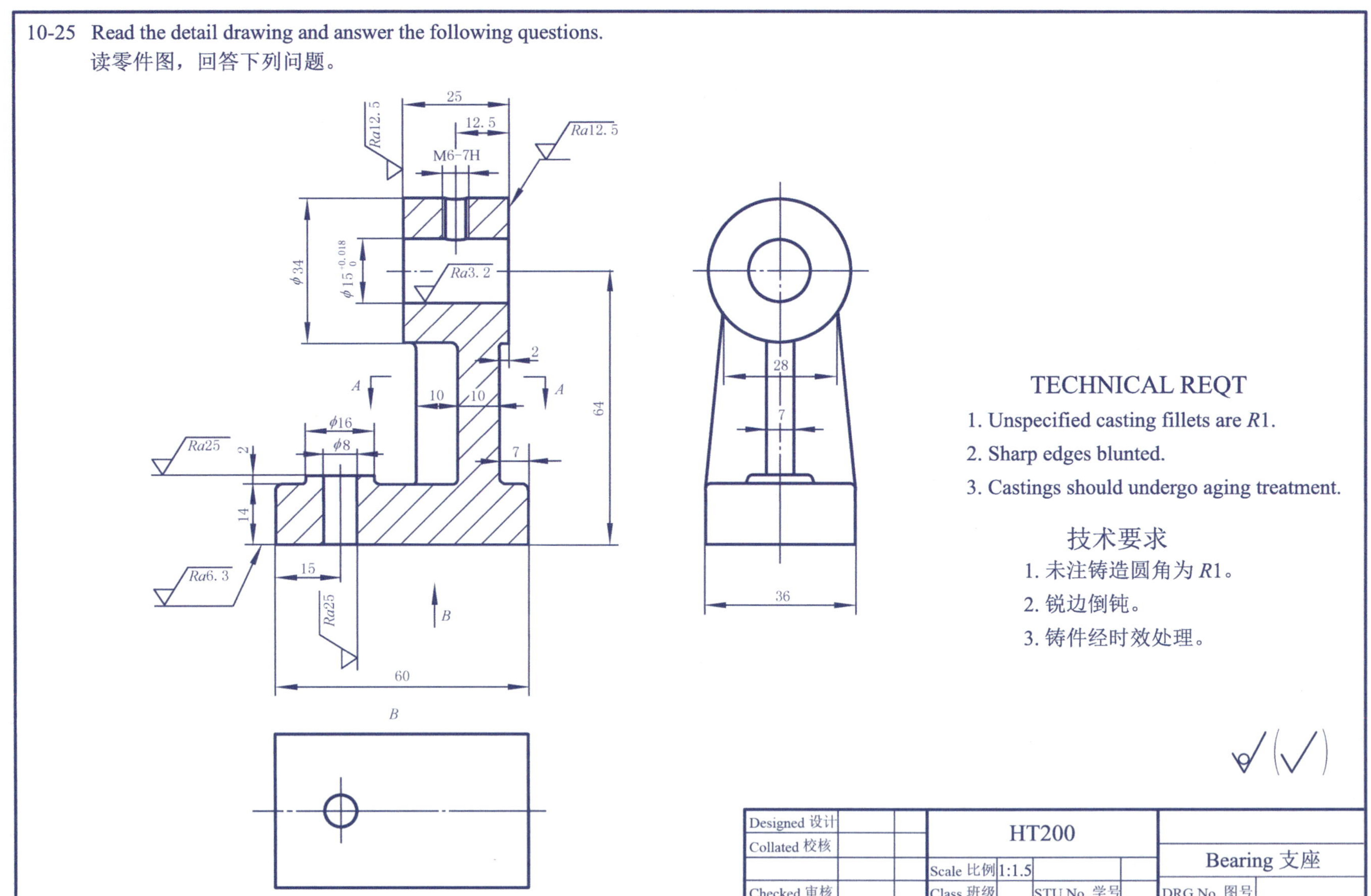

10-25 Read the detail drawing and answer the following questions.
读零件图，回答下列问题。

TECHNICAL REQT

1. Unspecified casting fillets are R1.
2. Sharp edges blunted.
3. Castings should undergo aging treatment.

技术要求

1. 未注铸造圆角为 R1。
2. 锐边倒钝。
3. 铸件经时效处理。

Class 班级： Name 姓名： ID Number 学号：

10-25 Continued.（续）

(1) The name of the part is ___________, the material is ______________, the scale is ______________, belongs to _______________.
该零件的名称是_________，材料是_________，比例是________，属于___________。

(2) In size M6-7H, M represents ___________, 6 is __________, 7H is _________________________, the rotation direction is _______________, and the positioning dimension is ______________________.
尺寸M6-7H中，M表示_____________，6是__________，7H是___________________，旋向是_____________，其定位尺寸是____________。

(3) The nominal size of dimension $\phi15^{+0.018}_{0}$ is ______ , the upper limit size is _______ , the lower limit size is _______ , the upper deviation is ________ , the lower deviation is __________ , and the tolerance value is _______________.
尺寸$\phi15^{+0.018}_{0}$的公称尺寸是_____，上极限尺寸是______，下极限尺寸是______，上偏差是______，下偏差是______，公差值是_______。

(4) Among all the machined surfaces of the part, the smoothest surface roughness code is ________ ; the roughest surface roughness code is ____________.
零件所有加工表面中，最光滑的表面粗糙度代号是__________________；最粗糙的表面粗糙度代号是__________________。

(5) Draw the *A-A* sectional view.
画出*A-A*剖视图。

10-26 Read the detail drawing and answer the following questions.
读零件图，回答下列问题。

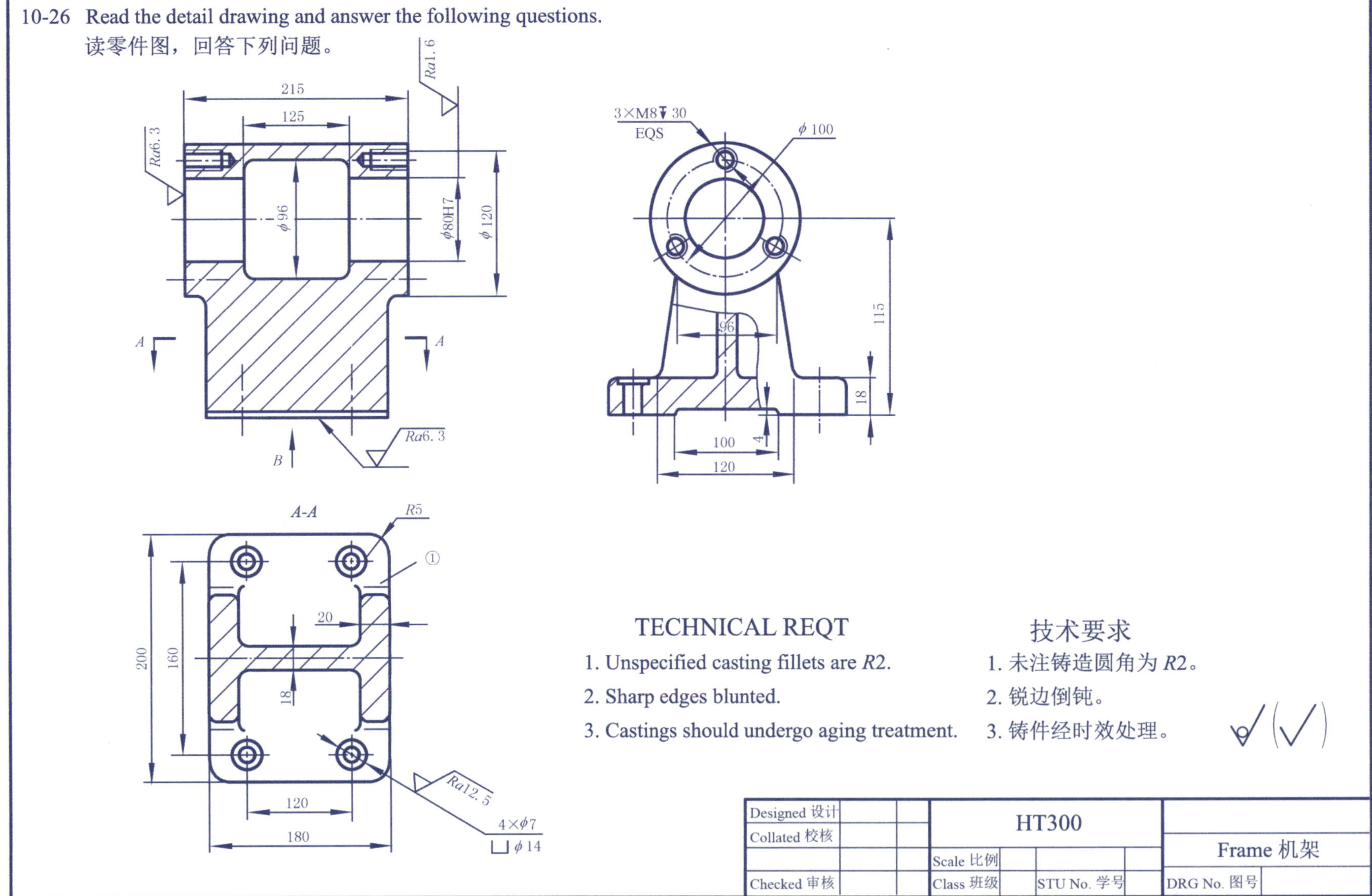

Class 班级： Name 姓名： ID Number 学号：

10-26 Continued. （续）

(1) In a set of diagrams expressing the frame, the front view adopts the _______ view, the left view adopts the _______ view, and the continuous thin line referred to by ① in the top view is the _______ line.

表达机架的一组图中，主视图采用了_______图，左视图采用了_______图，俯视图中①所指细实线是_______线。

(2) The nominal size of ϕ80H7 is _______, the tolerance zone code is _______, the basic deviation code is _______ , and the tolerance grade is _______.

ϕ80H7的公称尺寸是_______，公差带代号为_______，基本偏差代号是_______，公差等级是_______。

(3) In the left view, which size type (shaping, positioning) does the following size belong to ?

左视图中下列尺寸属于哪种尺寸类型（定形、定位）？

100 _______ 4 _______

115 _______ 18 _______

(4) Among all the machined surfaces of the part, the smoothest and roughest surface roughness codes are ___________.

该零件所有加工表面中，最光滑和最粗糙的表面粗糙度代号分别是___________。

(5) The roughness code of the non-machined surface of the part is ___________.

该零件非加工表面的粗糙度代号是___________。

(6) Draw the *B*-direction view.

画出*B*向视图。

10-27 Read the detail drawing and answer the following questions.
读零件图，回答下列问题。

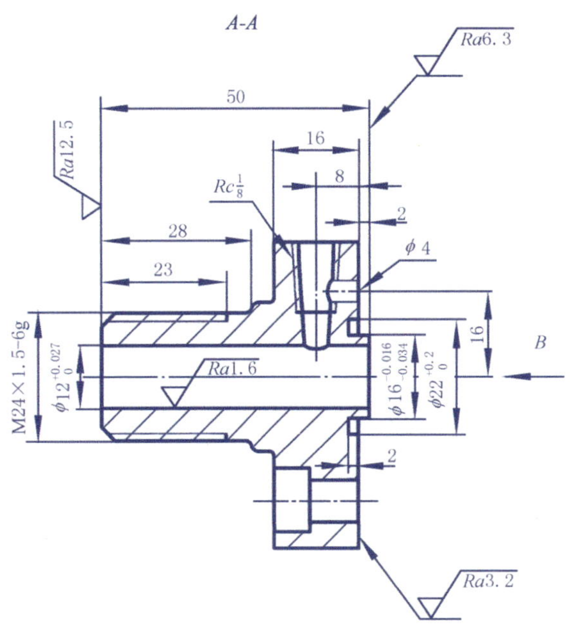

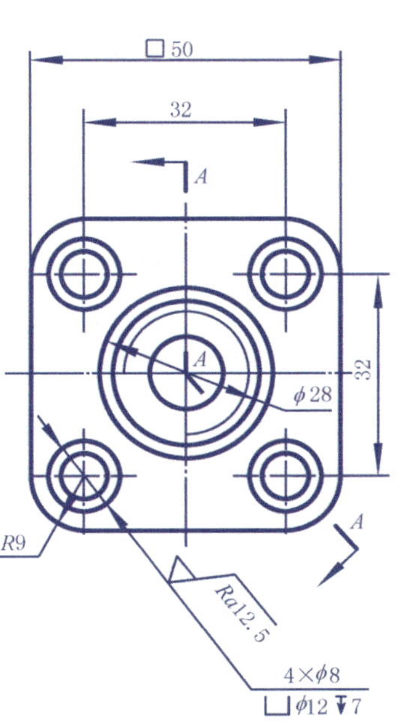

TECHNICAL REQT

1. Unspecified casting fillets are $R1$.
2. Sharp edges blunted.
3. Castings should undergo aging treatment .

技术要求

1. 未注铸造圆角为 $R1$。
2. 锐边倒钝。
3. 铸件经时效处理。

Designed 设计			HT150			
Collated 校核						Joint 管接头
			Scale 比例	1:1.5		
Checked 审核			Class 班级		STU No. 学号	DRG No. 图号

 Class 班级： Name 姓名： ID Number 学号：

10-27 Continued.（续）

(1) The name of the part is ____, the material is ____, and the scale is ____, which belongs to ______ scale.
该零件的名称是____，材料是____，比例是____，属于______比例。

(2) The front view is a ____ sectional view, and its cutting method is ______.
主视图是____剖视图，其剖切方法是______。

(3) The nominal size of $\phi16^{-0.016}_{-0.034}$ is ____, the upper deviation is ______, the lower deviation is ______, and the tolerance is ______.
$\phi16^{-0.016}_{-0.034}$的公称尺寸是____，上偏差是______，下偏差是______，公差是______。

(4) The maximum limit size of $\phi12^{-0.027}_{0}$ is _____, and the minimum limit size is ______.
$\phi12^{-0.027}_{0}$的最大极限尺寸是____，最小极限尺寸是______。

(5) Among all the machined surfaces of the part, the smoothest and roughest surface roughness codes are ______.
该零件所有加工表面中，最光滑和最粗糙的表面粗糙度代号分别是______。

(6) The roughness code of the non-machined surface of the part is ______.
该零件非加工表面的粗糙度代号是______。

(7) Draw the *B*-direction view.
画出*B*向视图。

10-28 Read the detail drawing and answer the following questions.
读零件图，回答下列问题。

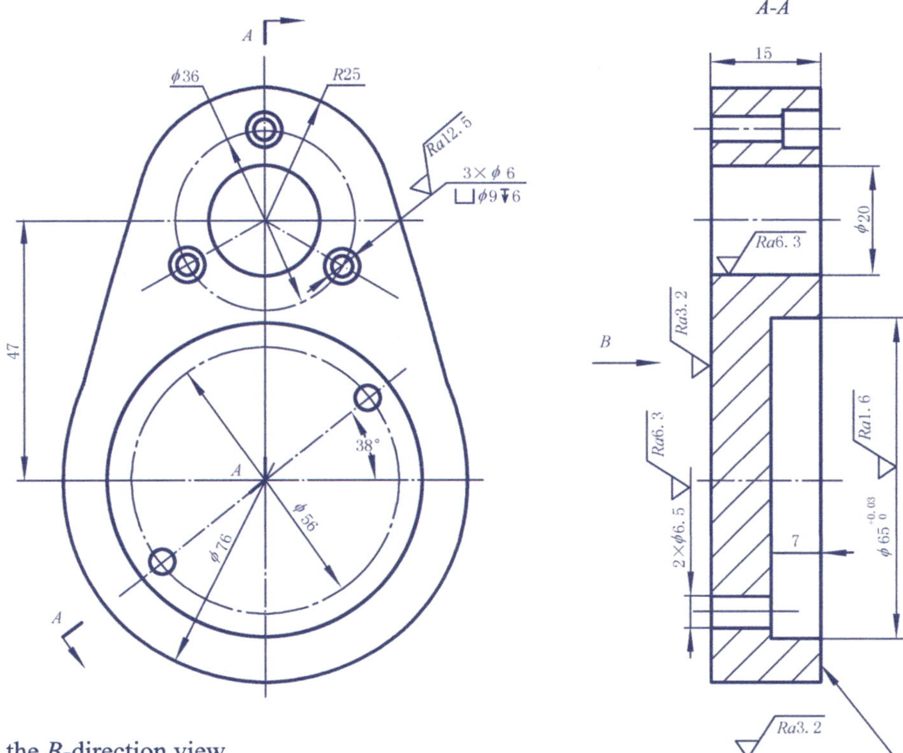

(1) Draw the *B*-direction view.
补画*B*向视图。

(2) Size 2 × ϕ6.5 indicates that there are _____ holes with a size of _____, and its positioning dimension is _____.
尺寸2 × ϕ6.5表示有_____个尺寸是_____的孔，其定位尺寸是______。

(3) Among all the machined surfaces of the part, the smoothest surface roughness code is ______, and the roughest surface roughness code is ______.
零件所有加工表面中，最光滑的表面粗糙度代号是________；
最粗糙的表面粗糙度代号是________。

Designed 设计			HT200				
Collated 校核							Bearing 支座
			Scale 比例				
Checked 审核			Class 班级		STU No. 学号		DRG No. 图号

　Class 班级：　Name 姓名：　ID Number 学号：

10-29 Read the detail drawing and answer the following questions.
读零件图，回答下列问题。

(1) This part is a ____ type part. The front view conforms to the _______ position of the part.
此零件是___类零件，主视图符合零件的__________位置。

(2) There is/are ____ ϕ9 hole(s) on the right end surface of the part, which are matched with the bolt with diameter _______.
零件右端面上有___个ϕ9的孔，是与直径_______的螺栓相配的。

(3) ϕ90 belongs to __________ dimension.
ϕ90属于___________尺寸。

(4) ϕ70d11 indicates that the nominal size is _________, the standard tolerance grade is _________, and the basic deviation code is_________.
ϕ70d11表示公称尺寸为_________，标准公称等级为_________，基本偏差代号为_________。

(5) Draw the *B-B* sectional view and *C*-direction view.
画出*B-B*剖视图和*C*向视图。

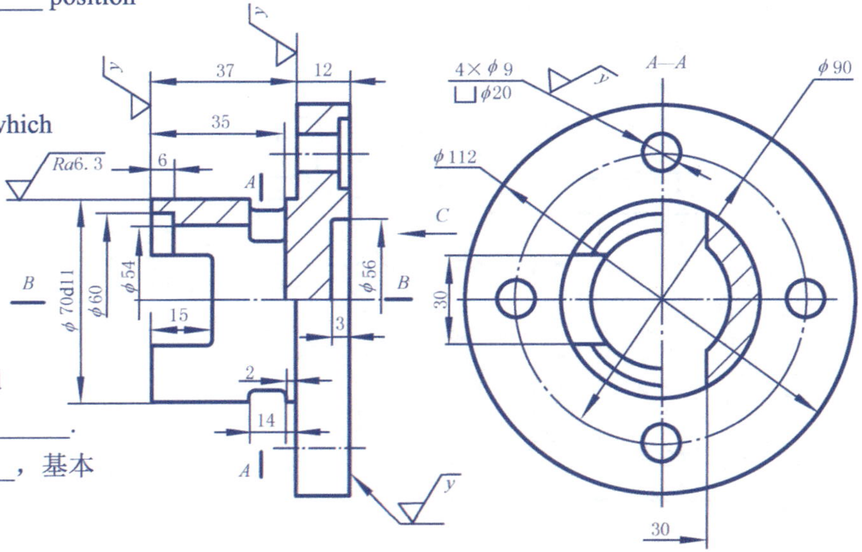

TECHNICAL REQT

1. The casting should not have defects such as pores and cracks.
2. Unspecified casting fillets are *R*3.

技术要求

1. 铸件不得有气孔、裂纹等缺陷。
2. 未注铸造圆角为*R*3。

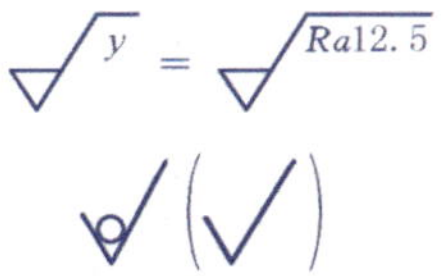

y = Ra12.5

Designed 设计			45			
Collated 校核						Cap 轴承盖
			Scale 比例			
Checked 审核			Class 班级		STU No. 学号	DRG No. 图号

Class 班级： Name 姓名： ID Number 学号：

10-30 Read the detail drawing, analyze the shape of the part, and complete the *B*-direction view and the *C*-*C* sectional view.
读零件图，分析零件形状，补画 *B* 向视图并补画 *C*-*C* 剖视图。

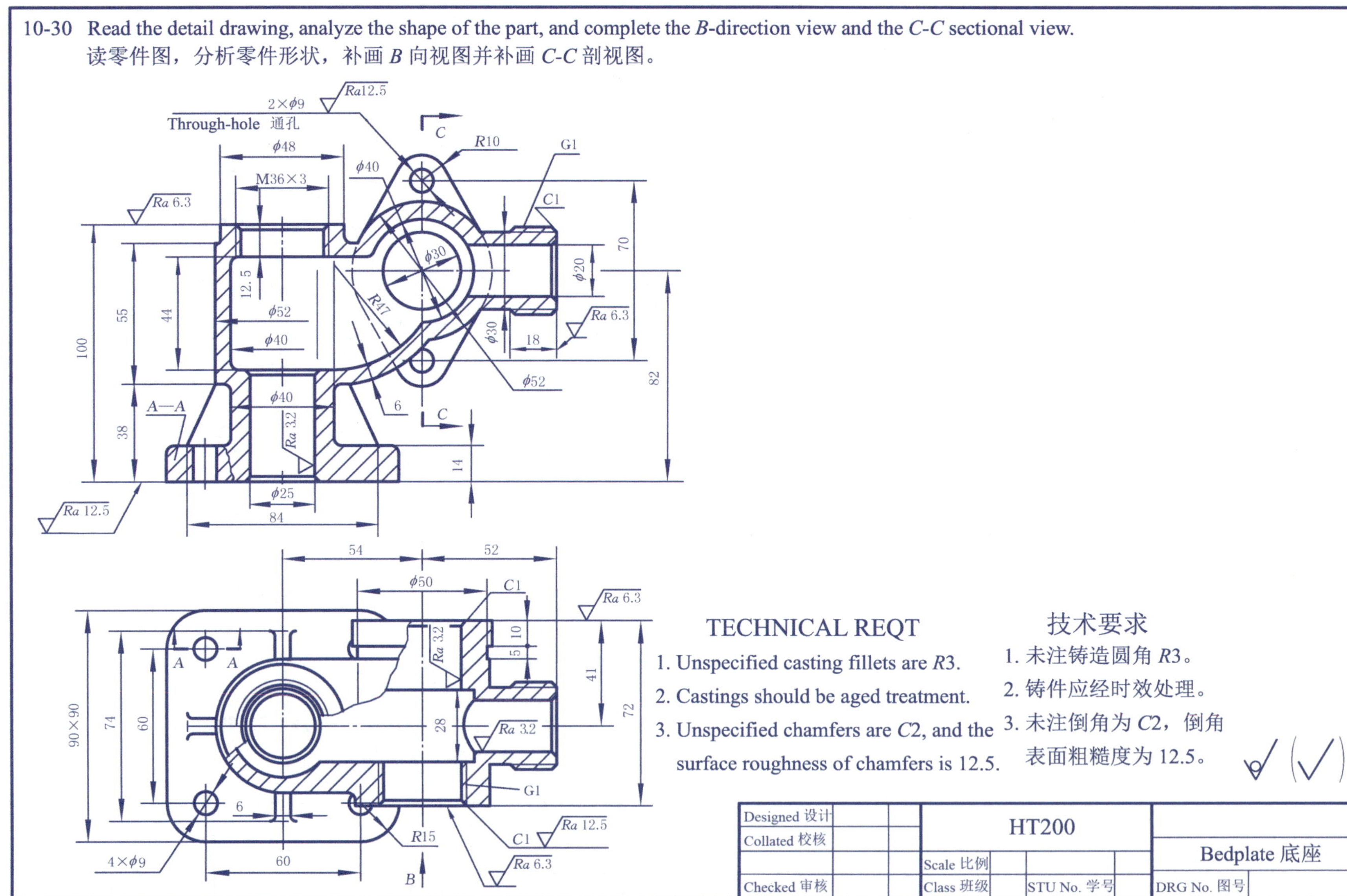

 Class 班级： Name 姓名： ID Number 学号：

11-1 Draw an assembly drawing of the shafting. 画轴系装配图。

Schematic diagram of the shafting
轴系装配示意图

Drawing sequence: Shaft 5 → Gear 3 → Key 4 → Shaft sleeve 2 → Bearing 1(left) → Bearing 1 (right) → Pulley 7 → Pin 6.

画图顺序：轴5→齿轮3→键4→轴套2→轴承1(左)→轴承1(右)→带轮7→销6。

No. 序号	Code 代号	Description 名称	No.off 数量	MATL 材料	PW 单重	TW 总重	Remark 备注
7		Pulley 带轮	1	HT200			
6	GB/T 119.1—2000	Pin 销 6×35	1	35			
5		Shaft 轴	1	45			
4	GB/T 1096—2003	Key 键 10×22	1	45			
3		Gear 齿轮	1	40Cr			
2		Shaft sleeve 轴套	1	Q235			
1	GB/T 276—2013	Bearing 轴承 6206	2	组件			

Designed 设计				
Collated 校核				Shafting assembly 轴系装配
		Scale 比例		
Checked 审核		Class 班级	STU No. 学号	DRG No. 图号

Designed 设计			40Cr	
Collated 校核				Gear 齿轮
		Scale 比例		
Checked 审核		Class 班级	STU No. 学号	DRG No. 图号

Class 班级：　　Name 姓名：　　ID Number 学号：

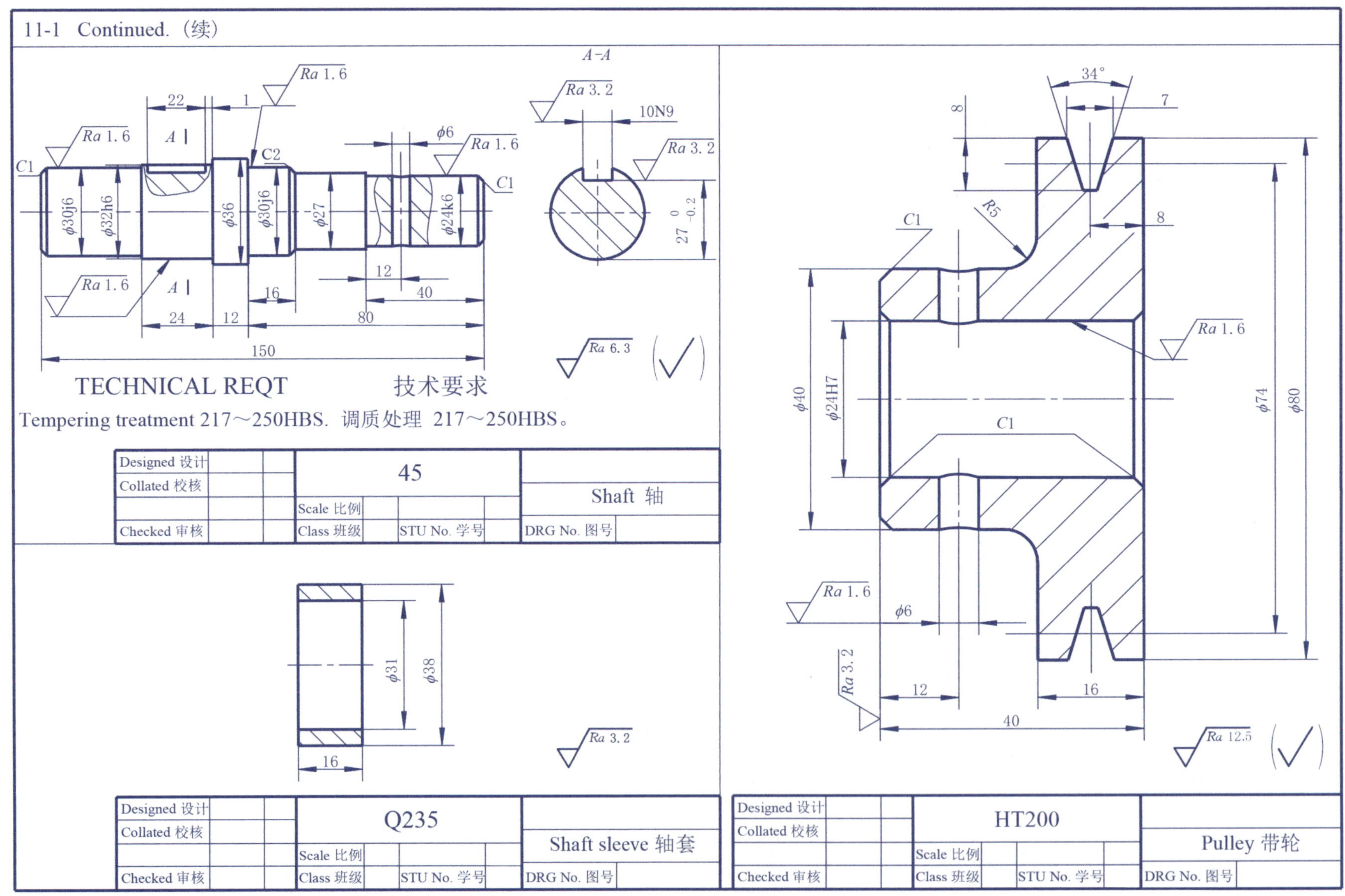
11-1 Continued.（续）
A-A
Ra 1.6
Ra 3.2
10N9
22
1
A
C1
C2
ϕ6
ϕ30j6
ϕ32h6
ϕ36
ϕ30j6
ϕ27
ϕ24k6
27 0 -0.2
12
16
40
24
12
80
150
Ra 6.3
TECHNICAL REQT
技术要求
Tempering treatment 217～250HBS. 调质处理 217～250HBS。
Designed 设计
Collated 校核
Checked 审核
45
Scale 比例
Class 班级
STU No. 学号
Shaft 轴
DRG No. 图号
ϕ31
ϕ38
16
Ra 3.2
Designed 设计
Collated 校核
Checked 审核
Q235
Scale 比例
Class 班级
STU No. 学号
Shaft sleeve 轴套
DRG No. 图号
34°
7
8
R5
C1
8
Ra 1.6
ϕ40
ϕ24H7
C1
ϕ74
ϕ80
Ra 1.6
ϕ6
Ra 3.2
12
16
40
Ra 12.5
Designed 设计
Collated 校核
Checked 审核
HT200
Scale 比例
Class 班级
STU No. 学号
Pulley 带轮
DRG No. 图号

11-2 Draw an assembly drawing of the jack. 画千斤顶装配图。

Working principle: The jack is a component for lifting heavy objects. When in use, simply turn the rotating rod 3 counterclockwise, and the lifting screw 2 will move upwards to lift the heavy object.

工作原理：千斤顶是顶起重物的部件，使用时，只需逆时针方向转动旋3，起重螺杆2就向上移动，并将重物顶起。

Specific requirements: Select A3 drawing sheets, customize the scale, fully and clearly express the working principle and assembly relationship of the jack, mark necessary dimensions, annotate part numbers, and fill in the item lists and title block.

具体要求：选用A3的图幅，自定比例，完整清晰地表达千斤顶的工作原理和装配关系，标注必要的尺寸，编注零件序号，填写明细栏和标题栏。

Drawing sequence (recommended): Base 1→ Lifting screw 2→ Rotating rod 3→ Top cover 5→ Screw 4 (drawn from outside to inside).

画图顺序(建议)：底座1 →起重螺杆2 →旋转杆3 →顶盖5 →螺钉4（由外向内画）。

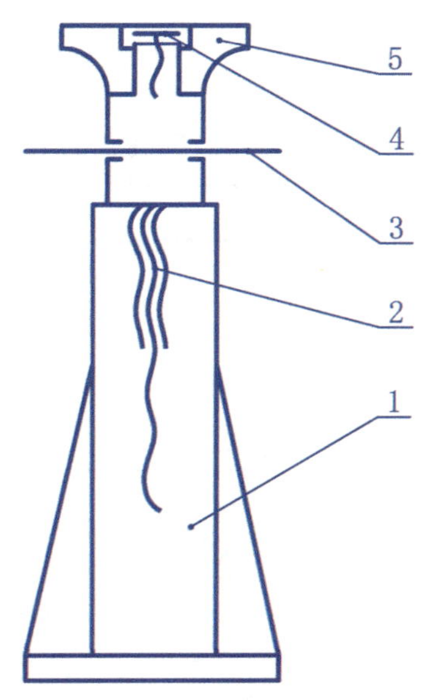

Schematic diagram of the Jack
千斤顶示意图

Technical requirements:

1. After the installation of the jack, apply lubricating grease to the lifting screw.
2. Brush red anti rust paint on the surface of the jack base.

技术要求：

1. 千斤顶安装后，起重螺杆涂润滑脂。
2. 千斤顶底座表面刷红色防锈油漆。

5		Top cover 顶盖	1	45			
4		Screw 螺钉	1	30			
3		Rotating rod 旋转杆	1	45			
2		Lifting screw 起重螺杆	1	45			
1		Base 底座	1	HT300			
No. 序号	Code 代号	Description 名称	No.off 数量	MATL 材料	PW 单重	TW 总重	Remark 备注

Designed 设计						
Collated 校核						Jack 千斤顶
			Scale 比例			
Checked 审核			Class 班级		STU No. 学号	DRG No. 图号

11-2 Continued.（续）

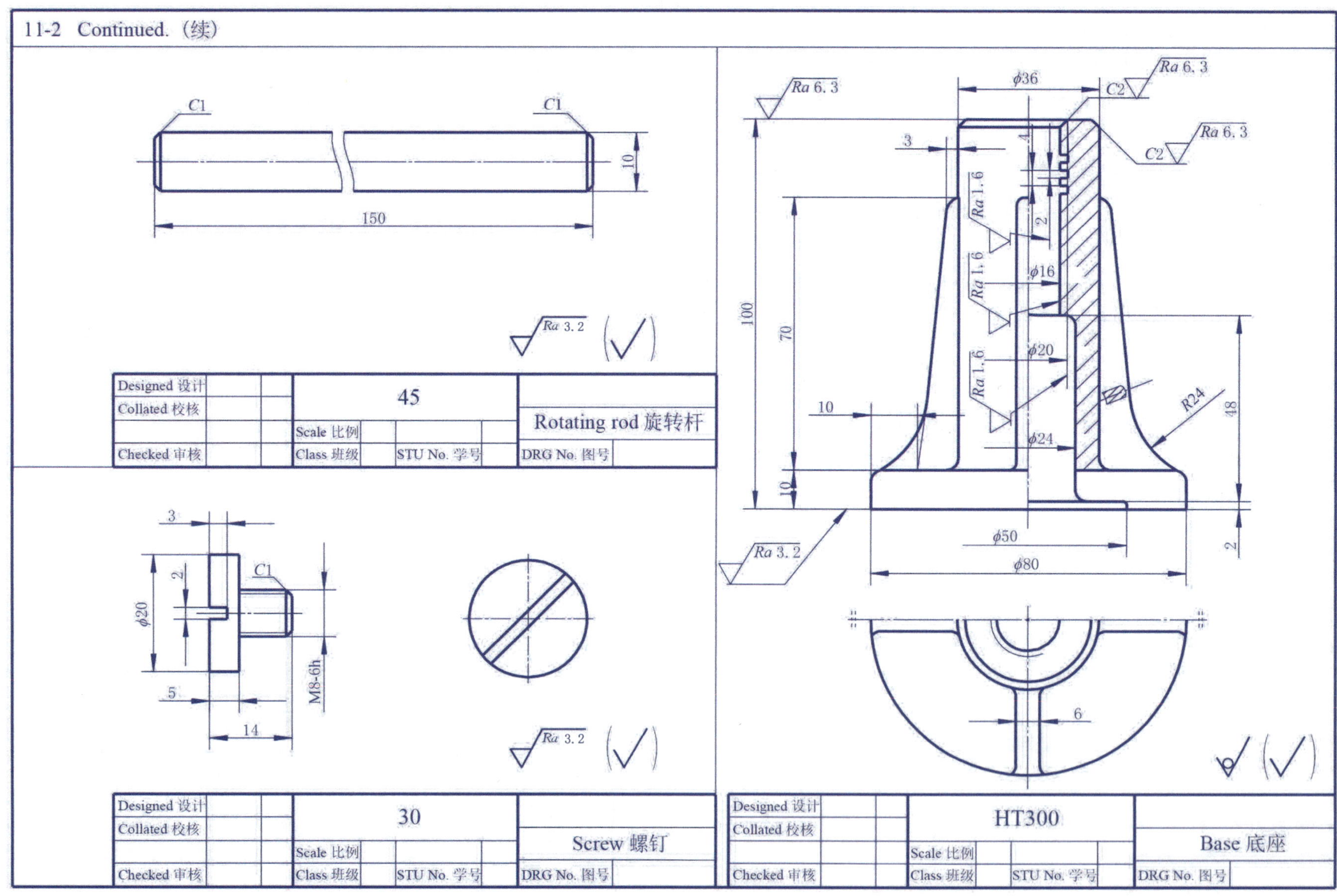

Class 班级： Name 姓名： ID Number 学号：

11-2 Continued.（续）

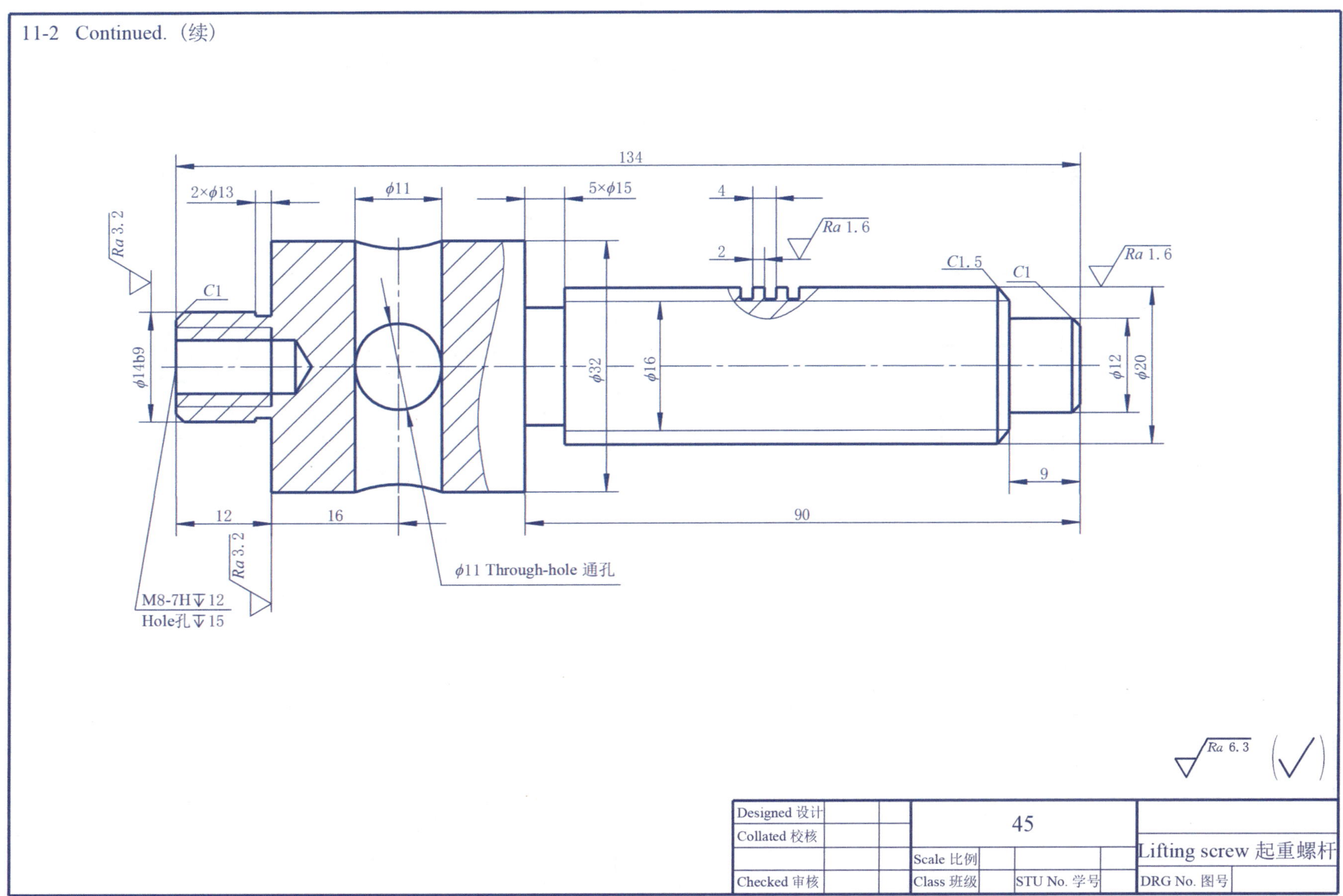

Designed 设计			45				Lifting screw 起重螺杆	
Collated 校核								
			Scale 比例					
Checked 审核			Class 班级		STU No. 学号		DRG No. 图号	

11-2 Continued.（续）

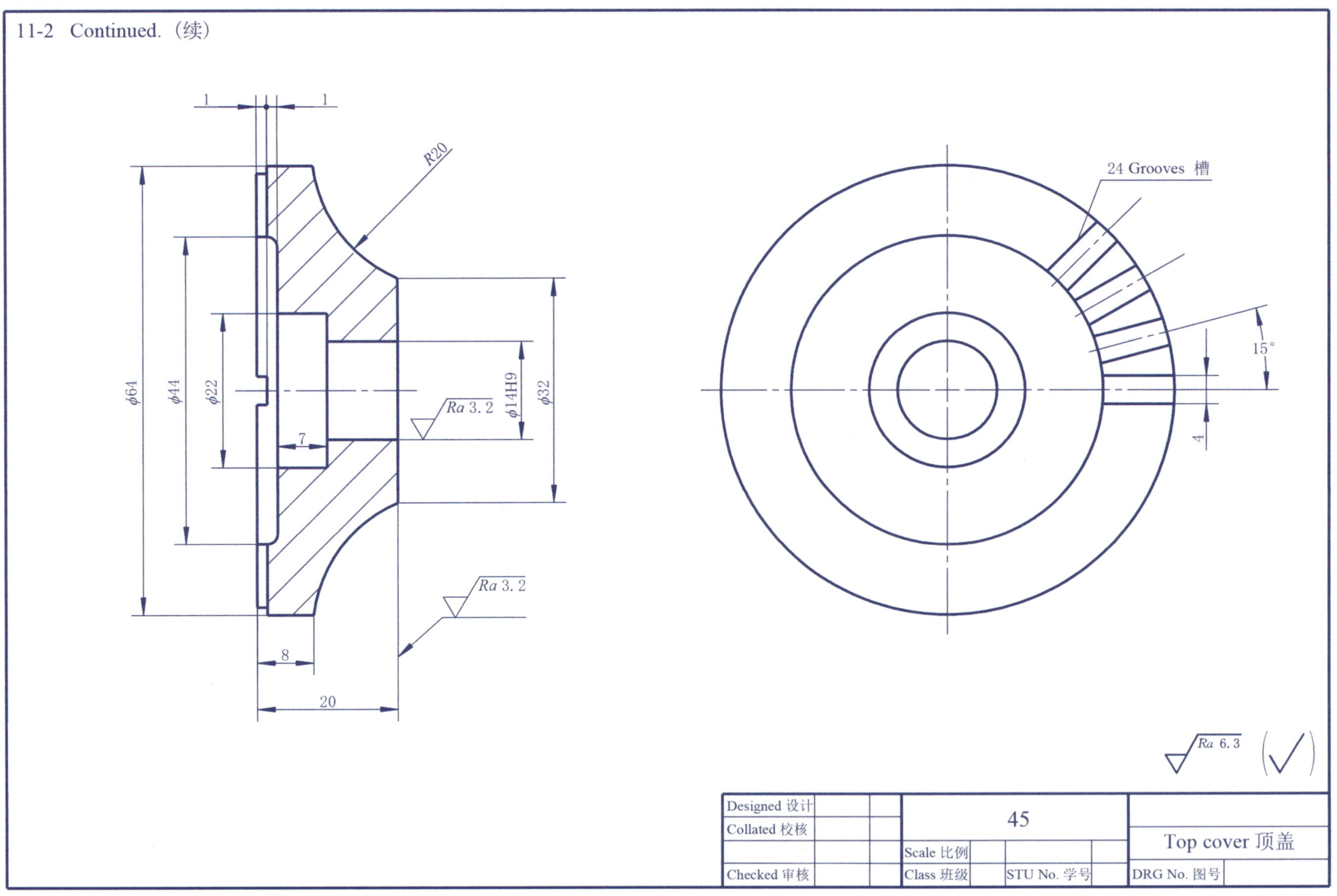

 Class 班级： Name 姓名： ID Number 学号：

11-3 Draw an assembly drawing of the cock. 画旋塞装配图。

Working principle: The cock is a switch in the pipeline, which is characterized by rapid switch action. Its flange is connected to the external pipeline with bolts. Move the plug 90 ° with a wrench to fully open the pipeline.

工作原理：旋塞是管路中的一种开关，特点是开关动作比较迅速。它的法兰用螺栓与外管道连接，用扳手将塞子搬动90°，就可全部打开管路。

Fill the asbestos packing between the conical plug and the housing, and then install the cover, and then tighten the nuts on the double end studs to compress the packing to prevent leakage.

在锥形塞与壳体之间填满石棉盘根，再装上压盖，然后拧动双头螺柱上的螺母，使压紧填料，用以防止泄漏。

Specific requirments: Select A3 drawing sheet, customize the scale, fully and clearly express the working principle and assembly relationship of the cock, mark necessary dimensions, annotate part numbers and fill in the item lists and title block.

具体要求：选用A3的图幅，自定比例，完整清晰地表达旋塞的工作原理和装配关系，标注必要的尺寸，编注零件序号，填写明细栏和标题栏。

Drawing sequence (recommended):

Housing → Plug → Cover → Stud and nut.

画图顺序(建议)：

壳体→塞子→压盖→螺柱螺母。

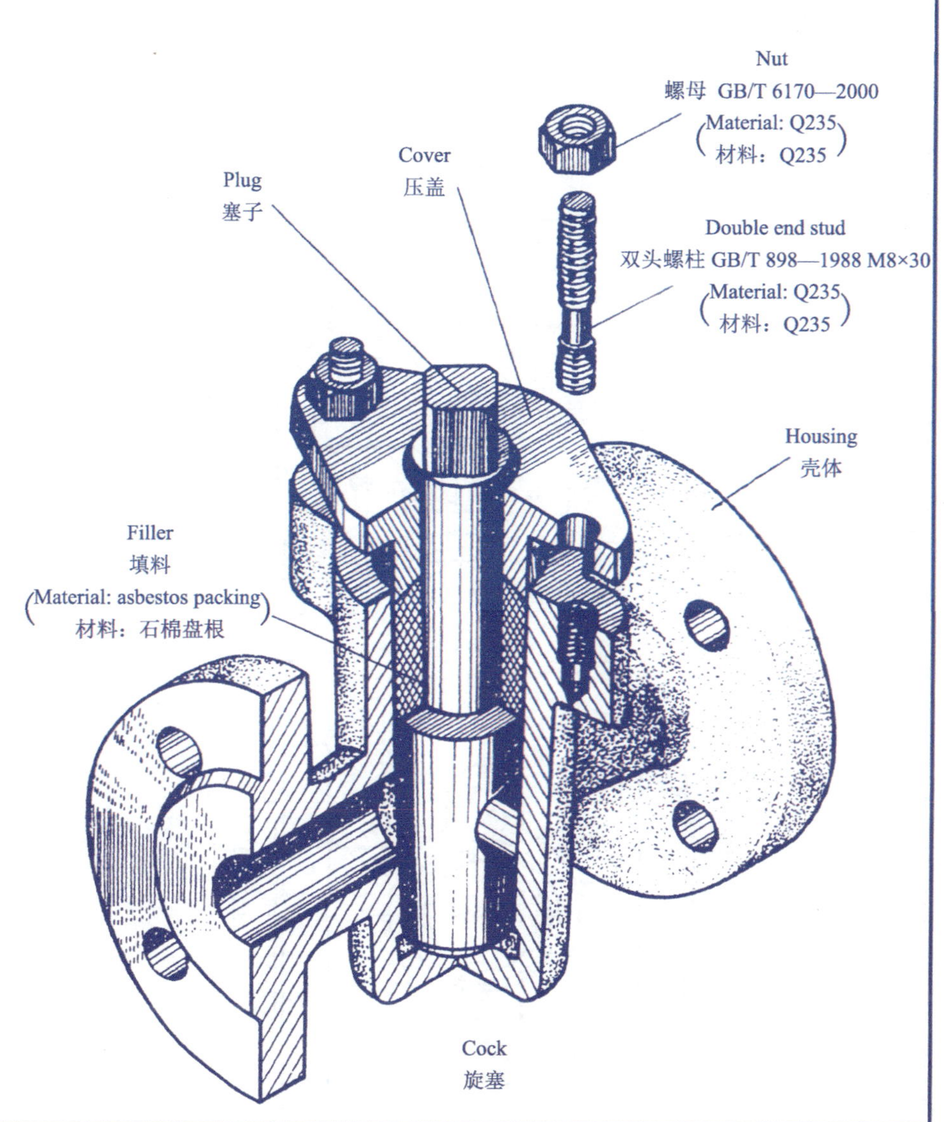

11-3 Continued. （续）

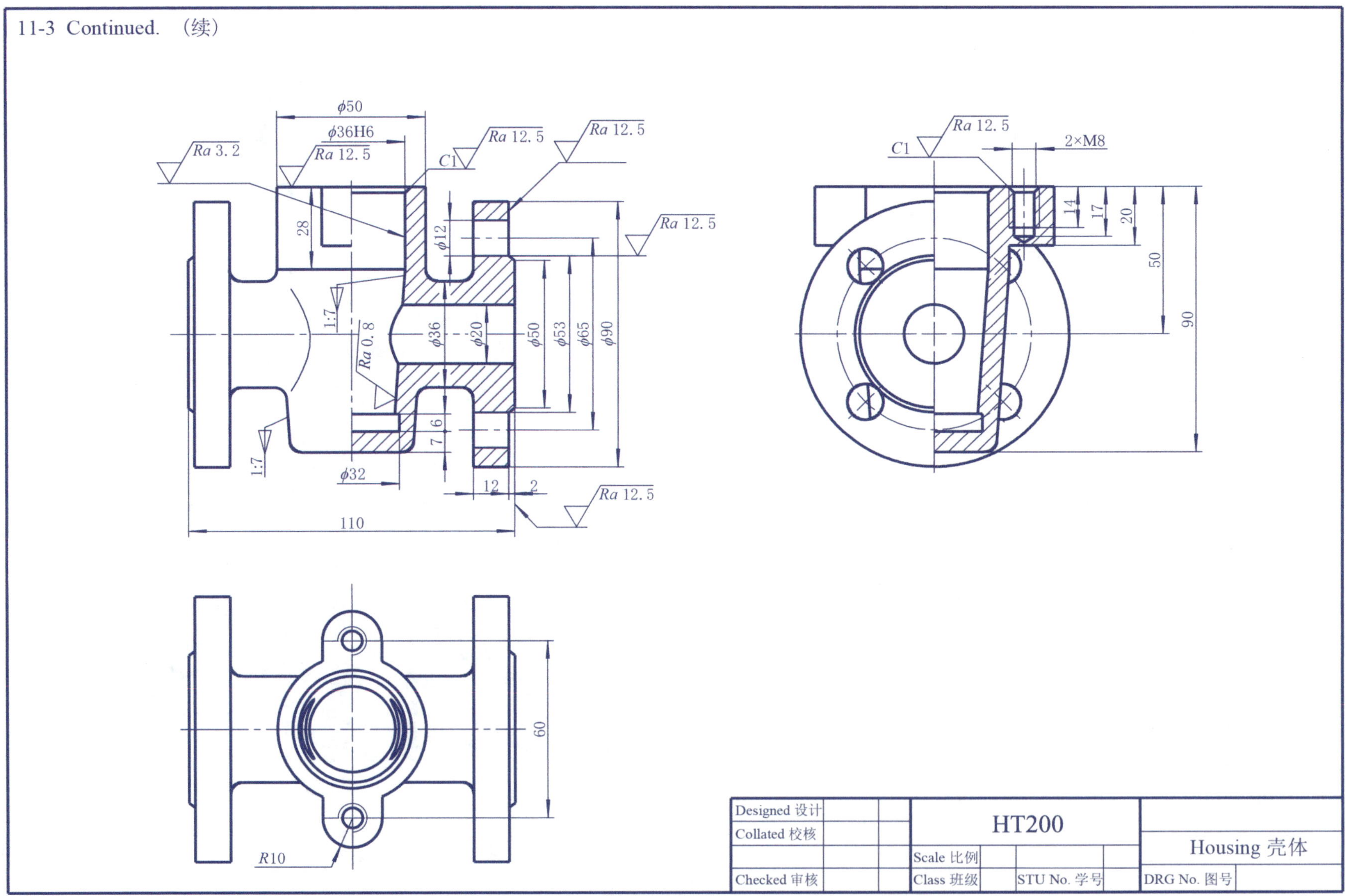

Class 班级：　　Name 姓名：　　ID Number 学号：

11-3 Continued. （续）

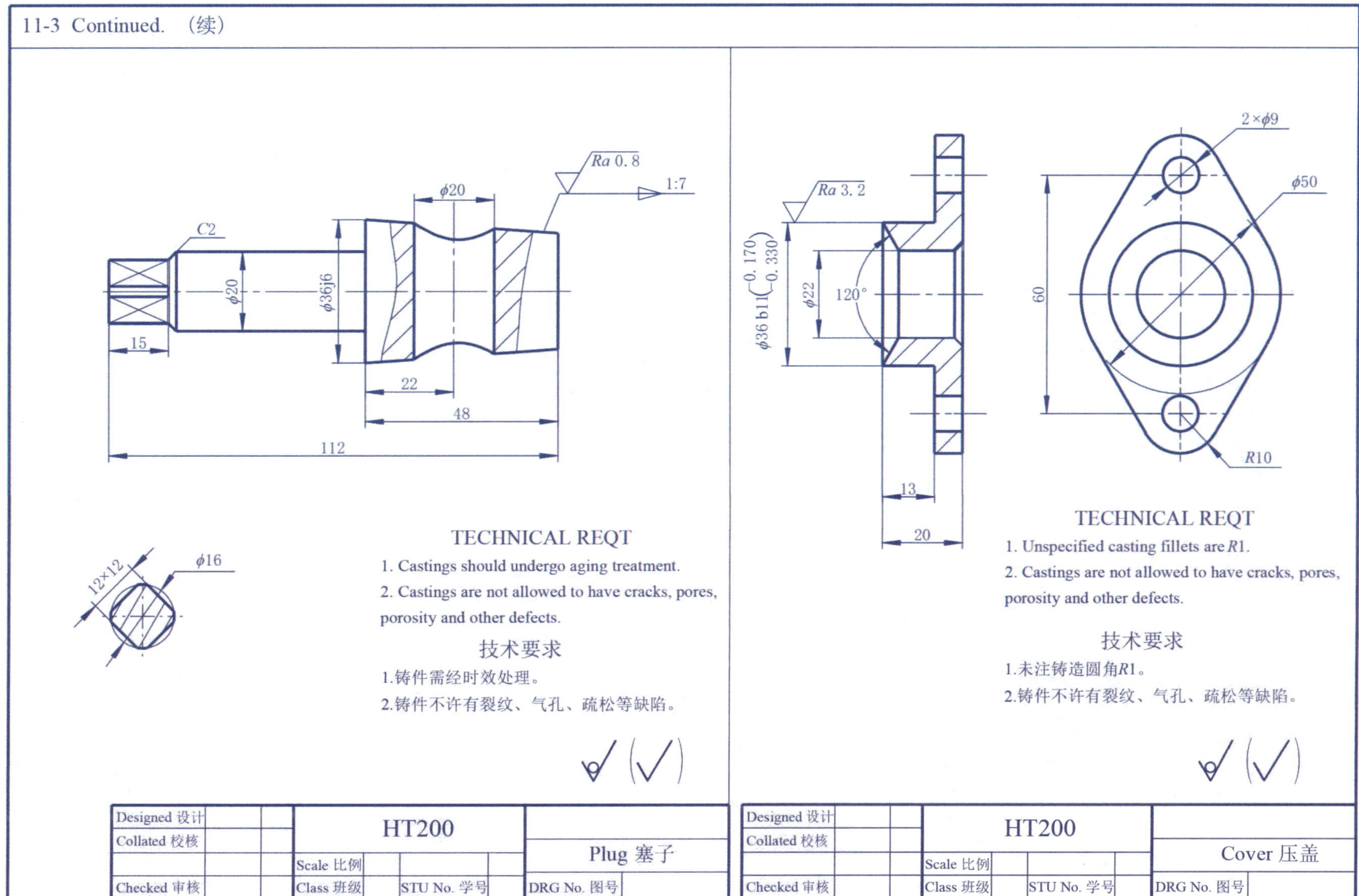

Class 班级： Name 姓名： ID Number 学号：

11-4 Read the assembly drawing of the drill jig, fill in the blanks, and dismantle the detail drawing of sheft 4.
读钻模装配图，完成填空，并拆画轴4的零件图。

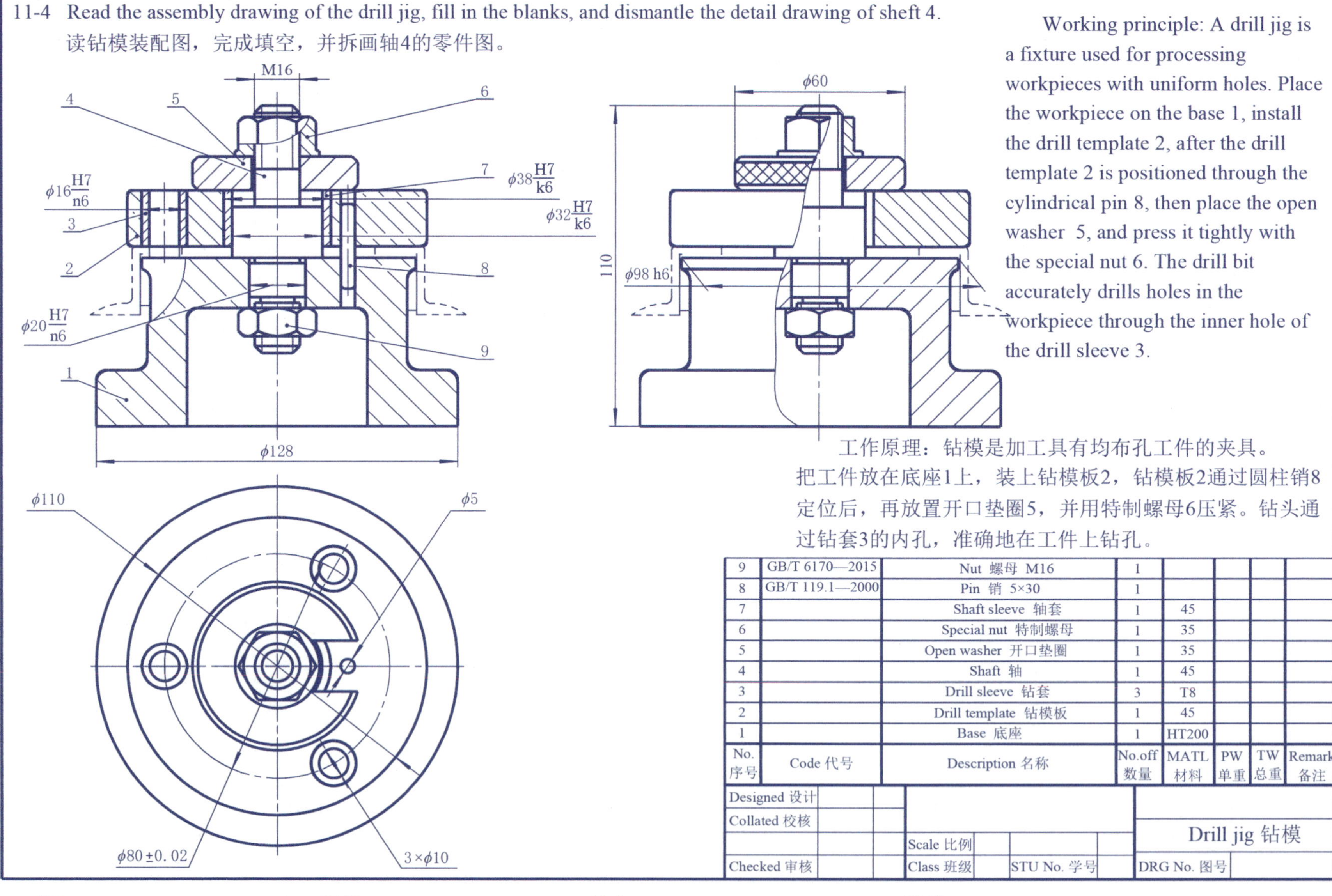

Working principle: A drill jig is a fixture used for processing workpieces with uniform holes. Place the workpiece on the base 1, install the drill template 2, after the drill template 2 is positioned through the cylindrical pin 8, then place the open washer 5, and press it tightly with the special nut 6. The drill bit accurately drills holes in the workpiece through the inner hole of the drill sleeve 3.

工作原理：钻模是加工具有均布孔工件的夹具。把工件放在底座1上，装上钻模板2，钻模板2通过圆柱销8定位后，再放置开口垫圈5，并用特制螺母6压紧。钻头通过钻套3的内孔，准确地在工件上钻孔。

9	GB/T 6170—2015	Nut 螺母 M16	1				
8	GB/T 119.1—2000	Pin 销 5×30	1				
7		Shaft sleeve 轴套	1	45			
6		Special nut 特制螺母	1	35			
5		Open washer 开口垫圈	1	35			
4		Shaft 轴	1	45			
3		Drill sleeve 钻套	3	T8			
2		Drill template 钻模板	1	45			
1		Base 底座	1	HT200			
No. 序号	Code 代号	Description 名称	No.off 数量	MATL 材料	PW 单重	TW 总重	Remark 备注

Designed 设计			Drill jig 钻模
Collated 校核		Scale 比例	
Checked 审核		Class 班级	STU No. 学号	DRG No. 图号

 Class 班级： Name 姓名： ID Number 学号：

11-4 Continued.（续）

(1) The drill jig consists of ________ types and a total of ________ parts. There is/are ________ standard part(s) in it.

钻模由________种共________个零件组成，其中标准件有________种。

(2) There is/are ________ $\phi16\frac{H7}{n6}$ drill sleeve hole(s) on the drill template 2, the positioning dimensions of its hole are ________. The material of drill sleeve 3 is ________.The double dotted lines in the drawing represent ________, belonging to the ________ drawing method.

钻模板2上有________个$\phi16\frac{H7}{n6}$配合的钻套孔，其孔的定位尺寸是________。钻套3的材料是________。图中双点画线表示________件，属于________画法。

(3) There is/are ________ circular groove(s) on base 1. The positioning dimensions between the base and the machined part are ________.

底座1上有________个圆弧槽。底座与被加工件的定位尺寸是________。

(4) From the drawing, it can be seen that the machined part needs to be drilled ________ holes with a diameter of ________.

从图中可以看出，被加工件需钻________个直径为________的孔。

(5) Dimension $\phi38\frac{H7}{k6}$ is the________size of part ________ and part ________. It belongs to the________fit of basic ________ system.

尺寸$\phi38\frac{H7}{k6}$是件________和件________的________尺寸，属于________制的________配合。

(6) The overall dimensions of the drill jig are ____________.

钻模的总体尺寸为____________。

(7) After drilling the holes, part ________ should be loosened first, then part ________ and part ________ should be removed, so that the machined part can be removed.

被加工件钻完孔后，应先旋松件________，再取下件________和件________，被加工件便可取下。

11-5 Read the assembly drawing of the pipe wrench, fill in the blanks, and dismantle the detail drawing of part 1 clamp seat.
读管钳装配图，完成填空，并拆画件1钳座的零件图。

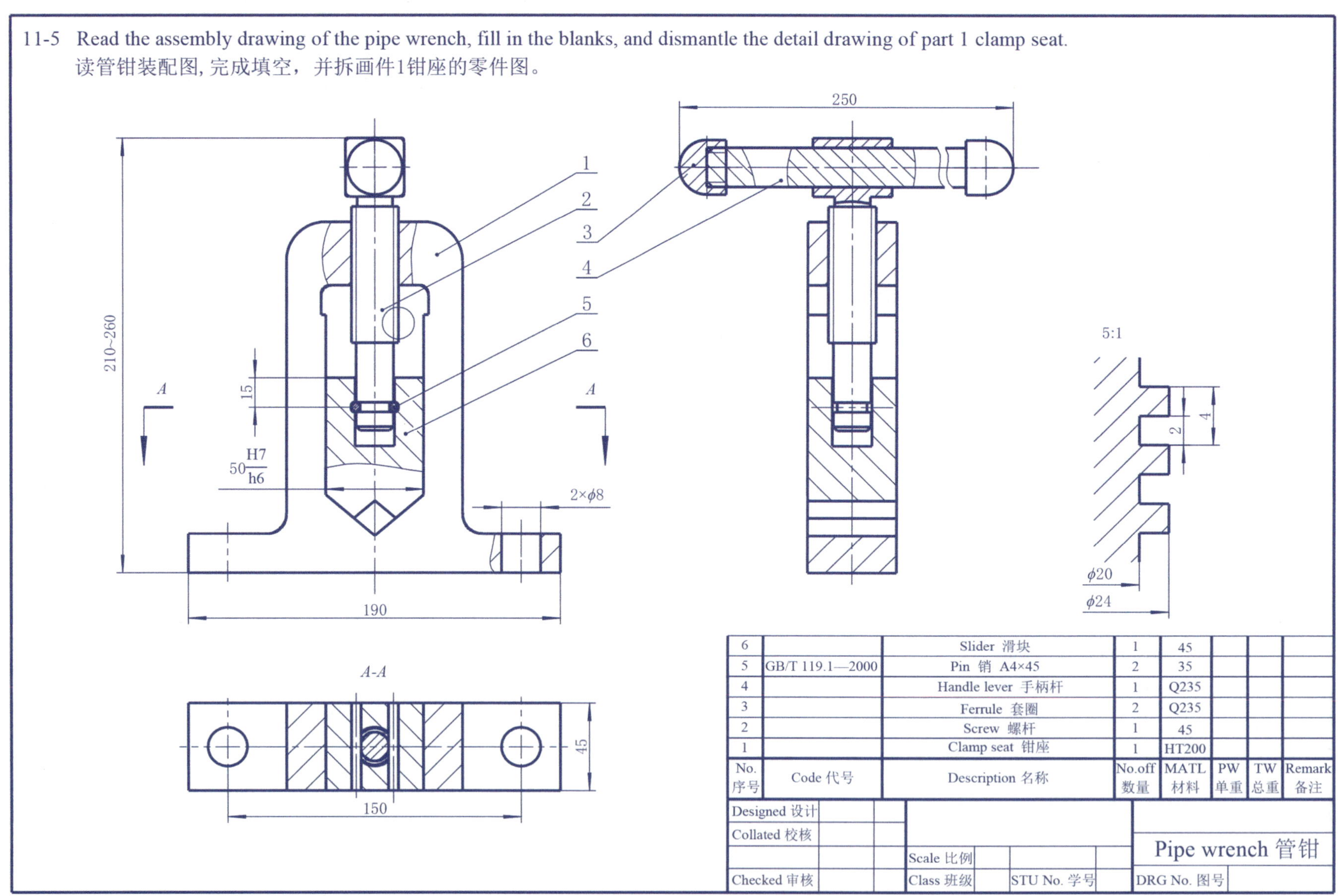

6		Slider 滑块	1	45			
5	GB/T 119.1—2000	Pin 销 A4×45	2	35			
4		Handle lever 手柄杆	1	Q235			
3		Ferrule 套圈	2	Q235			
2		Screw 螺杆	1	45			
1		Clamp seat 钳座	1	HT200			
No. 序号	Code 代号	Description 名称	No.off 数量	MATL 材料	PW 单重	TW 总重	Remark 备注

Designed 设计			Pipe wrench 管钳
Collated 校核		Scale 比例	
Checked 审核		Class 班级	STU No. 学号	DRG No. 图号

Class 班级： Name 姓名： ID Number 学号：

11-5 Continued.（续）

(1) The name of this assembly is_________ , consisting of _________ types and __________ parts.
该装配体的名称是__________，由__________种共_________个零件组成。

(2) In the front view of the assembly, there is/are _________ part(s) that has/have been _________ . The top view adopts a __________ drawing method. The left view adopts a __________ drawing method and a __________ drawing method. The handle lever 4 adopts a ___________ drawing method. In addition, there is a __________ view.
该装配体的主视图中有_________处作了__________，俯视图采用了_________，左视图采用了__________和_________，件4手柄杆采用了__________画法，另外还有一个__________图。

(3) When the screw 2 rotates, the slider 6 will move in _________ , with a lifting range of _________ mm.
当件2螺杆转动时，件6滑块将作__________运动，其升降范围是________mm。

(4) The screw 2 and the slider 6 belong to _______________connection, the screw 2 and the clamp seat 1 belong to ______________ connection, and the handle lever 4 and the ferrule 3 belong to ______________ connection.
件2螺杆与件6滑块属于____________连接，件2螺杆和件1钳座属于_____________连接， 件4手柄杆和件3套圈属于_______________。

(5) The size $50\frac{H7}{h6}$ is the __________ size for part __________ and part ___________, which belongs to the __________ fit of basic __________ system. The maximum outer diameter of the pipe clamped by this pipe wrench is ____________.
尺寸 $50\frac{H7}{h6}$ 是件__________和件__________的__________尺寸，属于基___________制的___________配合，该管钳所夹管子的最大外圆直径是__________。

(6) The overall sizes of the pipe wrench are: __________ in length, __________ in width, and__________ in height. The installation size is __________.
管钳的总体尺寸是：长___________，宽____________，高___________。安装尺寸是___________。

(7) To remove the slider 6, the __________ should be removed first and then the ___________ should be rotated out to remove the slider.
欲取下件6滑块，必须先取出___________，再旋出____________，滑块便可取出。

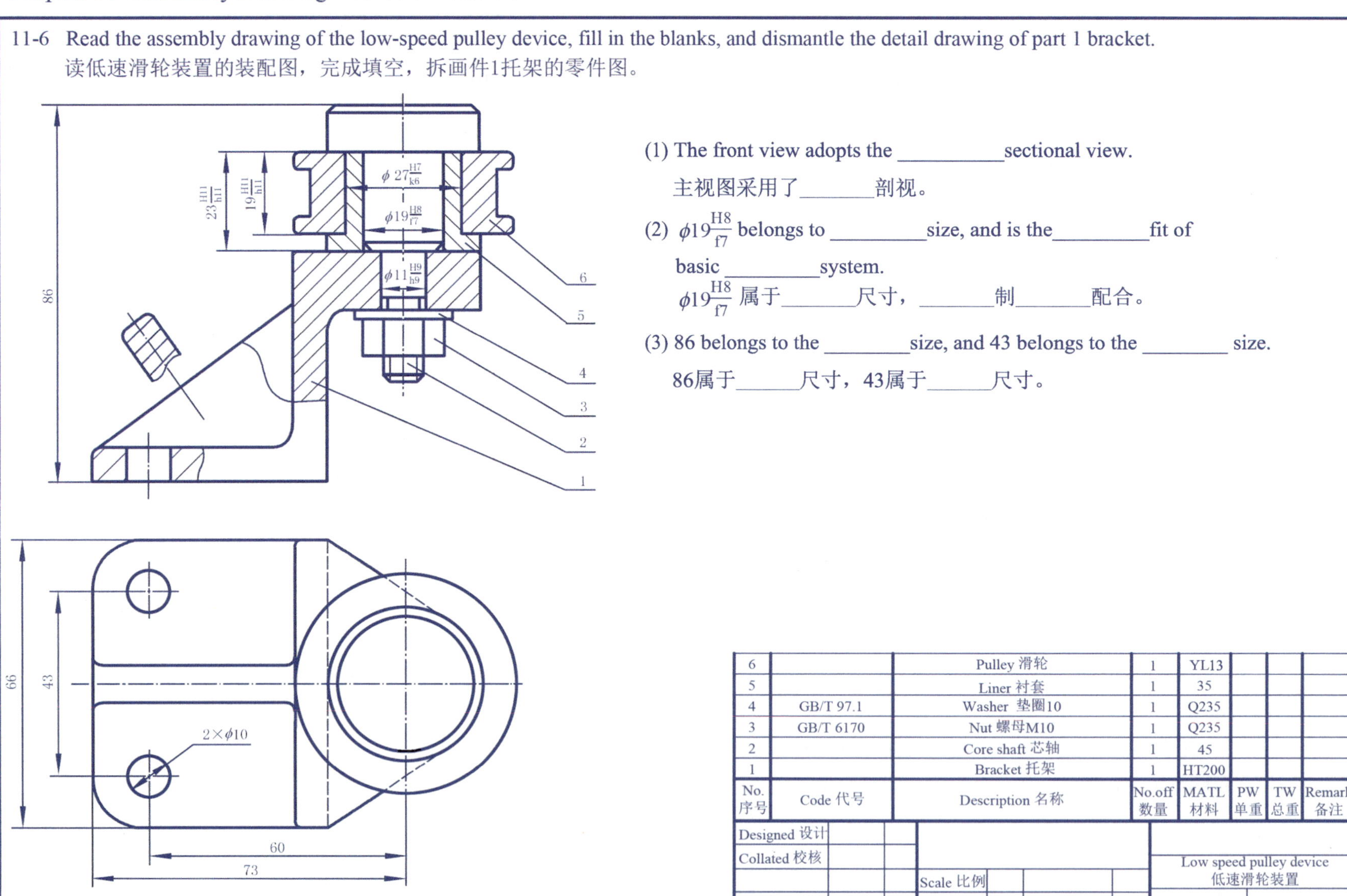

11-6 Read the assembly drawing of the low-speed pulley device, fill in the blanks, and dismantle the detail drawing of part 1 bracket.
读低速滑轮装置的装配图，完成填空，拆画件1托架的零件图。

(1) The front view adopts the ________ sectional view.

主视图采用了______剖视。

(2) $\phi 19\frac{H8}{f7}$ belongs to ________ size, and is the ________ fit of basic ________ system.

$\phi 19\frac{H8}{f7}$ 属于______尺寸，______制______配合。

(3) 86 belongs to the ________ size, and 43 belongs to the ________ size.

86属于_____尺寸，43属于_____尺寸。

No. 序号	Code 代号	Description 名称	No.off 数量	MATL 材料	PW 单重	TW 总重	Remark 备注
6		Pulley 滑轮	1	YL13			
5		Liner 衬套	1	35			
4	GB/T 97.1	Washer 垫圈10	1	Q235			
3	GB/T 6170	Nut 螺母M10	1	Q235			
2		Core shaft 芯轴	1	45			
1		Bracket 托架	1	HT200			

Designed 设计			Low speed pulley device 低速滑轮装置
Collated 校核		Scale 比例	
Checked 审核		Class 班级	STU No. 学号	DRG No. 图号

Class 班级： Name 姓名： ID Number 学号：

11-7 Read the assembly drawing of the valve, and dismantle the detail drawing of part 3 valve body. 读阀的装配图，拆画件3阀体的零件图。

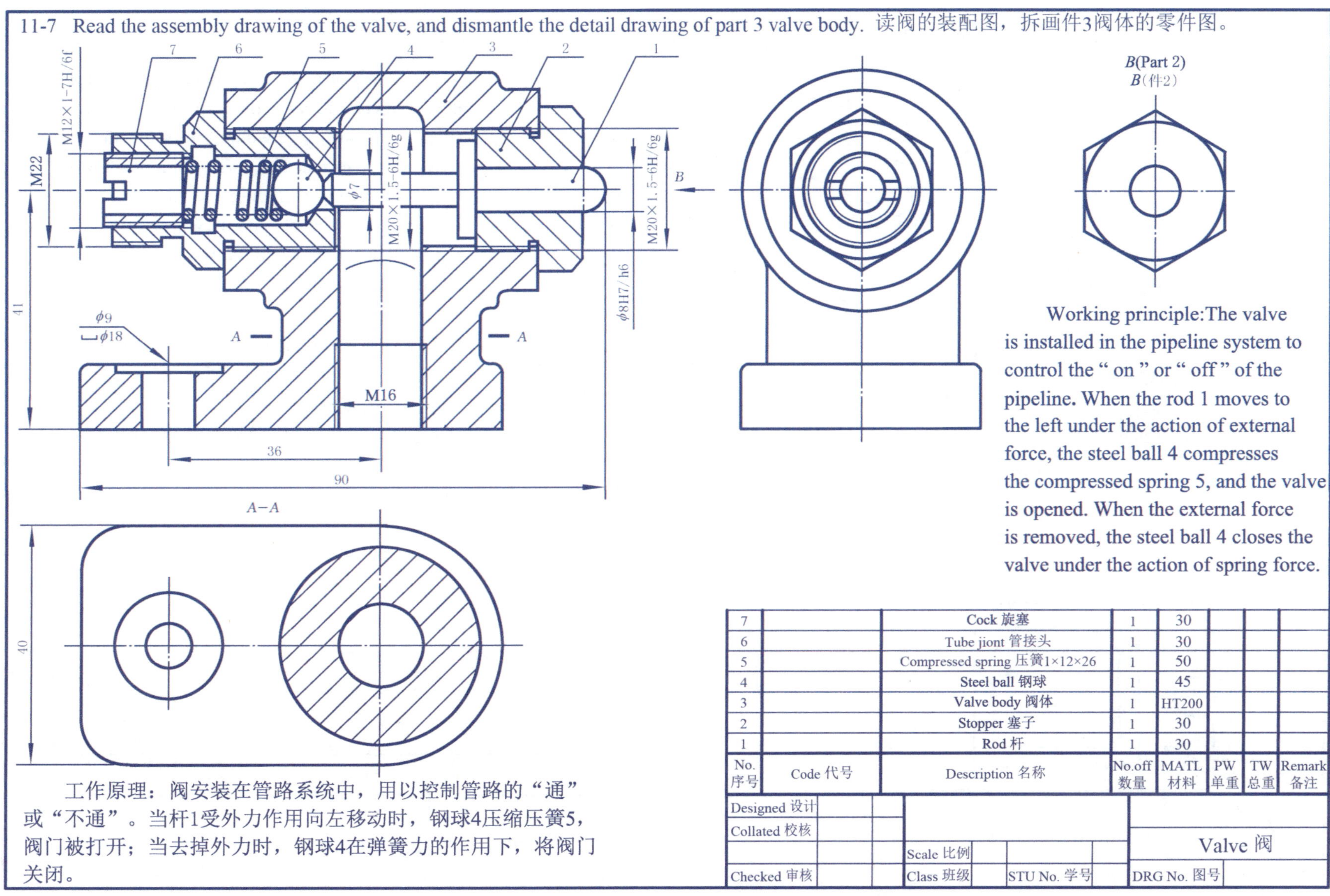

Working principle:The valve is installed in the pipeline system to control the “ on ” or “ off ” of the pipeline. When the rod 1 moves to the left under the action of external force, the steel ball 4 compresses the compressed spring 5, and the valve is opened. When the external force is removed, the steel ball 4 closes the valve under the action of spring force.

工作原理：阀安装在管路系统中，用以控制管路的“通”或“不通”。当杆1受外力作用向左移动时，钢球4压缩压簧5，阀门被打开；当去掉外力时，钢球4在弹簧力的作用下，将阀门关闭。

No. 序号	Code 代号	Description 名称	No.off 数量	MATL 材料	PW 单重	TW 总重	Remark 备注
7		Cock 旋塞	1	30			
6		Tube jiont 管接头	1	30			
5		Compressed spring 压簧1×12×26	1	50			
4		Steel ball 钢球	1	45			
3		Valve body 阀体	1	HT200			
2		Stopper 塞子	1	30			
1		Rod 杆	1	30			

Designed 设计			
Collated 校核		Scale 比例	Valve 阀
Checked 审核		Class 班级 / STU No. 学号	DRG No. 图号

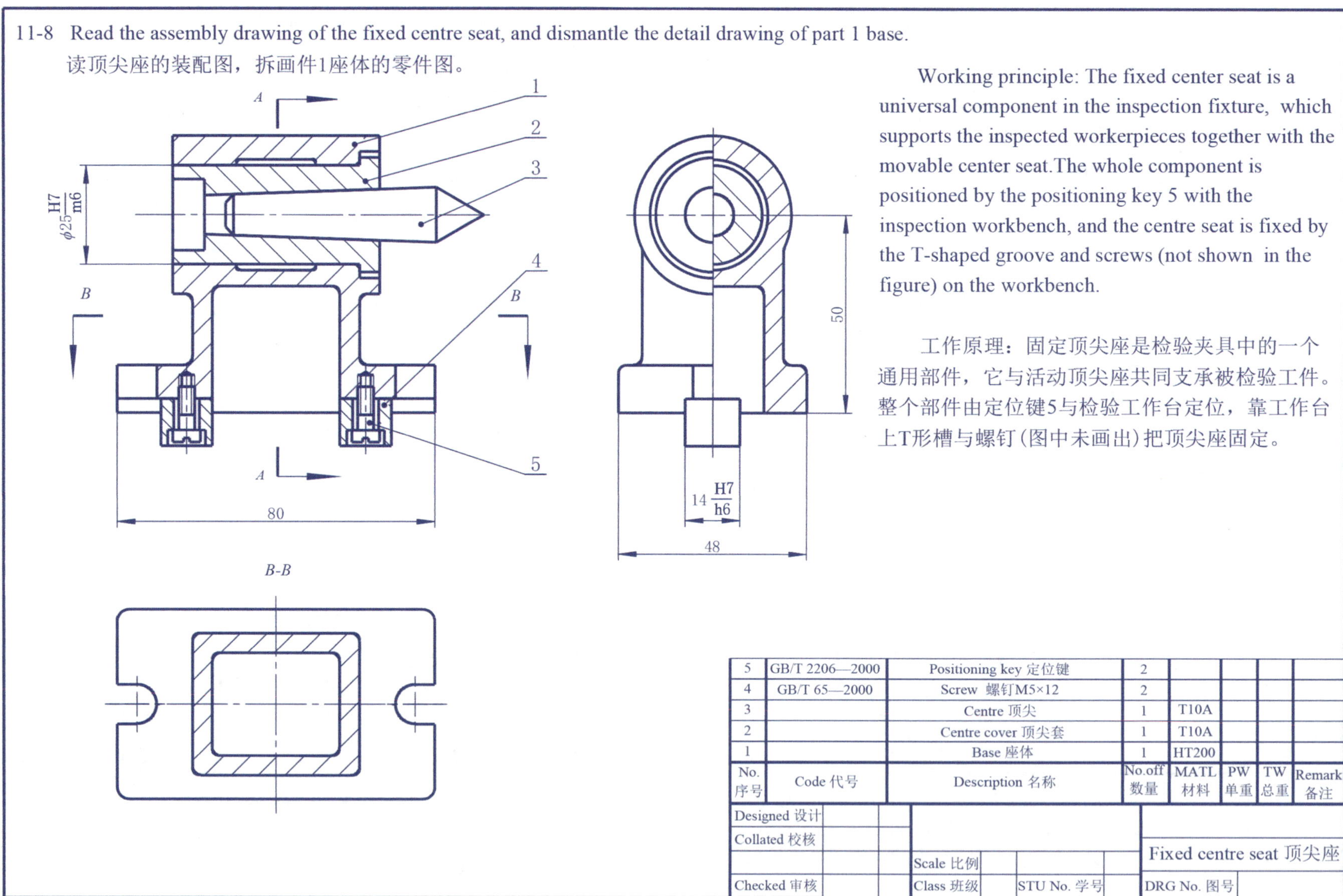

No. 序号	Code 代号	Description 名称	No.off 数量	MATL 材料	PW 单重	TW 总重	Remark 备注
5	GB/T 2206—2000	Positioning key 定位键	2				
4	GB/T 65—2000	Screw 螺钉M5×12	2				
3		Centre 顶尖	1	T10A			
2		Centre cover 顶尖套	1	T10A			
1		Base 座体	1	HT200			

Designed 设计				Fixed centre seat 顶尖座
Collated 校核				
		Scale 比例		
Checked 审核		Class 班级	STU No. 学号	DRG No. 图号

Class 班级： Name 姓名： ID Number 学号：

11-9 Read the assembly drawing of the broaching machine tool holder, and dismantle the detail drawing of part 1 tool holder seat.

读拉床托刀架的装配图，拆画件1刀座架的零件图。

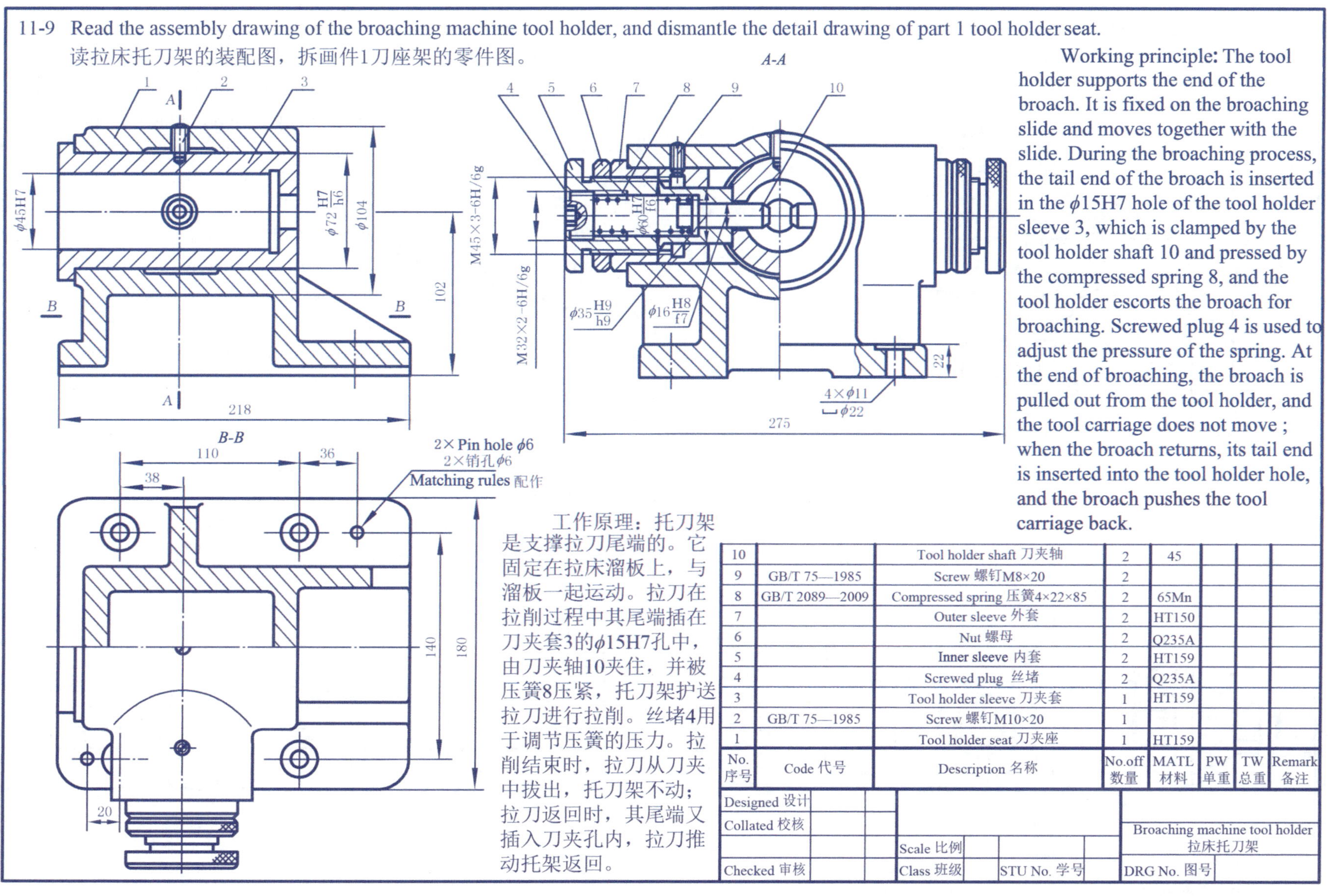

Working principle: The tool holder supports the end of the broach. It is fixed on the broaching slide and moves together with the slide. During the broaching process, the tail end of the broach is inserted in the ϕ15H7 hole of the tool holder sleeve 3, which is clamped by the tool holder shaft 10 and pressed by the compressed spring 8, and the tool holder escorts the broach for broaching. Screwed plug 4 is used to adjust the pressure of the spring. At the end of broaching, the broach is pulled out from the tool holder, and the tool carriage does not move ; when the broach returns, its tail end is inserted into the tool holder hole, and the broach pushes the tool carriage back.

工作原理：托刀架是支撑拉刀尾端的。它固定在拉床溜板上，与溜板一起运动。拉刀在拉削过程中其尾端插在刀夹套3的ϕ15H7孔中，由刀夹轴10夹住，并被压簧8压紧，托刀架护送拉刀进行拉削。丝堵4用于调节压簧的压力。拉削结束时，拉刀从刀夹中拔出，托刀架不动；拉刀返回时，其尾端又插入刀夹孔内，拉刀推动托架返回。

No. 序号	Code 代号	Description 名称	No.off 数量	MATL 材料	PW 单重	TW 总重	Remark 备注
10		Tool holder shaft 刀夹轴	2	45			
9	GB/T 75—1985	Screw 螺钉M8×20	2				
8	GB/T 2089—2009	Compressed spring 压簧4×22×85	2	65Mn			
7		Outer sleeve 外套	2	HT150			
6		Nut 螺母	2	Q235A			
5		Inner sleeve 内套	2	HT159			
4		Screwed plug 丝堵	2	Q235A			
3		Tool holder sleeve 刀夹套	1	HT159			
2	GB/T 75—1985	Screw 螺钉M10×20	1				
1		Tool holder seat 刀夹座	1	HT159			

Designed 设计			Broaching machine tool holder 拉床托刀架
Collated 校核		Scale 比例	
Checked 审核		Class 班级	STU No. 学号	DRG No. 图号

11-10 Read the assembly drawing of the reversing valve and dismantle the detail drawing of part 1 valve body. 读换向阀装配图，拆画件1阀体的零件图。

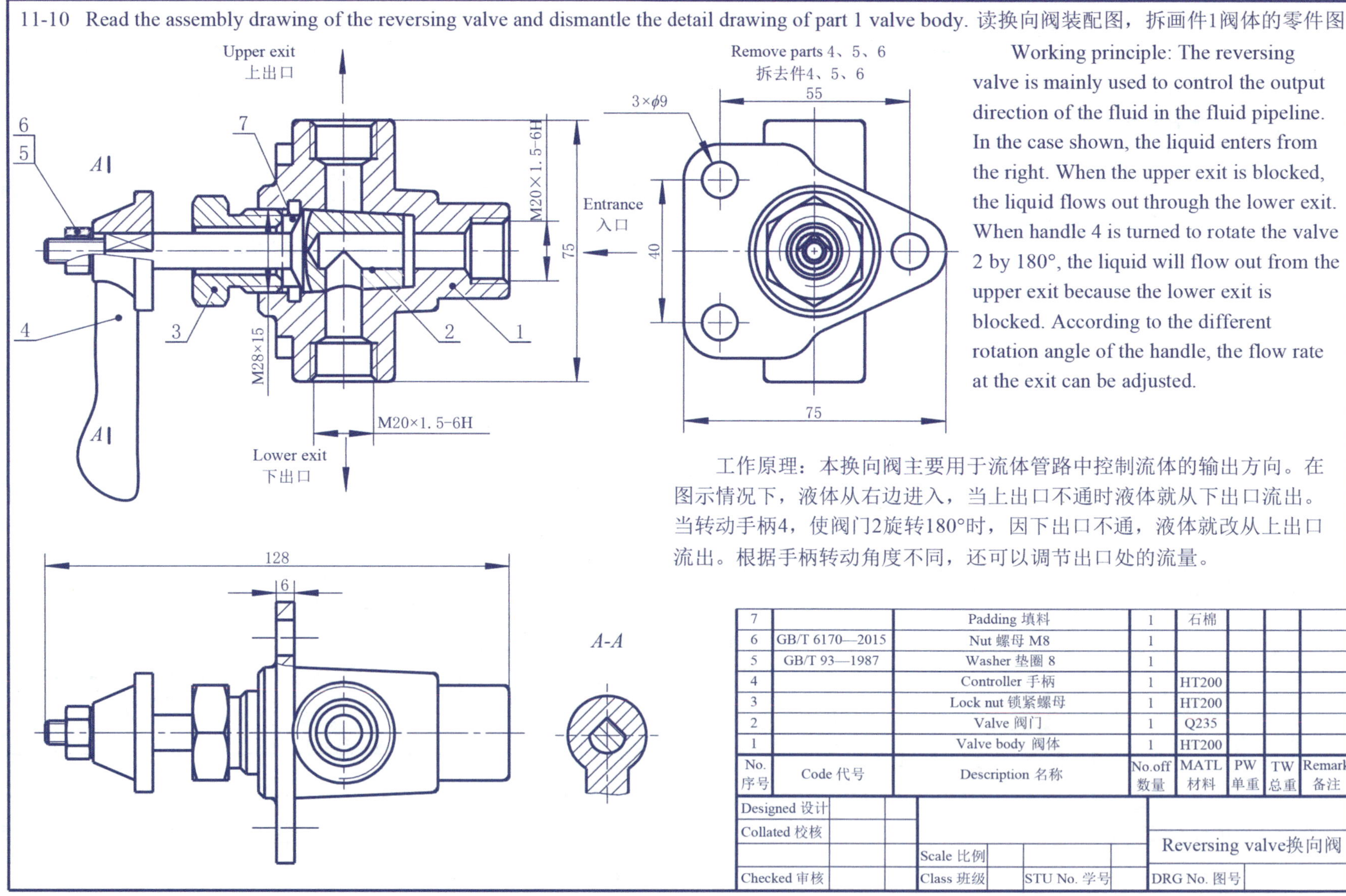

Working principle: The reversing valve is mainly used to control the output direction of the fluid in the fluid pipeline. In the case shown, the liquid enters from the right. When the upper exit is blocked, the liquid flows out through the lower exit. When handle 4 is turned to rotate the valve 2 by 180°, the liquid will flow out from the upper exit because the lower exit is blocked. According to the different rotation angle of the handle, the flow rate at the exit can be adjusted.

工作原理：本换向阀主要用于流体管路中控制流体的输出方向。在图示情况下，液体从右边进入，当上出口不通时液体就从下出口流出。当转动手柄4，使阀门2旋转180°时，因下出口不通，液体就改从上出口流出。根据手柄转动角度不同，还可以调节出口处的流量。

No. 序号	Code 代号	Description 名称	No.off 数量	MATL 材料	PW 单重	TW 总重	Remark 备注
7		Padding 填料	1	石棉			
6	GB/T 6170—2015	Nut 螺母 M8	1				
5	GB/T 93—1987	Washer 垫圈 8	1				
4		Controller 手柄	1	HT200			
3		Lock nut 锁紧螺母	1	HT200			
2		Valve 阀门	1	Q235			
1		Valve body 阀体	1	HT200			

Designed 设计			Reversing valve换向阀
Collated 校核		Scale 比例	
Checked 审核		Class 班级	STU No. 学号	DRG No. 图号

Class 班级： Name 姓名： ID Number 学号：

11-11 Read the assembly drawing of the instrument lathe tailstock, and dismantle the detail drawing of part 1 tailstock body.

读仪表车床尾座装配图，拆画件 1 尾座体的零件图。

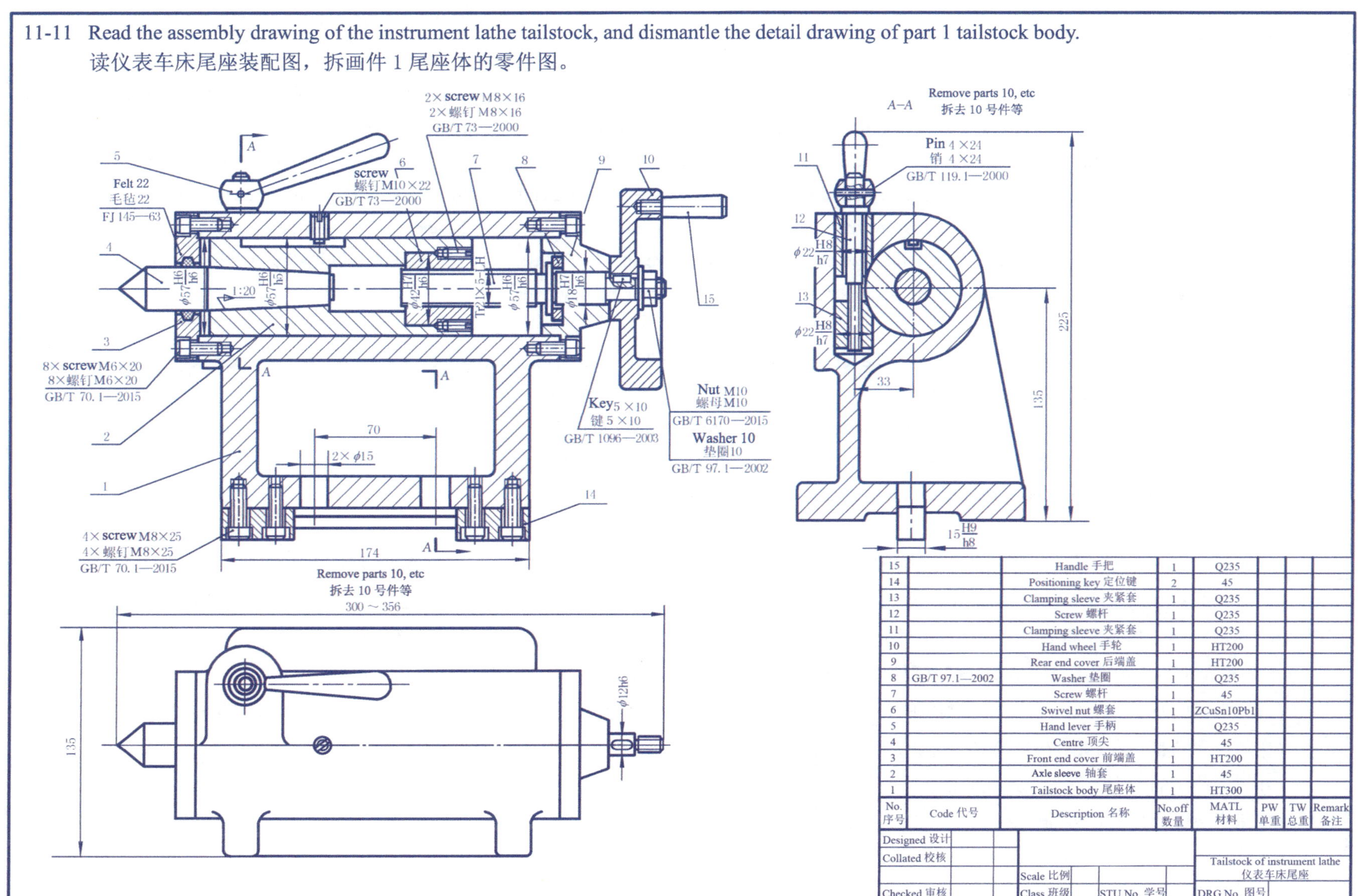

No. 序号	Code 代号	Description 名称	No.off 数量	MATL 材料	PW 单重	TW 总重	Remark 备注
15		Handle 手把	1	Q235			
14		Positioning key 定位键	2	45			
13		Clamping sleeve 夹紧套	1	Q235			
12		Screw 螺杆	1	Q235			
11		Clamping sleeve 夹紧套	1	Q235			
10		Hand wheel 手轮	1	HT200			
9		Rear end cover 后端盖	1	HT200			
8	GB/T 97.1—2002	Washer 垫圈	1	Q235			
7		Screw 螺杆	1	45			
6		Swivel nut 螺套	1	ZCuSn10Pb1			
5		Hand lever 手柄	1	Q235			
4		Centre 顶尖	1	45			
3		Front end cover 前端盖	1	HT200			
2		Axle sleeve 轴套	1	45			
1		Tailstock body 尾座体	1	HT300			

Designed 设计			Tailstock of instrument lathe 仪表车床尾座
Collated 校核		Scale 比例	
Checked 审核		Class 班级	STU No. 学号	DRG No. 图号

Class 班级： Name 姓名： ID Number 学号：

11-11 Continued.（续）

Working principle: Tailstock is a device used for tightening on instrument lathes for machining shaft parts , which is driven by the screw mechanism to move the center. The centre 4 of the component is installed in the axle sleeve 2, the swivel nut 6 is fixed with the axle sleeve 2 using two screws M8 × 16, and the screw M10 × 22 limits the axial movement of the axle sleeve 2. When the hand wheel 10 rotates, the screw 7 is rotated by the key 5 × 10, and then the centre 4 is moved axially with the axle sleeve 2 through the action of the swivel nut 6. When the centre 4 moves to the required position, it rotates and clamps on the hand lever 5 and the screw 12, so that the axle sleeve 2 is locked by the clamping sleeve 11 and 13. The tailstock is embedded in the T-shaped groove of the lathe by the positioning key 14 , and the distance between the centre and the headstock is adjusted by longitudinal sliding to adapt to parts of different lengths. When the adjustment is finished, the tailstock is locked on the lathe bed by bolts.

Read the picture requirements :

(1) What mechanism does the centre 4 achieve the left and right movement? If rotating the hand wheel clockwise (right), is the centre 4 in or out? Can the centre 4 rotate? Why?

(2) When the centre is adjusted to the right position, what mechanism is used to clamp?

(3) Can the drawing of the up-looking direction of the tailstock body 1 be figured out?

(4) Point out the meaning of each fit dimension symbol.

工作原理：尾座是用于仪表车床上加工轴类零件作顶紧用的装置，是借助于螺旋机构推动顶尖运动。部件中顶尖 4 装在轴套 2 内，螺套 6 用两个螺钉 M8 × 16 与轴套 2 固定，螺钉 M10 × 22 限制轴套 2 作轴向移动。当手轮 10 转动时，通过键 5 × 10 使螺杆 7 转动，再通过螺套 6 的作用，使轴套 2 带着顶尖 4 作轴向移动。当顶尖 4 移动到需要位置时，再旋转夹紧手柄 5 与螺杆 12，使夹紧套 11、13 将轴套 2 锁紧。尾座靠导向定位键 14 嵌入车身 T 形槽内，用纵向滑动来调整顶尖与床头箱的距离，适应不同长度的零件。当调整好后，用螺栓锁紧在床身上。

(1) 顶尖 4 的左右移动是靠什么机构实现的？如果顺时针（右转）转动手轮，顶尖 4 进还是退？顶尖 4 能转动吗？为什么？

(2) 当顶尖调整到合适位置时，靠什么机构夹紧？

(3) 尾座体 1 的仰视方向的图形能想出来吗？

(4) 指出各配合尺寸符号的含义。

* 12-1 Using the method of changing plane, find the real length of line segment AB and the inclination angles α and β to plane H and plane V.
用换面法求线段AB的实长及对H面、V面的倾角α、β。

* 12-2 Find the distance CD from point C to line AB and its projection using the method of changing plane.
用换面法求点C到直线AB的距离CD及其投影。

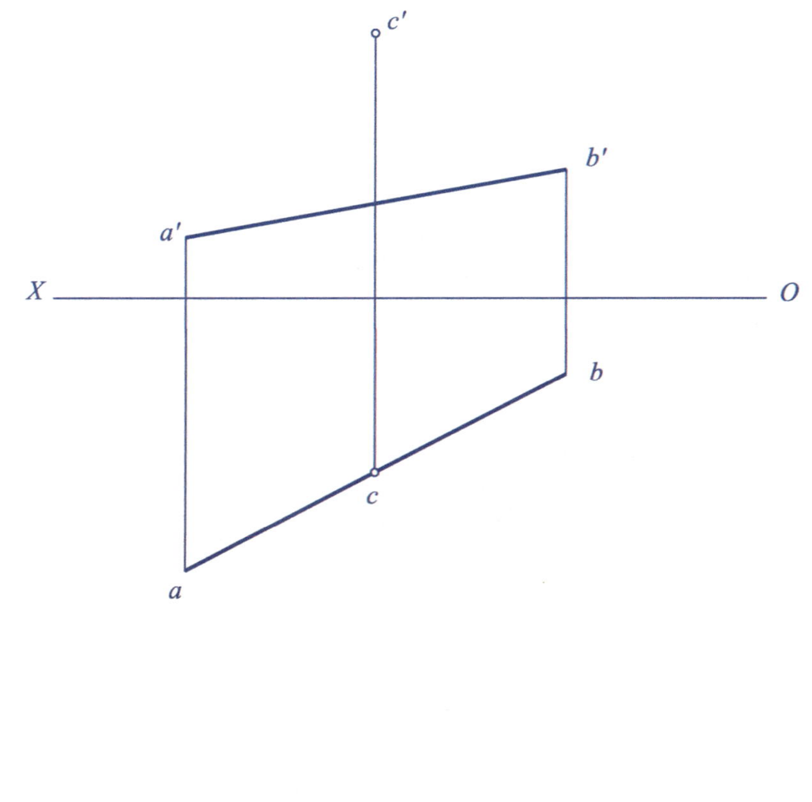

* 12-3 Find the distance from point D to plane $\triangle ABC$ and its projection using the method of changing plane.

用换面法求点D到$\triangle ABC$平面的距离及其投影。

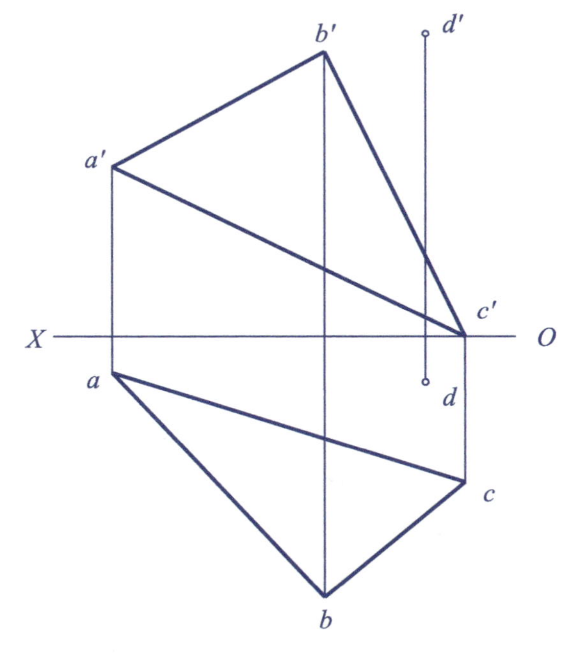

* 12-4 Complete the horizontal projection of an isosceles $\triangle ABC$ plane with AB as the base using the method of changing plane.

用换面法补全以AB为底边的等腰$\triangle ABC$平面的水平投影。

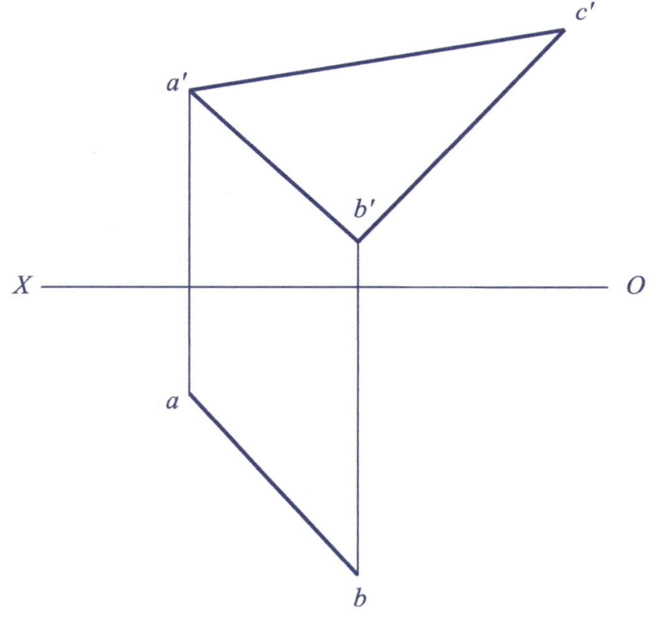

 Class 班级： Name 姓名： ID Number 学号：

*12-5 A given square *ABCD* has vertex *A* on line *EF* and vertex *C* on line *BG*, find the two projections of the square by the method of changing plane.

已知正方形*ABCD*的顶点*A*在直线*EF*上，顶点*C*在直线*BG*上，用换面法求该正方形的两面投影。

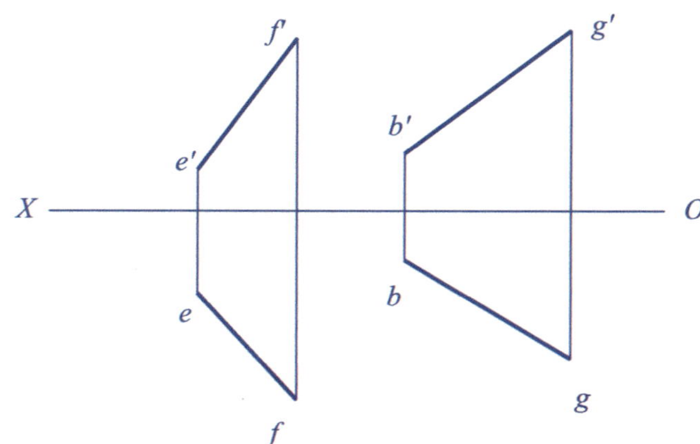

*12-6 The distance from point *K* to △*ABC* is 15mm. Find a line *KN* through point *K*, parallel to △*ABC* and intersecting with line *LM*, by the method of changing plane.

点*K*到△*ABC*的距离为15mm，用换面法过点*K*作一直线*KN*平行于△*ABC*并与直线*LM*相交。

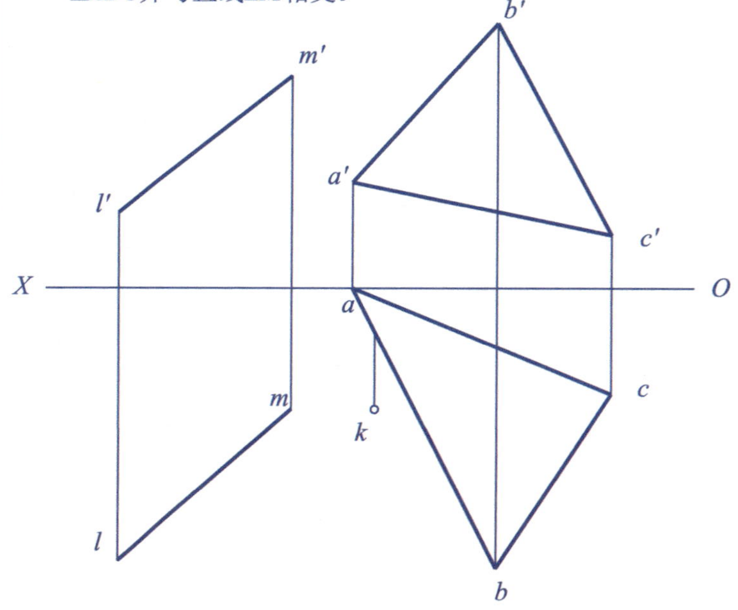

*12-7 Find the real length and projection of the common perpendicular *MN* between cross lines *AB* and *CD* using the method of changing plane.
用换面法求*AB*、*CD*两交叉直线间的公垂线*MN*的实长及其投影。

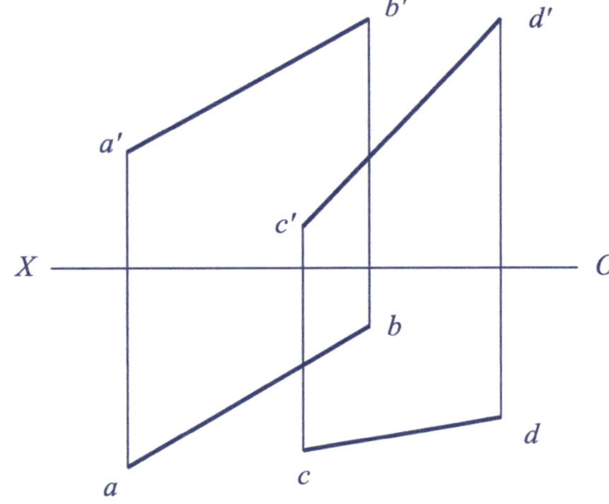

 Class 班级： Name 姓名： ID Number 学号：

*12-8 The angle between $\triangle ABC$ and $\triangle ABD$ is 60°. Find the frontal projection of $\triangle ABC$.

$\triangle ABC$与$\triangle ABD$的夹角为60°，求$\triangle ABC$的正面投影。

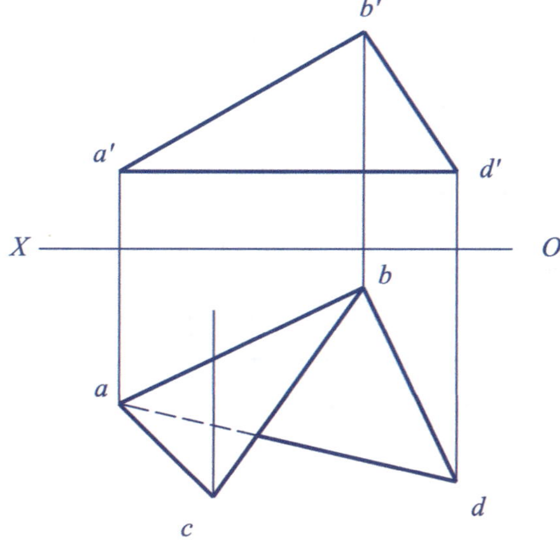

13-1 Draw the isometric projection of the part.
画出机件的正等轴测图。

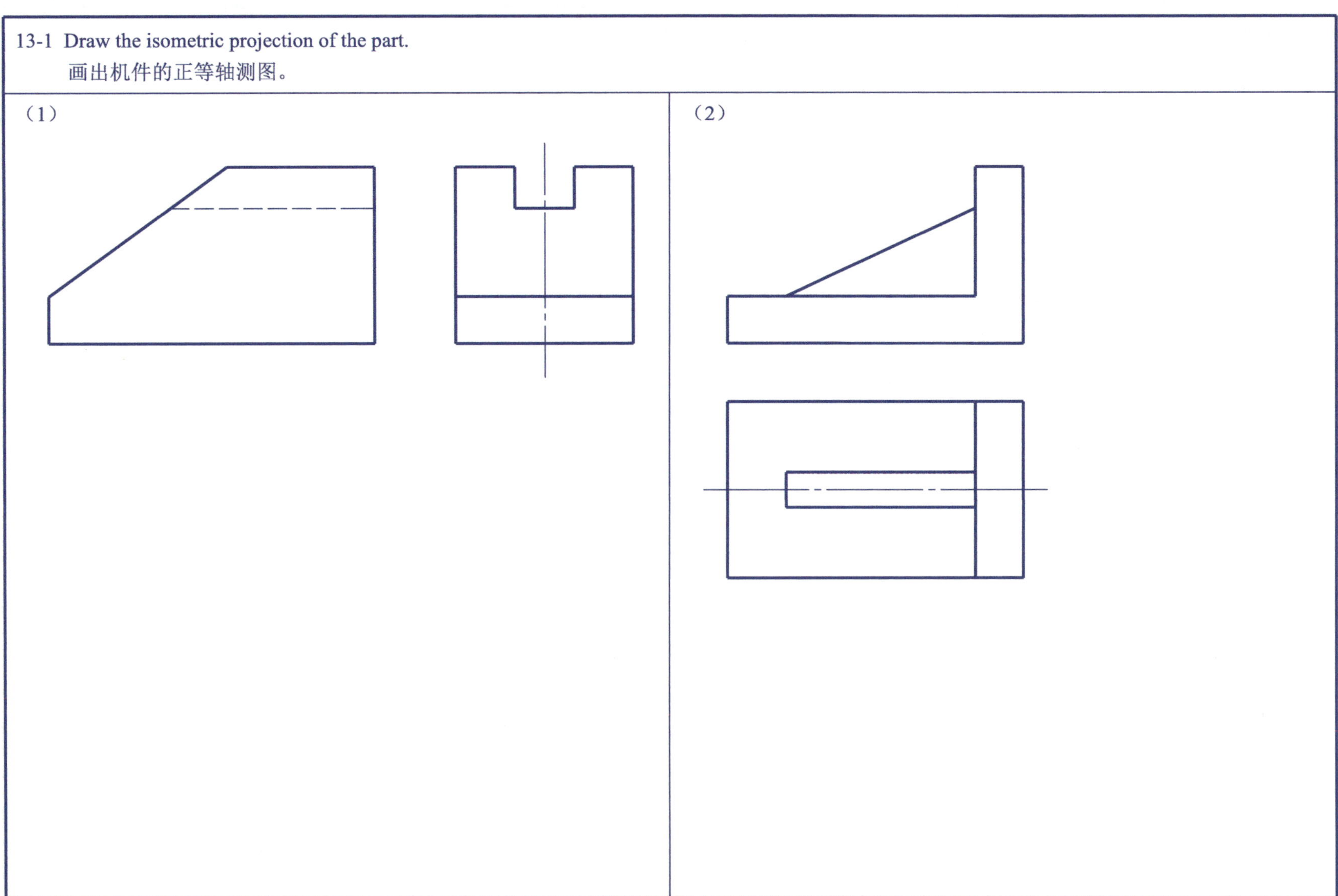

 Class 班级： Name 姓名： ID Number 学号：

13-2 Draw the isometric projection of the part.

画出机件的正等轴测图。

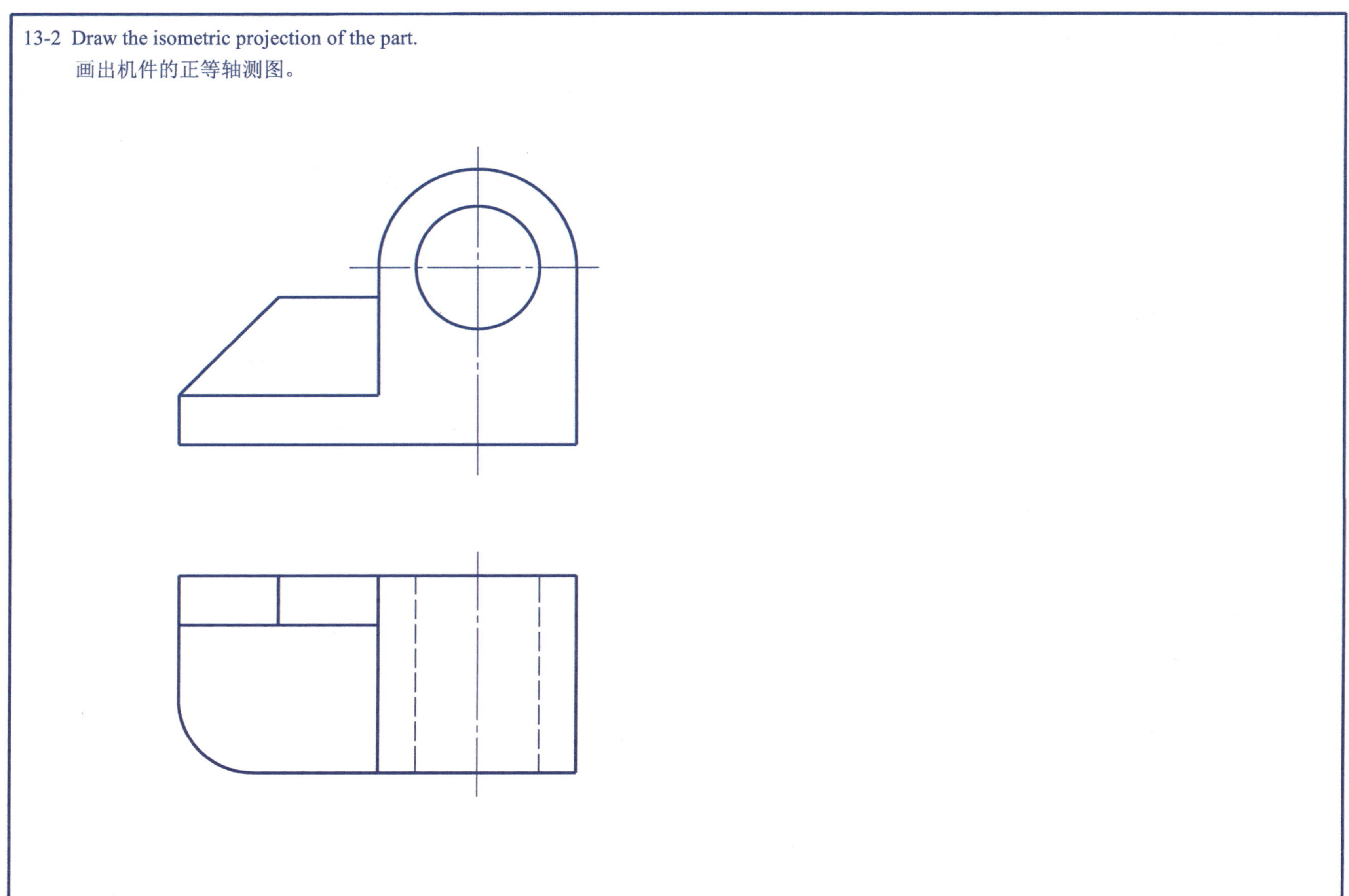

13-3 Draw the cabinet axonometry projection of the part.
画出机件的斜二轴测图。

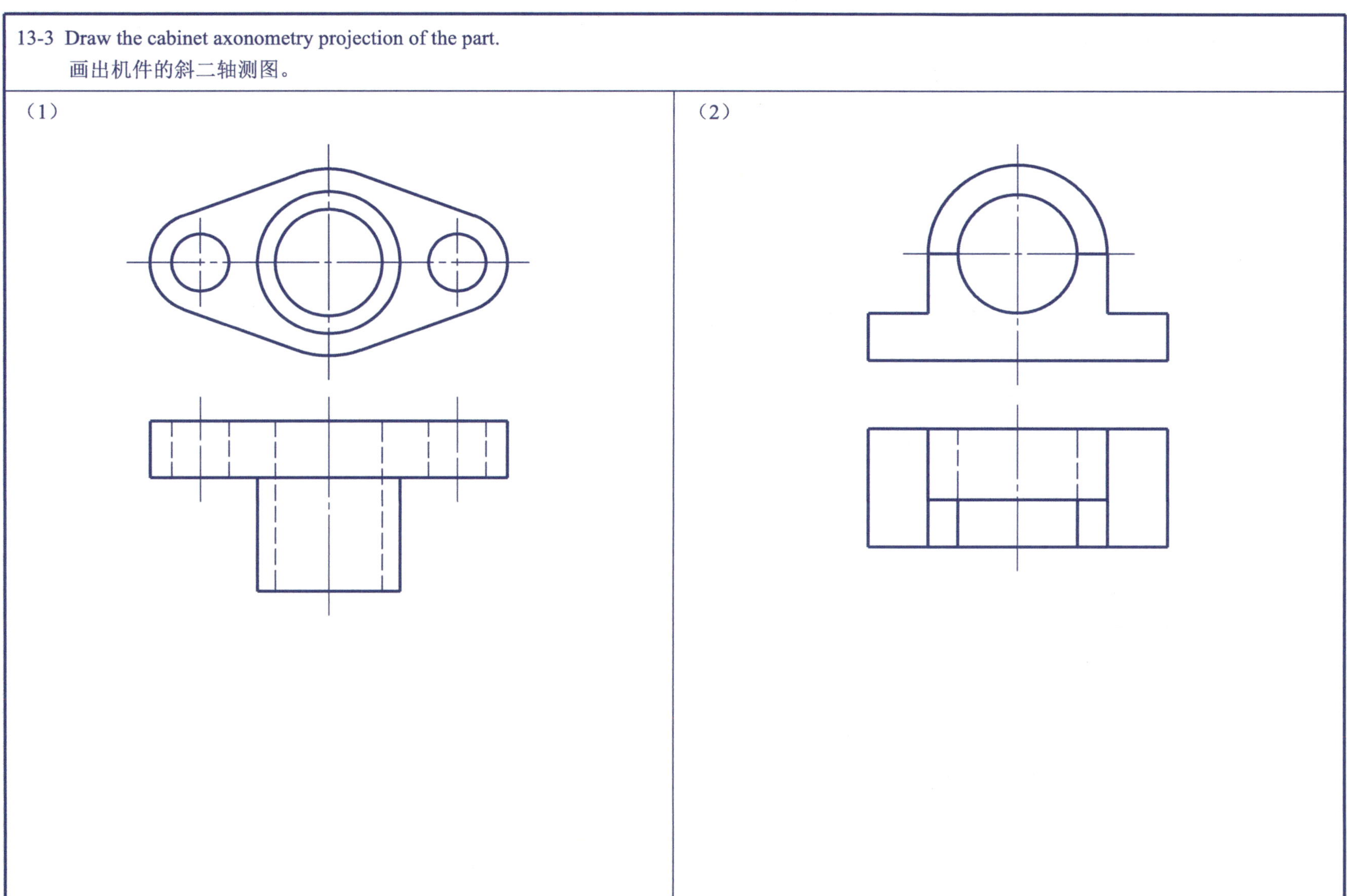

13-4 Based on the projections, draw （1） as a cabinet axonometry projection and （2） as a isometric projection.
根据投影图，将（1）画成斜二轴测图，将（2）画成正等轴测图。

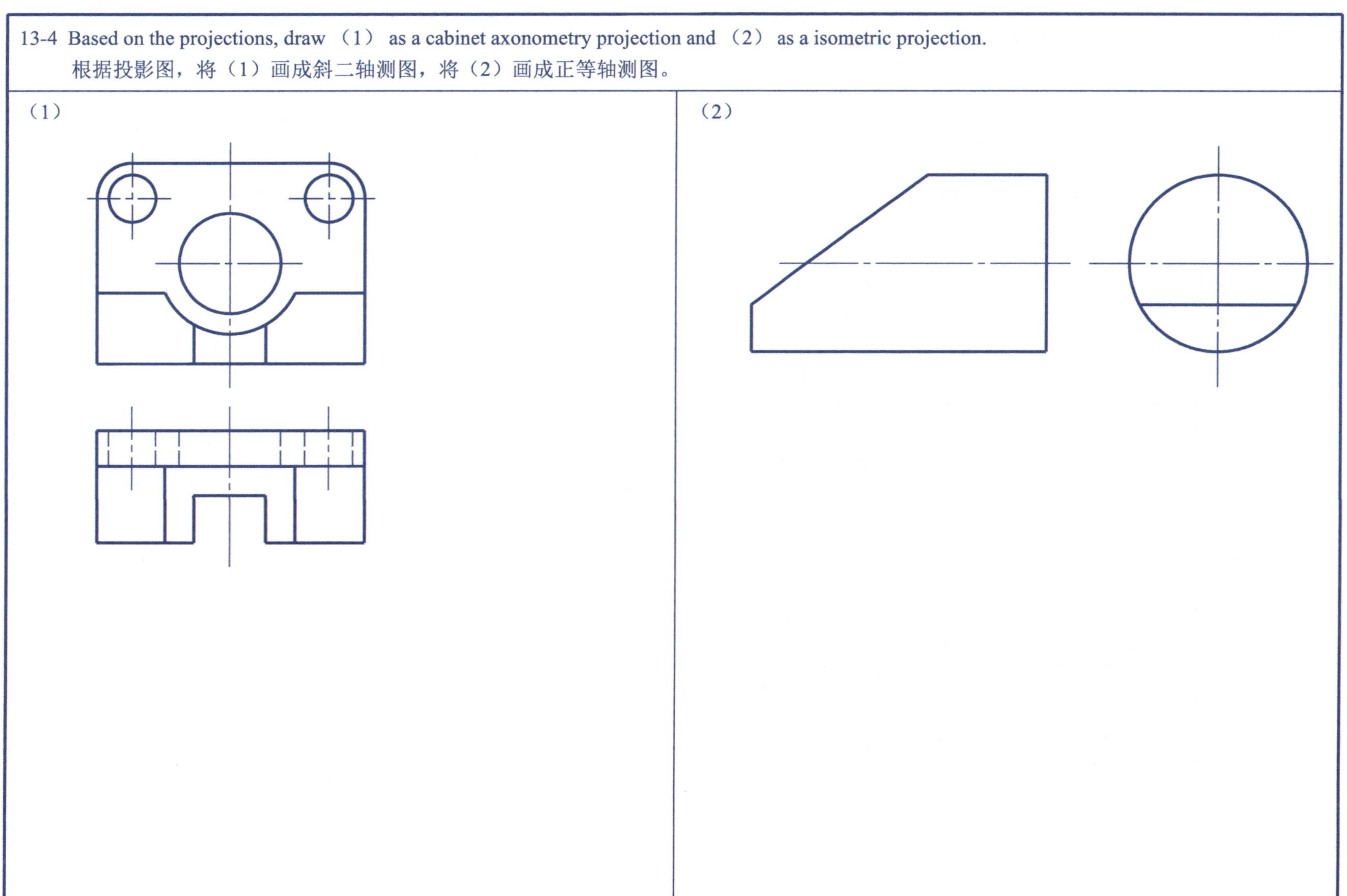

13-5 Draw the isometric projection of the part.
画出机件的正等轴测图。

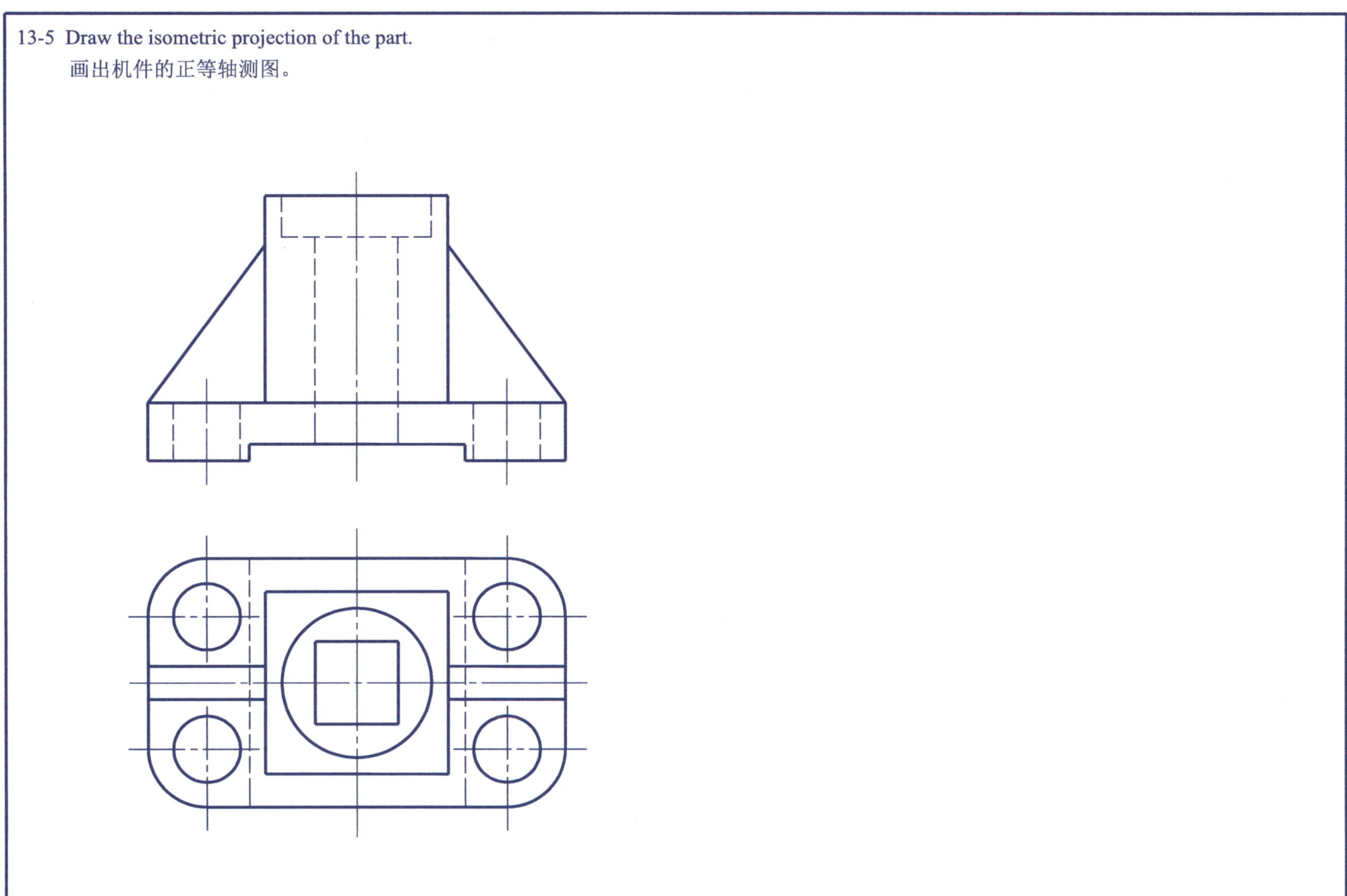

 Class 班级： Name 姓名： ID Number 学号：

13-6 Draw the isometric projection of the part.
画出机件的正等轴测图。

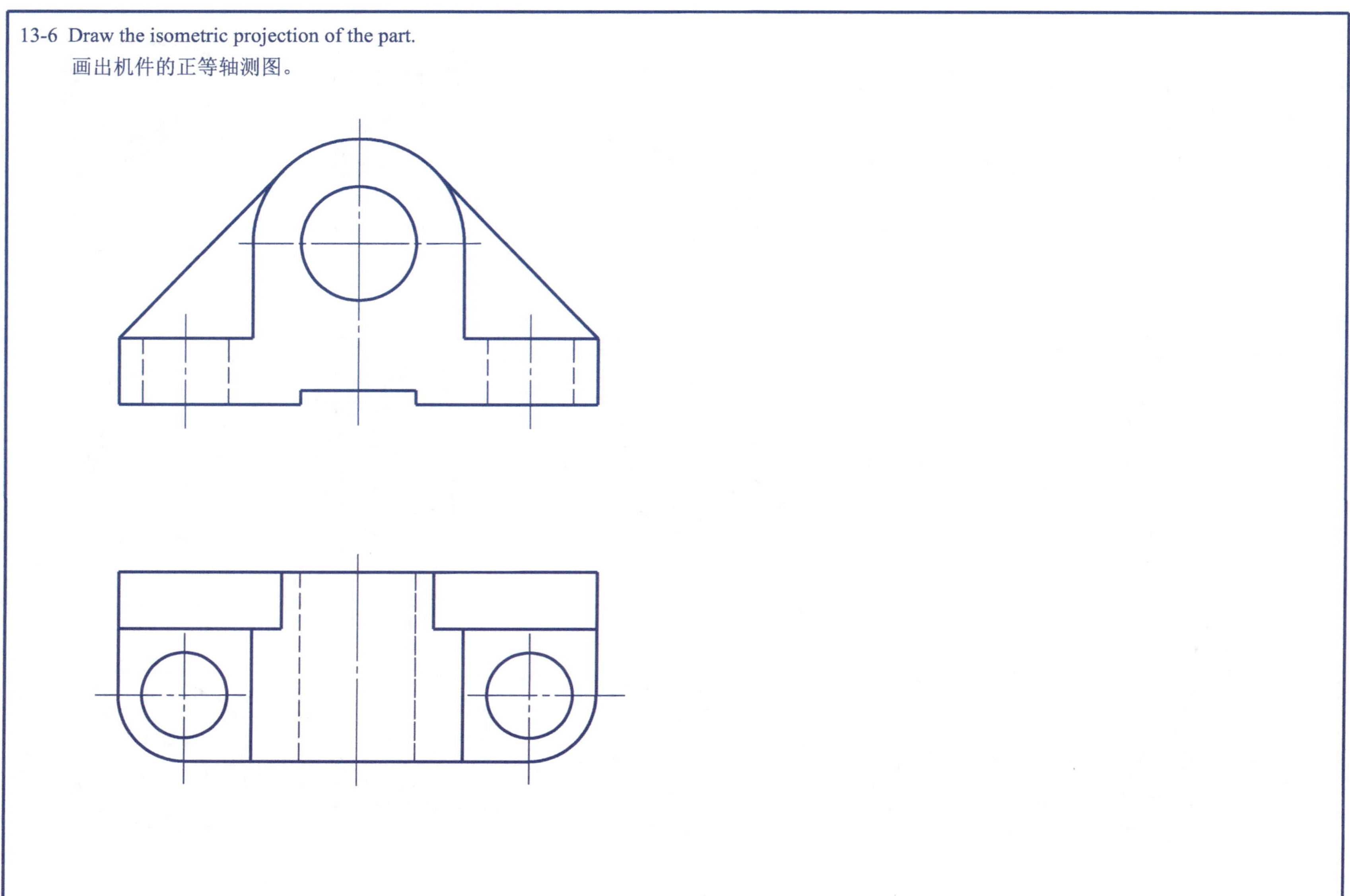

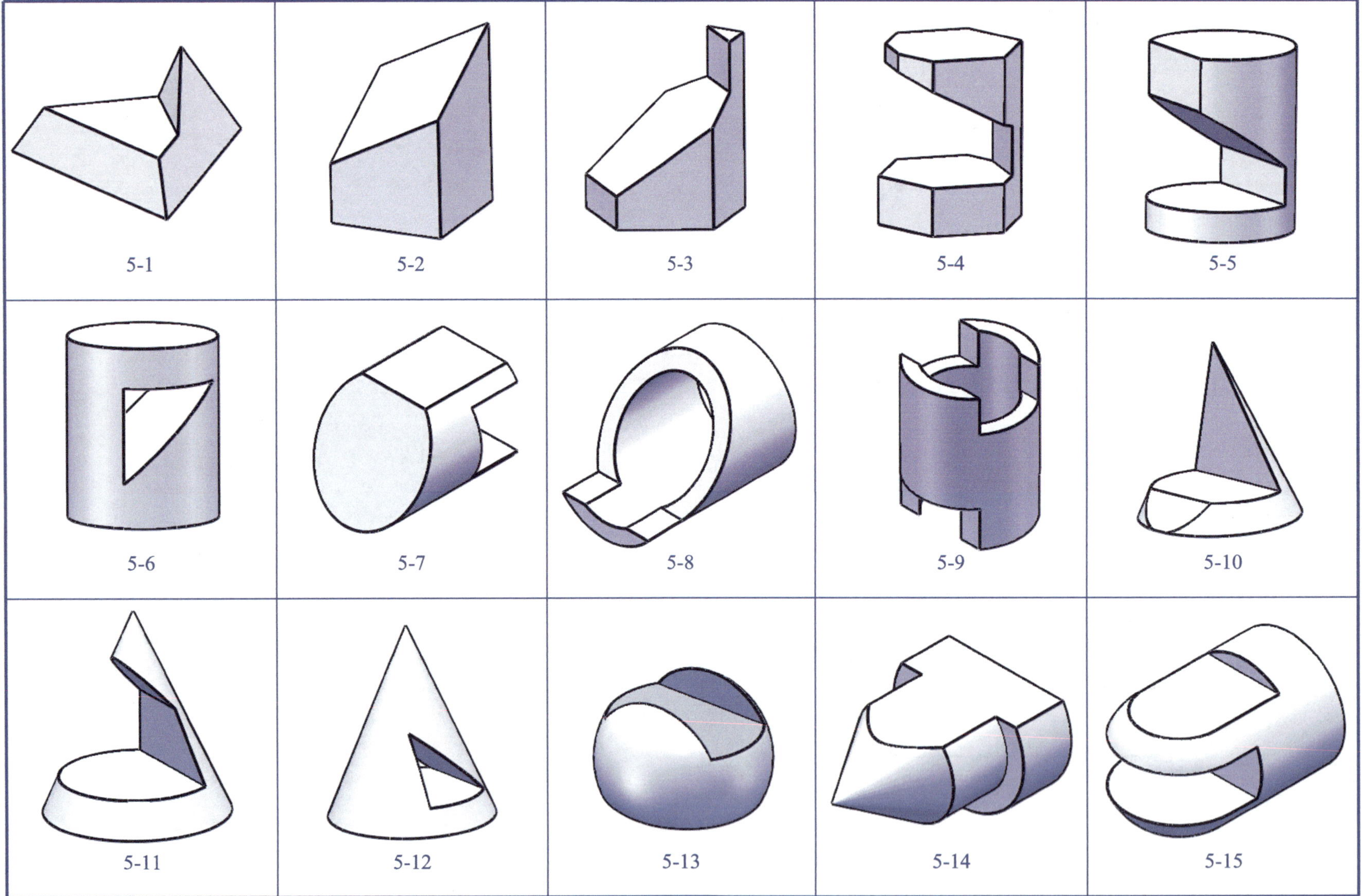
5-1
5-2
5-3
5-4
5-5
5-6
5-7
5-8
5-9
5-10
5-11
5-12
5-13
5-14
5-15

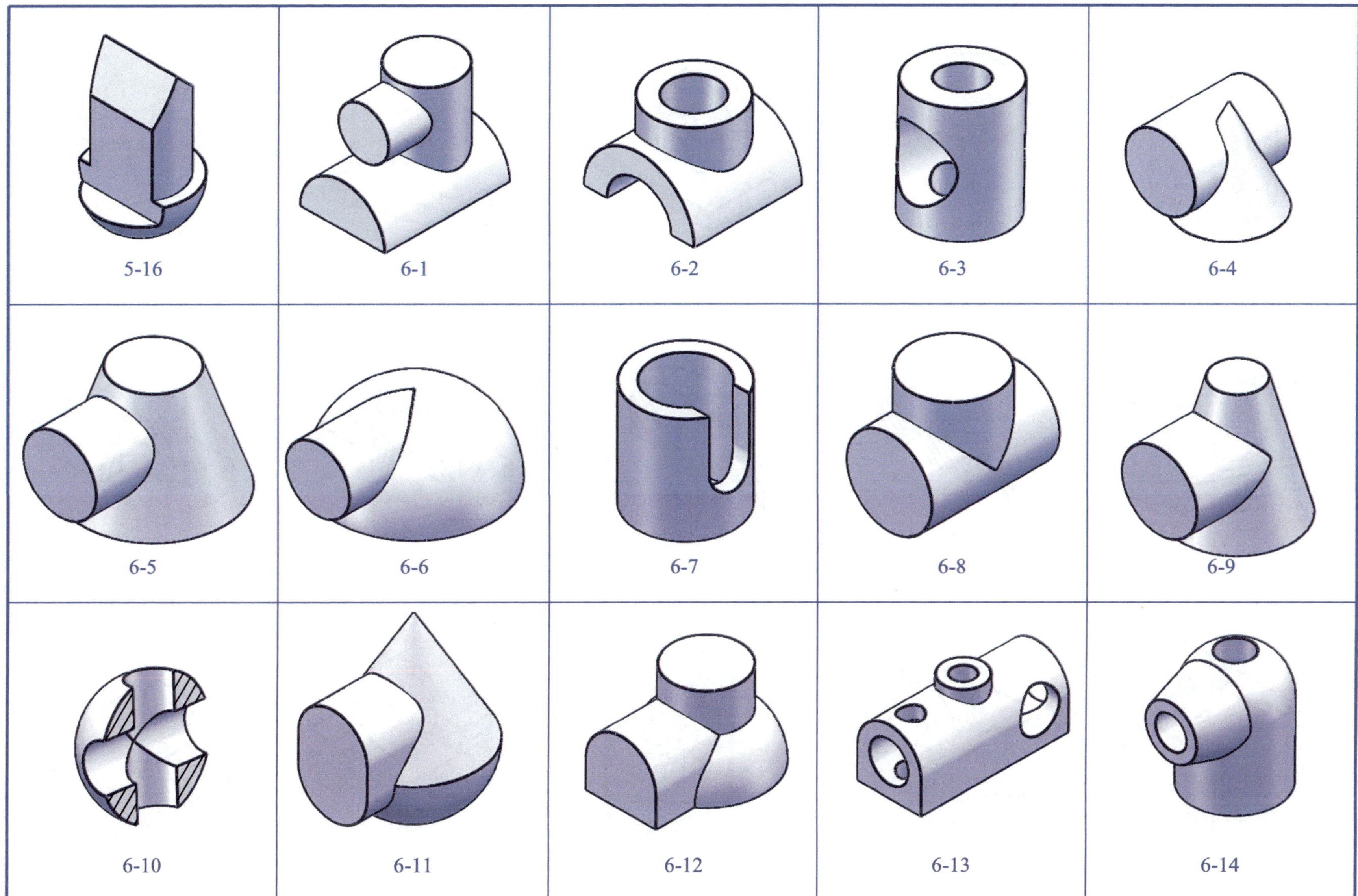
5-16
6-1
6-2
6-3
6-4
6-5
6-6
6-7
6-8
6-9
6-10
6-11
6-12
6-13
6-14

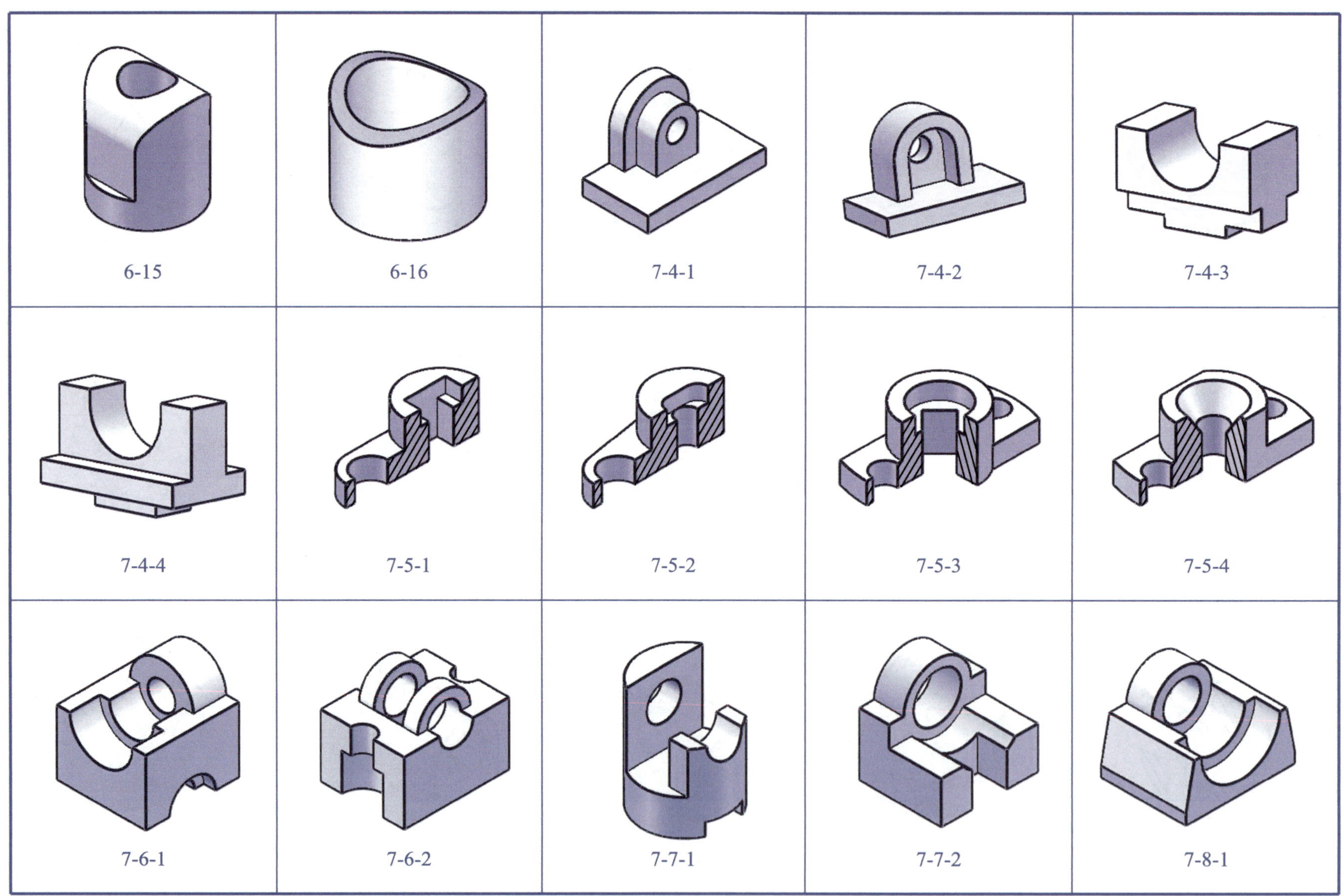
6-15
6-16
7-4-1
7-4-2
7-4-3
7-4-4
7-5-1
7-5-2
7-5-3
7-5-4
7-6-1
7-6-2
7-7-1
7-7-2
7-8-1

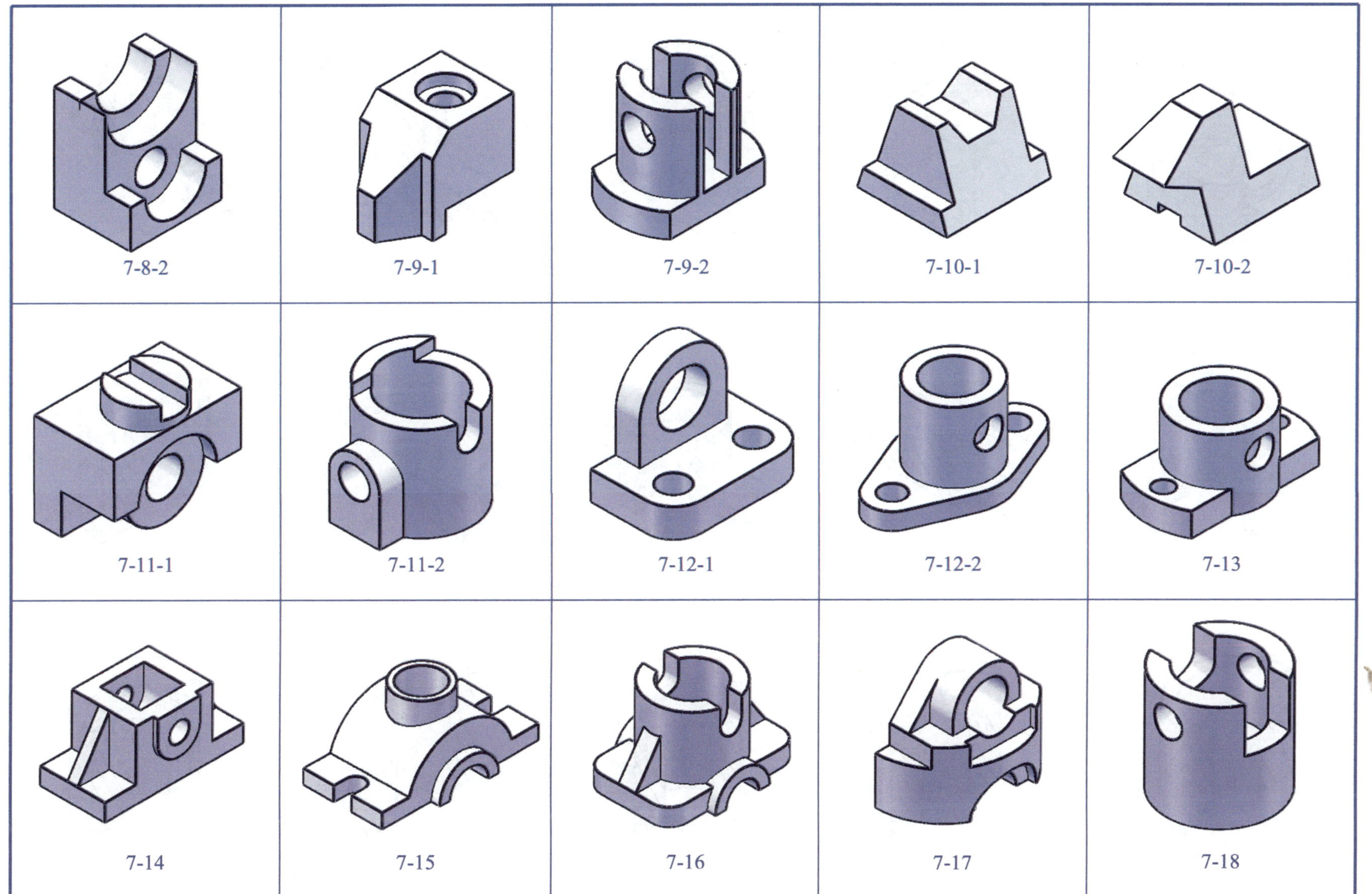
7-8-2
7-9-1
7-9-2
7-10-1
7-10-2
7-11-1
7-11-2
7-12-1
7-12-2
7-13
7-14
7-15
7-16
7-17
7-18

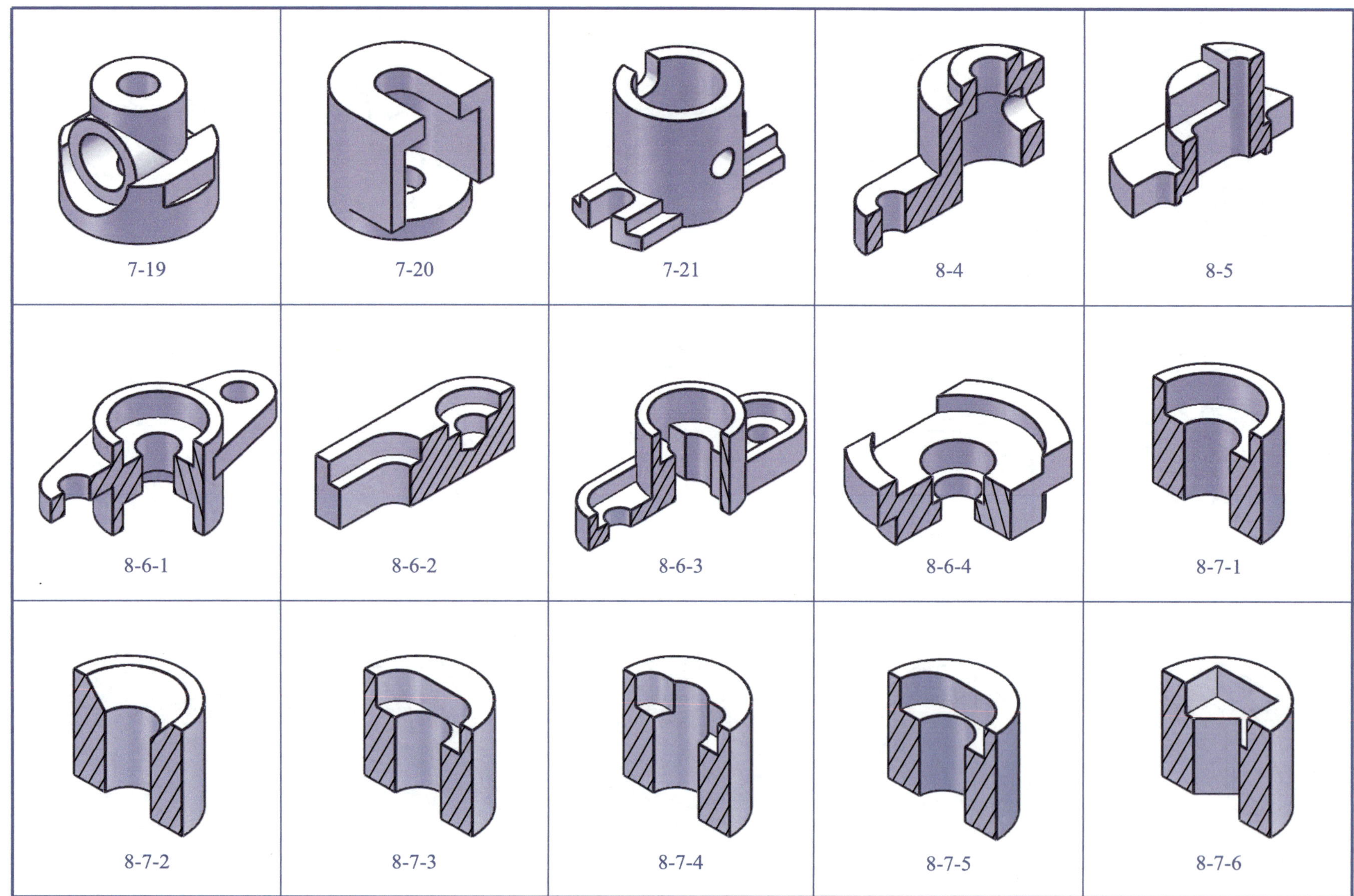
7-19
7-20
7-21
8-4
8-5
8-6-1
8-6-2
8-6-3
8-6-4
8-7-1
8-7-2
8-7-3
8-7-4
8-7-5
8-7-6

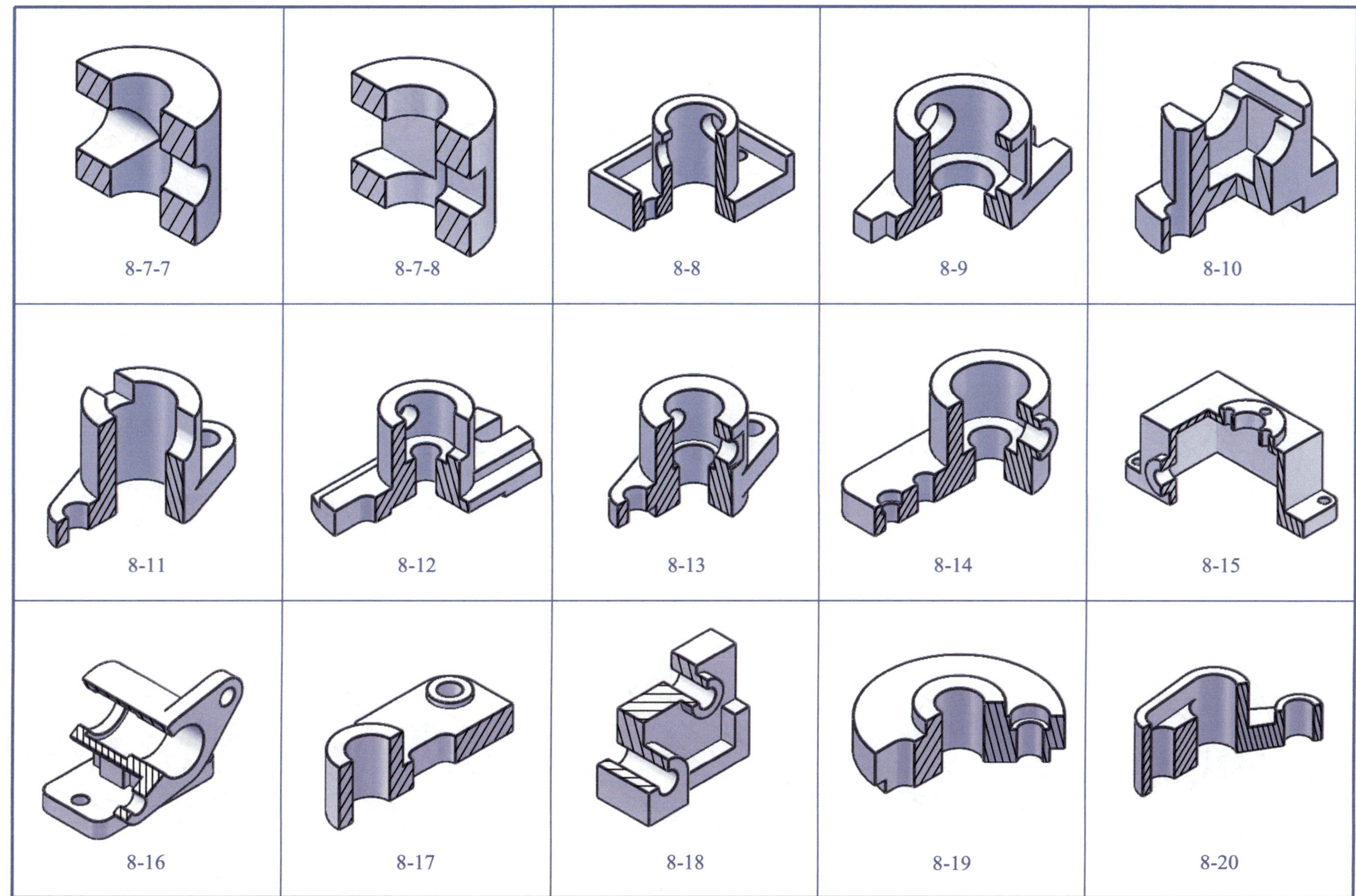
8-7-7
8-7-8
8-8
8-9
8-10
8-11
8-12
8-13
8-14
8-15
8-16
8-17
8-18
8-19
8-20

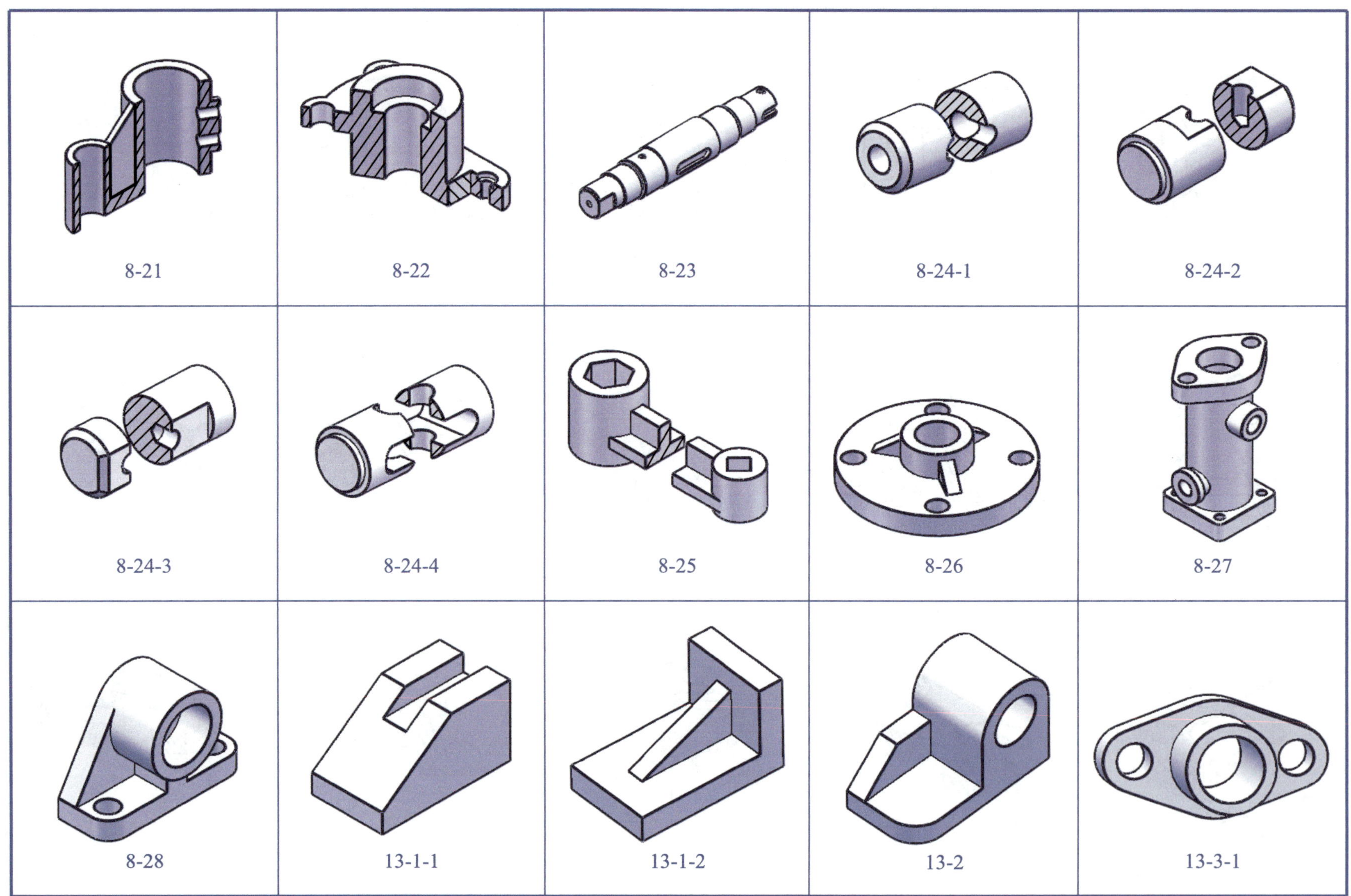
8-21
8-22
8-23
8-24-1
8-24-2
8-24-3
8-24-4
8-25
8-26
8-27
8-28
13-1-1
13-1-2
13-2
13-3-1

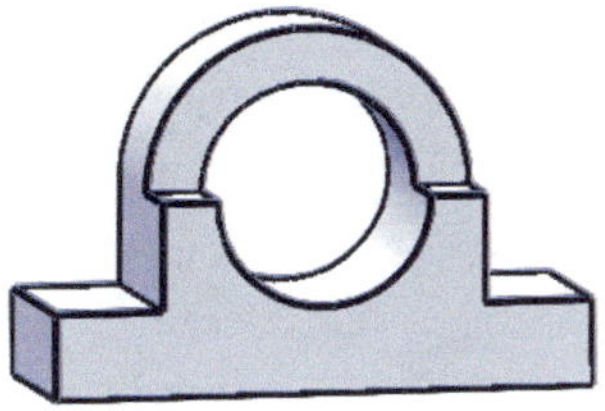

13-3-2	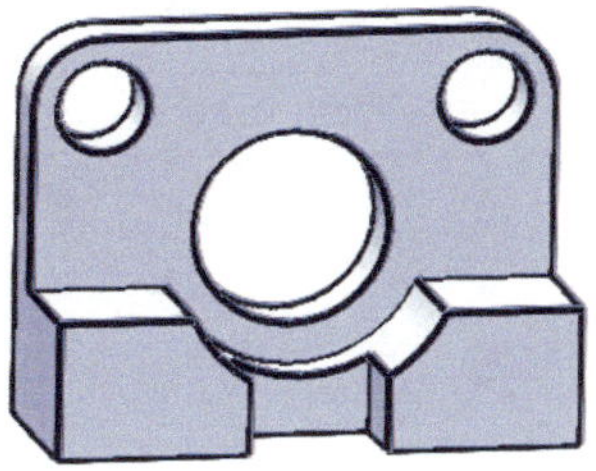 13-4-1	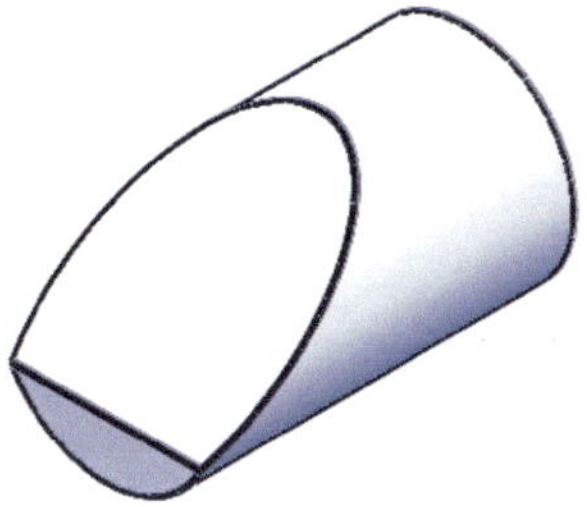 13-4-2	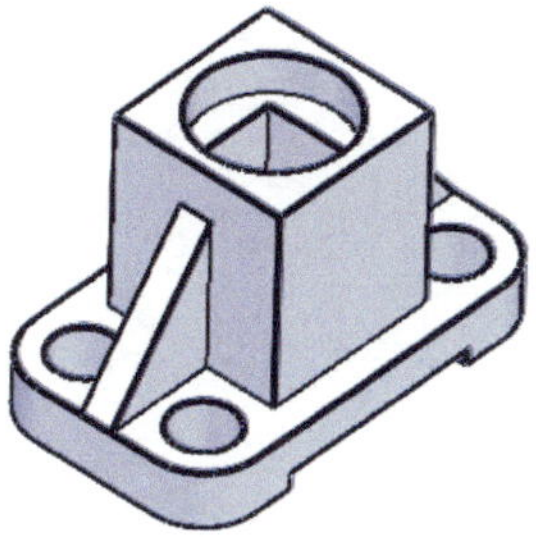 13-5	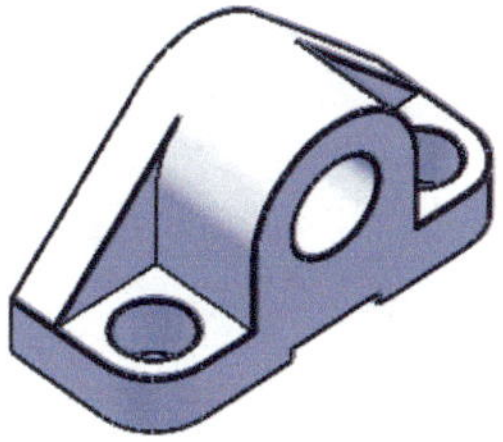 13-6